Heavy Metals
in the
Environment

Using Wetlands
for Their Removal

Heavy Metals

in the

Environment

Using Wetlands
for Their Removal

Howard T. Odum
Wlodzimierz Wójcik
Lowell Pritchard, Jr.
Shanshin Ton
Joseph J. Delfino
Malgorzata Wójcik
Slawomir Leszczynski
Jay D. Patel
Steven J. Doherty
Jacek Stasik

**Center for Environmental Policy and
Center for Wetlands
Environmental Engineering Sciences
University of Florida
Gainesville, Florida**

CRC Press
Taylor & Francis Group
Boca Raton London New York

CRC Press is an imprint of the
Taylor & Francis Group, an **informa** business

CRC Press
Taylor & Francis Group
6000 Broken Sound Parkway NW, Suite 300
Boca Raton, FL 33487-2742

First issued in paperback 2019

ISBN-13: 978-1-56670-401-4 (hbk)
ISBN-13: 978-0-367-39862-0 (pbk)

Library of Congress Cataloging-in-Publication Data

Heavy metals in the environment : using wetlands for their removal / Howard T. Odum
... [et al.].
       p. cm.
    Includes bibliographical references and index.
    ISBN 1-56670-401-4 (alk. paper)
    1. Heavy metals—Environmental aspects. 2. Lead—Environmental aspects. 3. Wetland ecology. 4. Bioremediation. 5. Ecological engineering. I. Odum, Howard T., 1924–

TD196.M4 H434 2000
628.5′2—dc21

99-089022
CIP

Library of Congress Card Number 99-089022

Visit the Taylor & Francis Web site at
http://www.taylorandfrancis.com

and the CRC Press Web site at
http://www.crcpress.com

Foreword

This is a book to deepen our understanding of the way nature realigns and reorganizes to accept the burdens associated with metal as it streams through society. Tadeusz and Berthe Sendzimir, industrial pioneers, first in Poland and later in the U.S., left a legacy to fit metals with environment and society by supporting the work which led to this book.

Much of the appearance and possibilities of modern life reside in sheet metal, the cowling shield of most machines. Tadeusz Sendzimir designed some of the first machines that made sheet metal affordable, mills that rolled red hot slabs into miles of thin metal to house the refrigerators and washing machines of expanding suburbia. Like so many of the creative eddies spun off a river of fossil fuel in this century, Tadeusz Sendzimir made sheet metal something common, and appliances became the altars of modern-day convenience. However, the load that convenience takes off human muscle has to be taken up elsewhere, and some of this has been borne as a series of burdens added to the Earth during the extraction and refinement of metal. In a sense, the true cost of convenience is emerging; and it is only fitting that the profits derived from metal's convenience should in turn be fed back to understand some of the impacts underlying the exploitation of metal. This balancing of cause and effect is part of the noble legacy of pioneers.

Many of us awakened to the true cost of convenience as the momentum of a century of industrialization finally surpassed the capacity of nature to assimilate it. Society's final buffer against uncontrolled pollution is to filter the news from state and corporate news media. When the official silence on environmental degradation was breached in Poland by the Solidarity Movement in the mid-1980s, Tadeusz and Berthe Sendzimir were shocked to learn of the effects of 50 years of short-sighted exploitation of natural systems. The bleak aspect of Polish rivers, lakes, and forests had testified to this for decades, and now the verdict of statistics left no doubt. Forests were acidified by rain, rivers were made corrosive with salt and acid, waters were polluted by sewage, and air in many cities was laden with the exhaust of cars, furnaces, and factories. Many places in the U.S. were also impacted.

With difficult choices and years of hard work, new progress can grow from these ashes. Tadeusz and Berthe decided that they could help by supporting small steps in understanding what might culminate in useful tools when the political will and the economic means to mitigate the damage to the environment had matured. This book describes the first project they launched. It had the goal of understanding how natural systems use wetlands to adapt to wastes. Lead was chosen because a capable group of investigators had incisive questions to apply and situations were available where human developments had inadvertently saturated wetland ecosystems with this heavy metal.

The key people in question were ecologists, Howard T. Odum and Lowell Pritchard, Jr.; chemists, Joseph Delfino and Shanshin Ton; and an engineer, Wlodzimierz Wójcik and associates. Breakthrough questions stem from ecological engineering, a discipline that probes how the design of natural systems can be employed to engineer resource flows in ways more efficient than fossil fuel-driven machine systems. Can systems powered by sunlight handle toxics more effectively than systems running on fossil fuel? At what scale and by what means does one usefully define efficiency? The locations where these studies occurred were a North Florida cypress swamp loaded with lead from a spill from a battery reprocessing plant and an herbaceous wetland in Silesia which has received the waste effluent from a lead mine for 400 years.

The questions raised in these studies resonate ever more strongly with a number of global challenges that have become more prominent since this study has been completed. Airborne deposition of toxics is no longer a regional phenomenon that can be avoided by moving elsewhere. It appears to be as global and inescapable as climate change. If the time comes when there is less use of fossil fuel-based technology, what means remain with which to clean up the aftermath of toxic misadventures? Ecological engineering appears to be one of the most promising avenues with which to explore answers to that question, and this book is an excellent step down that path.

Jan Sendzimir

Foreword

This is a book to deepen our understanding of the way nature realigns and reorganizes to accept the burdens associated with metal as it streams through society. Tadeusz and Berthe Sendzimir, industrial pioneers, first in Poland and later in the U.S., left a legacy to fit metals with environment and society by supporting the work which led to this book.

Much of the appearance and possibilities of modern life reside in sheet metal, the cowling shield of most machines. Tadeusz Sendzimir designed some of the first machines that made sheet metal affordable, mills that rolled red hot slabs into miles of thin metal to house the refrigerators and washing machines of expanding suburbia. Like so many of the creative eddies spun off a river of fossil fuel in this century, Tadeusz Sendzimir made sheet metal something common, and appliances became the altars of modern-day convenience. However, the load that convenience takes off human muscle has to be taken up elsewhere, and some of this has been borne as a series of burdens added to the Earth during the extraction and refinement of metal. In a sense, the true cost of convenience is emerging; and it is only fitting that the profits derived from metal's convenience should in turn be fed back to understand some of the impacts underlying the exploitation of metal. This balancing of cause and effect is part of the noble legacy of pioneers.

Many of us awakened to the true cost of convenience as the momentum of a century of industrialization finally surpassed the capacity of nature to assimilate it. Society's final buffer against uncontrolled pollution is to filter the news from state and corporate news media. When the official silence on environmental degradation was breached in Poland by the Solidarity Movement in the mid-1980s, Tadeusz and Berthe Sendzimir were shocked to learn of the effects of 50 years of short-sighted exploitation of natural systems. The bleak aspect of Polish rivers, lakes, and forests had testified to this for decades, and now the verdict of statistics left no doubt. Forests were acidified by rain, rivers were made corrosive with salt and acid, waters were polluted by sewage, and air in many cities was laden with the exhaust of cars, furnaces, and factories. Many places in the U.S. were also impacted.

With difficult choices and years of hard work, new progress can grow from these ashes. Tadeusz and Berthe decided that they could help by supporting small steps in understanding what might culminate in useful tools when the political will and the economic means to mitigate the damage to the environment had matured. This book describes the first project they launched. It had the goal of understanding how natural systems use wetlands to adapt to wastes. Lead was chosen because a capable group of investigators had incisive questions to apply and situations were available where human developments had inadvertently saturated wetland ecosystems with this heavy metal.

The key people in question were ecologists, Howard T. Odum and Lowell Pritchard, Jr.; chemists, Joseph Delfino and Shanshin Ton; and an engineer, Wlodzimierz Wójcik and associates. Breakthrough questions stem from ecological engineering, a discipline that probes how the design of natural systems can be employed to engineer resource flows in ways more efficient than fossil fuel-driven machine systems. Can systems powered by sunlight handle toxics more effectively than systems running on fossil fuel? At what scale and by what means does one usefully define efficiency? The locations where these studies occurred were a North Florida cypress swamp loaded with lead from a spill from a battery reprocessing plant and an herbaceous wetland in Silesia which has received the waste effluent from a lead mine for 400 years.

The questions raised in these studies resonate ever more strongly with a number of global challenges that have become more prominent since this study has been completed. Airborne deposition of toxics is no longer a regional phenomenon that can be avoided by moving elsewhere. It appears to be as global and inescapable as climate change. If the time comes when there is less use of fossil fuel-based technology, what means remain with which to clean up the aftermath of toxic misadventures? Ecological engineering appears to be one of the most promising avenues with which to explore answers to that question, and this book is an excellent step down that path.

Jan Sendzimir

Acknowledgments

This study was a joint project of the D.T. Sendzimir Family Foundation (Jan Sendzimir, Head), the Department of Environmental Engineering Sciences of the University of Florida in Gainesville (H.T. Odum, Principal Investigator), and the University of Mining and Metallurgy, Krakow, Poland (Wlodzimierz Wójcik, Principal Investigator). Joan Breeze was editorial assistant.

Joseph J. Delfino
Department of Environmental
  Engineering Sciences
University of Florida
Gainesville, Florida

Steven J. Doherty
Swedish University of Agricultural Science
Uppsala, Sweden

Slawomir Leszczynski
Institute of Enviroment Protection and
  Management
University of Mining and Metallurgy
Krakow, Poland

Howard T. Odum
Department of Environmental
  Engineering Sciences
University of Florida
Gainesville, Florida

Jay D. Patel
Department of Environmental
  Engineering Sciences
University of Florida
Gainesville, Florida

Lowell Pritchard, Jr.
Department of Environmental Studies
Emory University
Decatur, Georgia

Jacek Stasik
Institute of Environment Protection
  and Management
University of Mining and Metallurgy
Krakow, Poland

Shanshin Ton
Department of Environmental Science
Feng-Chia University
Tai-Chung, Taiwan

Malgorzata Wójcik
Institute of Environment Protection
  and Management
University of Mining and Metallurgy
Krakow, Poland

Wlodzimierz Wójcik
Institute of Environment Protection
  and Management
University of Mining and Metallurgy
Krakow, Poland

Contents

PART IV VALUE AND POLICY

To Tadeusz and Berthe Sendzimir
For their love of nature and the people of Poland
which returns again and again
in the new understanding and practice
toward which they set us.

PART I

Introduction and Background

After describing the contents of this book on lead and wetlands, the introductory chapter explains the *gaia* hypothesis for the role of wetlands in regulating toxic heavy metals. Chapter 2 identifies current questions and needs regarding lead and society. Chapter 3 reviews published literature on lead, wetlands, and society. Finally, Chapter 4 examines systems models of lead and the global cycle.

Introduction

CONTENTS

With restless energies of the sun and heat of the earth, the biogeosphere circulates materials on which human life depends. In the global cycle of the earth, the slow rise of the land is balanced by steady erosion as soils and sediments return to the sea carried by the cycle of water. In Figure 1.1 the ocean on the left supplies the rain to the land on the right, and the water carrying sediments drains back to the ocean. In the original landscape before economic development, much of the water runoff passed through wetlands which captured the fertile soils and toxic substances. The filtered material in peaty sediments was eventually converted into coal by geologic processes.

What the evidence of many studies now shows is that the wetlands all over the world evolved mechanisms that were mutually reinforcing life in the biosphere. Self-organization of life with the biogeochemistry of the earth has been called *gaia*, a word used in the popular press to mean "life operating the earth for itself." As the examples in these studies show, wetlands are gaia operations.

The development of human civilization was based on this earth cycle. Metals were mined from the earth, manufactured into swords and plowshares, used, and returned to the earth cycle as fragmented remains. Figure 1.2 shows the economy on the right using waters and earth resources. With accelerating growth of society and its demands, the flow of metals between the economy and the earth cycle increased, often distorted with waste accumulations and risks (waste tank on the right in Figure 1.2).

A sustainable civilization requires a harmonious fit between economic use of the earth materials and the environmental system. It makes sense to process and reuse concentrated materials in wastes, whereas the dilute wastes need to go back into the earth cycle in a way that promotes productivity and is not toxic. There are now many successful demonstrations that recycling the dilute wastes of cities and industries back through wetlands is a good way to connect the economy to nature. Especially where the natural wetlands were drained, new wetlands are being created to let the ecological systems adapt, becoming a low-cost contributor to the wealth of the landscape.

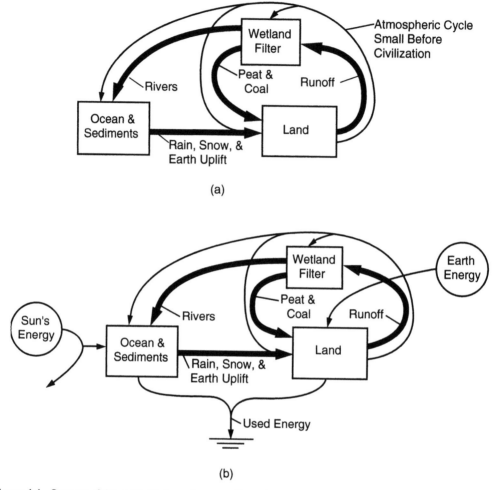

Figure 1.1 Concept of the cycle of atmosphere, earth, and water which moves the cycle of lead (thicker lines) from ocean to land, with runoff waters passing through wetland filters before returning to the sea. Swamp peats may eventually be turned into coal as shown. (a) Cycles of matter; (b) same with addition of the main energy flows.

This book is a story about scientific studies for achieving a better fit of the economy with the earth cycle. We call the development of partnership designs with nature *ecological engineering*. Others call it *industrial ecology*, making manufacturing and ecosystem processing mutually beneficial. This book is especially about the wetlands.

Starting in the last century, brilliant advances in the science and technology of chemistry made it possible to cover the earth with new chemical products in amazing variety as part of the new ways of farms and cities. At first little attention was paid to the questions of what could be sustained and how to return the leftovers and worn-out pieces of the new substances to the earth. As accumulations and toxic conditions developed threatening the health of humans, a new insight developed: perhaps the global earth system after millions of years of ecosystem self-organization was operating so as to support and protect the special chemistry of life and to filter out chemicals toxic to life.

New studies like those reported here showed that the *wetlands* were one of the main places where the toxic materials were filtered, immobilized, and returned to the geologic part of the earth cycle, mostly isolated from the living biosphere (Figure 1.2). The beds of peat in wetlands often were turned into coal by the pressures of the geologic processes, often binding the metals.

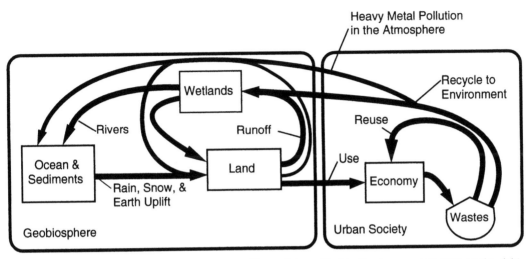

Figure 1.2 Cycle of earth, waters, and lead from Figure 1.1 modified by the human civilization on the right. The economy is shown using the waters and mined lead, generating by-product wastes (tank symbol). The most concentrated wastes are reused and dilute wastes are recycled to the environment by air and in waters that can be filtered by wetlands.

LEAD

One of the metals first used by developing civilization was *lead*, because of its low melting temperature and malleable properties, and which was easy for artisans to make into pipes, kitchenware, and bullets or to combine with other metals in alloys. Nriagu (1983) includes estimates of global lead processing, starting in ancient history with evidences of its toxicity affecting sustainability of both Greek and Roman civilizations. Patterson, in the 1980 National Academy Report, has a graph of lead production starting 5000 years ago. His minority statement warns of toxic accumulations. Later massive quantities were part of paint, gasoline anti-knock additives, and especially batteries still used by the billions. Unfortunately, too much lead in people is toxic, especially to brains and the nervous system. Soon there were enormous costs to society for treatment, lost work, and public health measures to eliminate lead from waters, foods, and dangerous waste disposal. See details in Chapter 3.

However, wetlands filter lead. This book is about lead, wetlands, and the relation of the global lead cycle to the industrial economy. In Figures 1.1 and 1.2 the cycle of lead as it accompanies atmosphere, waters, sediments, and rocks is shown as thick, dark arrows.

In small-scale science it is easy to set up experiments to study the effect of various factors on a system. Thus, there have been many laboratory-scale studies of the effect of lead on organisms. On this scale many replications could be made to ensure that the results are consistent. However, for the larger scale of whole ecosystems it is rarely possible to do such experimenting. Instead, one can look for situations where a test of a factor has been provided as a fortuitous experiment. In this study two situations were found where wetlands received and filtered large quantities of lead.

SAPP SWAMP IN FLORIDA

Field studies were made on the two wetlands which had received lead. The *Sapp site* in Jackson County, in Florida (Figure 1.3), is a 1-h drive west of Tallahassee. From 1970 to 1979 thousands

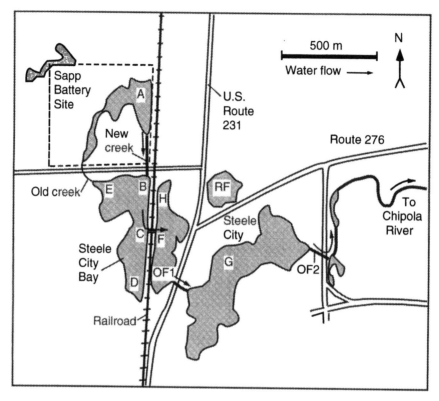

Figure 1.3 Map showing the site of the Sapp battery washing that drained into cypress-gum swamps in Steele City Bay in Jackson County, Florida. Letters are sampling stations.

of car batteries with their sulfuric acid were washed with more sulfuric acid to recover the lead which was returned to industry. The sulfuric acid and dissolved lead drained into a series of cypress-gum wetlands. In 1980 groundwaters developed 250 ppm manganese (Lynch, 1981; Trnovsky et al., 1988). Becker (1981) gathered photographic documentation.

This battery-washing area was designated a Superfund site and a $700,000 cleanup was contracted by the U.S. Environmental Protection Agency in 1981 to 1983. The battery-washing site and soils of the first wetland (within the dashed line at the top of Figure 1.3) were dug up and transported to Alabama. During 1990 to 1993 we studied the cypress swamps downstream that received the runoff from the Superfund site to determine the amount of lead present, its condition, long-term impacts, and values. Figures 1.4a to 1.4h show a selection of pictures of this site. Then economic evaluations were made and compared with an industrial battery recovery operation in Tampa, FL where wastes were not released to the environment.

BIALA RIVER MARSHES IN POLAND

The second lead-containing wetland studied was the Biala River Marsh (Figure 1.5), 60 mi from Krakow, Poland where lead and zinc from mining and processing had been running into grassy wetlands for 400 years. Lead and zinc and ecosystem characteristics were studied to understand long-range responses. In this site some of the metals washing out from the industry into the river were attached to particles that were captured and sedimented by the marsh vegetation. The complex of wetland vegetation is shown in the map in Figure 1.6 and a selection of pictures in Figures 1.7a to 1.7g.

Figure 1.4a   Lowell Pritchard, Jr. sampling water lilies *Nymphaea odorata* in Steele City Bay near Station D August 21, 1990 (Pritchard). See Figure 1.3.

Figure 1.4b   Area of dead blackgum trees with water lilies near Station B (Pritchard). See Figure 1.3.

Figure 1.4c   Lowell Pritchard, Jr. sampling sediments near Station B, with surviving swamp trees in less affected areas in the background March 15, 1991 (Pritchard). See Figure 1.3.

MODELS AND SIMULATION

After scientific studies have been made of an environmental system, ideas are developed about how the system works. To visualize a wetland system that captures lead requires some kind of simplified picture of the parts and processes. Such an overview is called a model. We have already used two models (Figures 1.1 and 1.2) to help the reader visualize the global cycles. For the human mind to understand, the model has to be kept simpler than the actual system, which usually has many elaborate details, not all of which are important to understanding the main processes. The models used in this study are shown with a standardized language of symbols (Appendix A1).

Like a machine, the parts of environmental systems have mathematical relationships that describe the mechanisms by which the ecosystem works. Research is directed at understanding these relationships. Systems diagrams are a way of representing the parts and processes in a more understandable way than comes from studying the equivalent mathematical equations.

If a model is calibrated (given numbers based on field measurements), the equations quantitatively relate each part and process. A calibrated model (calibrated equations), when programmed on a computer, simulates what the system would do given the arrangements and simplifications of the model. The simulation generates graphs of properties and processes such as the quantities of plants, lead, microbes, and productivity. Then the simulation results can be compared with what

Figure 1.4d Sampling sediments near Station B with Sapp Battery area (Superfund site) in the background March 15, 1991 (Pritchard). See Figure 1.3.

was observed in the field, a process called validation. If the simulations are not like what was actually observed, then the model and/or its calibrations have to be changed until there is better agreement. In this way the ideas about how a system works and what is most important are tested with reality and improved step by step.

In this study simulations were made of the lead absorption by wetlands and on a larger scale of the behavior of the lead in the economy. The wetlands were too large and lead too toxic for nearby people to purposefully add lead or change the biology. Instead, calibrated and validated simulation models were run with various changes of components and inputs, sometimes called "what if" experiments.

MICROCOSMS

Another way of studying large-scale ecosystem responses is to study miniaturized ecosystems, sometimes called microcosms. In small containers in a greenhouse, main components of a wetland such as soils, small trees, peaty soils, and a seeding of invertebrates and microbes are introduced and allowed to self-organize on a small scale. With smaller size, replications can be prepared. Although the microecosystems that develop differ in many ways from the large ecosystems, the

Figure 1.4e   View of gum and cypress trees in November 1991 after autumn leaf fall in the unimpacted reference area at Station RF (Pritchard). See Figure 1.3.

Figure 1.4f   View August 21, 1990 of damaged area near Station G where many trees survived (Pritchard). See Figure 1.3.

Figure 1.4g   View of area near Station F August 20, 1990 where some trees survived and were resprouting (Pritchard). See Figure 1.3.

Figure 1.4h   Sampling dissolved oxygen August 21, 1990 at Station C (Pritchard). See Figure 1.3.

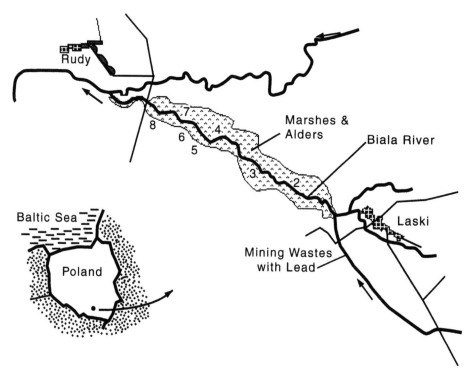

Figure 1.5 Map showing the industries with lead and zinc wastes that drained into marsh wetlands around the Biala River, near Krakow, Poland. Note location in the Poland inset.

essence of production, consumption, and material recycle is present. In this study microcosms were developed to resemble the Florida wetlands as much as possible, and then treated with lead to study the way lead was processed and its impact on ecosystems. Models for the wetlands were adapted to represent the observations from the microcosms as well.

On a global scale lead circulates as part of the general cycles of air, water, earth, and wetland processing. The industrial society changed these cycles with new pathways of circulation — from mines to factory to consumer to waste disposal. The flow of material such as lead through the network of pathways in the environment and economy is evaluated in the scientific field of biogeochemistry. To suggest the importance of wetlands to the global cycle, data on chemical uptake and impact by wetlands from the field study, simulations, and microcosms were extended to the planetary scale. In this study the potential for absorbing lead in the nation's wetlands was very roughly estimated by using the filtration ability of those systems studied. You need many more examples before drawing general conclusions. Still, gross extrapolations based on what you have measured are a starting point.

EVALUATION

At the end of the decade of operation of the Sapp battery-washing operation, Lynch (1981) made an economic evaluation of the enormous costs and losses caused by a poor fit of wastes and environment. He included costs of the battery operations and lead reuse, costs of Superfund treatment, cost of controlling toxic groundwaters, losses of fishing and recreation values in the downstream watershed, and losses of forest resources and ecosystems.

In this study we evaluate the swamps below the Superfund site and their role in lead retention 10 years later. We seek answers to the more general questions: How important is lead filtration to

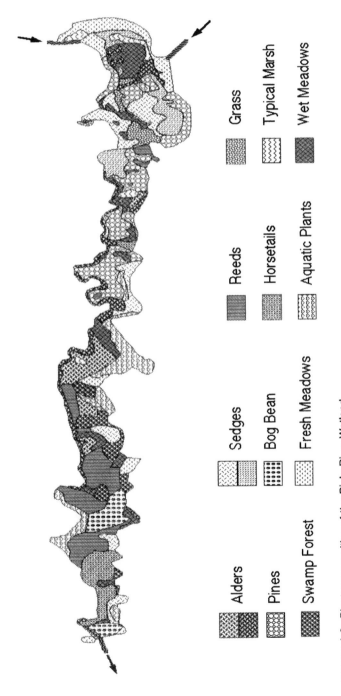

Figure 1.6 Plant communities of the Biala River Wetland.

Figure 1.7a   View of the sedge community, Station 3, Biala River, Poland.

Figure 1.7b   View of bog bean, sedges, and reeds, Station 9, Biala River, Poland.

Figure 1.7c   View of cattails and reeds, Station 4, Biala River, Poland.

Figure 1.7d   View of reeds and cattails, Station 7, Biala River, Poland.

Figure 1.7e   View of Station 3, Biala River, Poland.

Figure 1.7f   View of deschampsia grass with the piezometers for groundwater monitoring in the foreground, Station 2, Biala River, Poland.

Figure 1.7g   View of alder, reed, cattail, sedge, and deschampsia grass, Station 5, Biala River, Poland.

the whole economy of humanity and nature? How important are the wetlands and the lead absorptions to the rest of the landscape and to the whole world?

Many have sought perspective on environmental processes using market costs and prices based on human judgments of value such as willingness to pay money. However, since money is only paid to people, money only measures the human service contribution to the process. A method is required that evaluates the contribution of the wetlands *and* the people in the economy. Figure 1.8 shows the way contributions of work from the environment go into support of the economy and the wealth that circulating money buys.

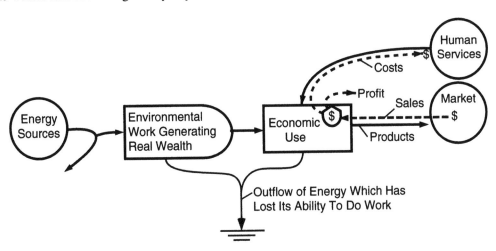

Figure 1.8 Diagram of environmental work (on the left) contributing real wealth to economic uses. The work contributed by human services is shown (from the right) paid for by dollars (dashed lines).

To *evaluate the impact and contributions of environment and the costs and benefits to people* on one common basis, a relatively new method, *EMERGY evaluation*, was used which measures the work of nature and of the human economy on a common basis. See Chapter 4 and recent book summary (Odum, 1996). Emergy is the sum of all types of work, expressed in units of one kind of energy. These values are expressed as *emdollars*, the dollars of equivalent Gross Economic Product. Figure 1.9 shows the total emergy contributed to support the global gross economic product and the ratio of emergy to money. We multiplied data on human labor and services in dollars by the emergy/money ratio for the year to obtain equivalent emergy contribution. The emergy method of evaluation measures both work contributions on a common basis.

Also included is a more traditional economic analysis of the environmental contribution. Economic methods evaluate environment inputs by what dollars people are willing to spend beyond costs. Humans are often willing to pay more than the cost for processing an environmental product. Thus, when the price is higher than the cost, the dollar amount a producer receives that is greater than the dollar cost is the "producer surplus," which is sometimes regarded as the value of the environmental input. The dashed line in Figure 1.8 shows money from sales going to pay for the costs of items purchased from the economy. Money flow larger than this is shown going into profit, eventually going to humans for additional services.

Included in Chapter 11, economic analysis uses the replacement cost method of evaluating producer surplus. The idea is that people place a value on environmental systems and their work, even if they are not made to pay money for it. One way to estimate this value is to measure how much people would pay to replace the environmental services they presently enjoy free if they were to be taken away. When free environmental services are lost, people will try to substitute services from the human economy, and they must pay money for them. The cost in terms of money for those substitute services is called replacement cost, and is an economic measure of how much benefit people were previously deriving free from natural systems. It is not a complete measure, since even the replacement services contain unevaluated environmental services.

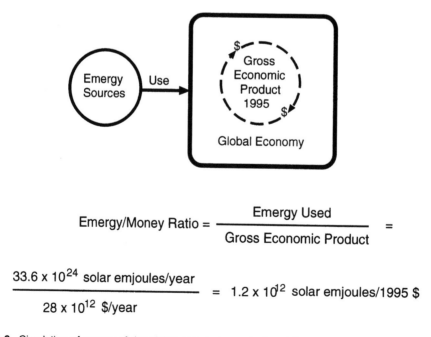

$$\text{Emergy/Money Ratio} = \frac{\text{Emergy Used}}{\text{Gross Economic Product}} =$$

$$\frac{33.6 \times 10^{24} \text{ solar emjoules/year}}{28 \times 10^{12} \text{ \$/year}} = 1.2 \times 10^{12} \text{ solar emjoules/1995 \$}$$

Figure 1.9 Circulation of money of the global economic product and the energy sources which are used in work processes to make real wealth. The emergy/money ratio is calculated by dividing the emergy used by the money circulated (Brown and Ulgiati, 1999).

With this method is a paradox which limits its validity. When the total environmental resource is abundant and its contribution therefore great, the replacement cost for a little bit of it seems very low, and the economic value is very low. When the total environmental contribution has almost disappeared, it would be economically very costly to replace even a little bit, so the economic value is very high. In Figure 1.8 the contribution of environmental work is shown coming from the left. The willingness to pay does not measure this contribution to the real wealth of the economy. In fact, it is the reverse. When the environment contributes most, its resource is taken for granted and the willingness to pay is least.

CHAPTER 2

Problems and Needs

CONTENTS

Originally, before human civilization, the cycle of heavy metals in the geobiosphere was compatible with the living cover of the earth, aided in large measure by filtering ecosystems, especially wetlands. Most of the metals were in deposits in geological strata, rarely interacting at high concentrations with life. Accelerating over several centuries, the heavy metal storage reserves in the geological strata were brought into industrial manufacture, with uses changing rapidly as technologies changed. What was most economical at first was to mine, to use, and to discard, letting environments absorb wastes and toxicities. What are the *problems* we face now due to this history of resource exploitation?

ENVIRONMENTAL HEALTH FOR HUMANITY

The condition of lead at the turn into the third millenium AD illustrates the serious problems for society with heavy metals out of place. The environmental lead burden is not smoothly distributed spatially, and the reduction in atmospheric lead pollution from leaded gasoline means that background levels will not grow as rapidly as before, focusing attention on more concentrated sources. Most studies of the physiological and economic costs of lead poisoning have involved cases of acute lead poisoning at high exposure levels, and more recently they have examined the effects of exposure to elevated levels of lead in urban environments, but not at lower levels in the general environment. It is estimated that the average burden of lead in the human body is now at least 300 times that of pre-industrial human populations (National Research Council, 1993). The effect of this on overall human health is not known precisely, although there is no known lower threshold for deleterious effects of lead.

While lead has long been known as a toxin, the last 20 years has confirmed the poisonous effects of lead for the neurological system, even at very low levels (Needleman et al., 1979, 1990). Lead levels in human blood as low as 0.1 ppm (10 micrograms per deciliter = 10 µg/dl) are now considered cause for intervention by public agencies such as the Centers for Disease Control (CDC) and Environmental Protection Agency (EPA) (National Research Council, 1993). Low-level lead exposure has been linked to lowered intelligence, diminished cognitive function, and delinquent behavior in children (Schwartz, 1994a; Jacobs, 1996b; Needleman et al., 1996). The sources of lead exposure have changed in recent years, with the decline of leaded gasoline and lead solder in food cans, but the largest remaining sources are especially risky for children — lead-based paint and lead-polluted soil (Jacobs, 1996b).

Lead pollution from leaded gasoline and food cans was much more easily addressed by environmental regulations than the residual stock of lead in house paint and soil. Manufacturers of leaded gasoline and tainted foodstuffs could be identified and regulations readily enforced. Although the use of lead-based paint in houses declined from 1940 onward (National Research Council, 1993), the danger remains as old paint flakes, crumbles, turns to dust, or is disturbed by renovation.

The U.S. Department of Housing and Urban Development (HUD) estimated the cost of completely removing lead-based paint from the nation's housing stock to be about $500 billion (HUD, 1990, cited in Jacobs, 1996a). (By comparison, EPA estimated the annual cost of removing lead from gasoline to be only about $500 million.) That cost is misleading, however, since interim control methods (to reduce exposure but not to remove all lead-based paint) would be much less expensive. Given the toxic effects of low-level lead exposure discussed above, there are clear economic benefits to such an enterprise. Schwartz (1994b) estimated a social benefit of $1.7 billion (1989 dollars) for a 0.1-µg/dl (1.0 ppb) drop in blood lead levels, using methods of neoclassical economics. Measuring the social benefit of reducing lead exposure using emergy has not been undertaken, but would be expected to be at least as dramatic.

ECOTOXICOLOGY

There is finally a general realization that humanity can be healthy only when its environment is also healthy. Understanding heavy metal toxicology also requires an understanding of these elements and their impact on the environment — the field of ecotoxicology. Some of the questions about heavy metals and environment are scientific issues about the way systems self-organize their cycles of toxic material use.

Twenty scientists gave their views on this field in a book entitled *Ecotoxicology, a Hierarchical Approach* (Newman and Jagoe, 1996). Starting with the very small scale of cells and molecular biology, chapters considered the processing of toxic substances by organisms and their behavior, populations, communities, ecosystems, lakes, and the earth as a whole. Newman, in his introduction, defines the field as: "knowledge about the fate and effects of toxicants in ecosystems based on explanatory principles." But mechanisms at a small scale of time and space (example: chemical reactions) cannot by themselves explain distributions at much larger scales. Principles were not included for relating separate chapters on phenomena of different scale. A discipline is sometimes defined as understanding phenomena of a particular scale (example: chemistry concerned with molecules and their processes), but it is unlikely that principles of one scale can suffice.

In Chapter 4 of this book we try to relate phenomena of different scales using energy hierarchy concepts. Emergy (spelled with an "m") is a quantitative measure of the work accumulated in concentrating valuable substances from one scale to another. Units and processes are given their hierarchical position by their unit emergy, which increases with scale and concentration and correlates with toxicity (Genoni and Montague, 1991). Systems models are required which combine what is important among the scales of interest. In this book a systems model of lead in a wetland is simulated in Chapter 8.

Much science in this century has approached each phenomenon of interest by studying component mechanisms one scale smaller or more. In *Ecotoxicology, a Hierachical Approach* most of the chapters discuss various mechanisms of toxic interaction at the smaller scales of stressed organisms. This is the traditional approach of trying to understand all the component mechanisms even though the larger-scale parts of the system and flow in the biogeochemical cycles are controlling the rates. If the questions and needs are environmental, is it really necessary to understand the DNA damage, molecular biology, membrane physiology, histology, organism behavior, population demography, ecosystem oscillations, biogeography, and community composition (the topics of the ecotoxicology book)? It is probably not practical to put that much detail from the parts in a model intended to explain the environmental role of heavy metals.

Minimodels are simplifications in the form of diagrams and mathematical equations relating the various parts and processes of a system. For example, Beyers and Odum (1993) simulated the role of a toxic metal that recirculates through the photosynthesis and respiration of an ecosystem, with a generalized minimodel named TOXICPR. Another model of a cycle of toxic substance (named TOXICOSM) simulated the stress on biodiversity. These are top-down models, highly aggregated to show only the most important components and processes. How useful are minimodels of this class in explaining heavy metal distribution in real cases?

Ecological microcosms are containers with living, miniaturized ecosystems (examples: aquaria and terraria). Perez et al. (1977) and Perez (1995) studied similar treatments in microcosms of three different sizes and found that the impacts were dependent on the size of the microcosm. Large-scale impact oscillations were excluded from the small scale. Questions were raised about using small microcosms to judge toxicity impact of large environmental systems. Beyers and Odum (1994) reviewed some of the extensive literature that used ecological microcosms to study cycles and toxicities. In this review we suggested transferring conclusions from small-scale microcosms to larger-scale systems by multiplying the dimensions of the observed phenomena's time, space, and unit emergy by the scaling factor relating microcosm to the full-scale system. Underlying this theory is the general systems idea that processes of all scales are similar because of the energy principles guiding self-organization.

HEAVY METALS IN ENVIRONMENT

Other questions concern the way the lead cycle is changing in its relationship to society. What do we need to know to set public policy on heavy metals?

1. *How can human society reorganize the toxic impacts of heavy metals?* People can avoid concentrations of the heavy metals such as those in acid drainages and hot-spring waters around volcanoes. The principal toxic exposures now are leftovers from unwise uses of heavy metals in close proximity to people. The lead use in pipes started in Roman times. The tetraethyl lead that was an "antiknock" gasoline additive spread lead with motor vehicle exhausts on all the roadsides. With passage of the Clean Air Act of 1972, lead emissions from automobiles began to decrease in the U.S., and the lead concentrations in transportation decreased dramatically (Eisenreich et al., 1986). The action needed to stop uses that are not compatible with life and human society is well under way.

2. *Can heavy metal accumulations be made safe from human interaction including the natural ore deposits and industrial waste sites?* Currently there are toxic distributions and accumulations from the rush to use heavy metals in the past. These include mines left open to the surface waters and aquatic ecosystems. They include Superfund sites like the Sapp Swamp in this study. Digging up the metal residues and soils and putting them in a dump somewhere else is not a solution. Getting the biogeochemical cycle on a track to become part of the geological cycle of slow and

more isolated processes is the appropriate course of action. Processing Superfund wastes through regular technological metal reprocessing was recommended by Staley (1996).

3. *When can we increase the fraction of heavy metals supplied from reuse?* Well under way are the collection and return of heavy metals for reprocessing and resupply. This saves the earth reserves and closes the cycles, preventing environmental accumulations (Chapter 3). Florida, for example, requires lead, nickel, and cadmium batteries to be recycled for reprocessing, not cast into general solid waste disposal systems.

4. *Can the rate of use be balanced with a sustainable rate of supply?* Although recycling for reuse was substantial in 1997, the rate of use of ores of lead and other heavy metals was not sustainable. As the more concentrated ores become increasingly scarce, prices will rise, and more and more of the metals will have to come from recycle and reuse at greater cost. Then the former great economic stimulus of cheap abundant raw materials will be over. Decreasing availability of cheap critical materials and the rising costs of fuels will contribute to the leveling and decline of the global economy.

5. *Can society eliminate the dangerous concentrations of heavy metals left over from earlier ill-advised uses?* For example, lead-based paint, crumbling into dust from old houses, is dangerous if ingested, especially by children playing on the ground or on the floors. Lead paints have been banned from use in houses since 1978, and laws now require checks and information to be supplied in real estate transactions. Public health recommendations call for checking for lead in children.

ORGANIZATION OF HEAVY METAL CYCLES

6. *Is there an energetic principle of chemical self-organization that accounts for the natural distribution of the heavy metals?* The tendency for heavy metals to circulate and become concentrated high in the energy hierarchy of the earth and of the economy is related to energy and emergy (spelled with an "m") in Chapter 4. Many of the heavy metals of the earth are concentrated into ore bodies, the result of converging processes of the energy hierarchy of the global system. According to this concept, position in the energy hierarchy determines real wealth value, and justifies costs in processing and the kind of reuse or recycle that is appropriate.

7. *How does the precivilization cycle of heavy metals compare with the present cycle?* Apparently, the earlier system kept lead and other heavy metals bound in cycles of the land, keeping levels in the atmosphere, lakes, and oceans low. Is global toxicity ultimately due to letting heavy metals get into atmospheric and aquatic circulation?

8. *How can the use and processing of heavy metals by the economy make use of the natural principles guiding scarce elements?* Public policies on mining and scarce minerals now are largely guided by short-range economic markets. Emergy and emdollars are used in Chapters 4 and 11 as a better basis for evaluating and planning society's use of scarce metals in the long range. Alternatives are recommended that maximize the production and use of real wealth.

9. *What incentives help consumers recycle and reuse heavy metals?* Already, deposits are being required to facilitate returns for recycling, as in war times and earlier times of less affluence. Tax exemptions can be offered to the public based on emdollar values of recycle and reuse. The high value to the public of keeping the metals in appropriate reuse and out of the air and water cycles justifies public subsidy. Emergy-emdollar evaluations indicate how much is appropriate.

Incentives can also be used to help the natural processes regulate the environmental recycling. Conservation and special-use easements can be purchased where lands and waters need to be managed for the public interest and safety. For example, a public easement can be purchased so that a wetland containing heavy metals is not excavated. Special restriction easements are less expensive than outright land purchase; they allow private interests to use lands and waters for compatible purposes.

EVALUATION OF ALTERNATIVES

10. *Can the real-wealth value (emergy) of heavy metals be given an equivalent economic basis?* *Emdollars* are the dollars of the gross economic product made possible by an emergy contribution. Emergy values in solar emjoules are divided by the emergy/money ratio from Figure 1.9 to obtain emdollars. Chapter 4 contains examples of emdollar evaluation of lead.

11. *Can we define what is the appropriate best use for scarce chemical elements?* For any need, a desirable policy selects alternatives that achieve objectives with the least use of the most valuable materials. Here the emergy value measures of the accumulated work of nature and humans are better than the market values for measuring which resources are most valuable and scarce, to be used sparingly. See Chapter 4.

12. *How does the worth of scarce metals compare to strategic needs and safety of nations? On what basis can policies be defined for conservation?* Resources essential to military prowess and security were stockpiled during World War II, but used to accelerate the economy after the war. Both practices need to be put on a comparative basis. Economic use and defense contributions can both be evaluated on an emergy-emdollar basis, thus representing the real-wealth value to the public. Emdollar values of heavy metals are much higher than market values. A student report by Pat Dalton in 1986 found emdollars of defense about twice the economic value, partly because of the high transformities of strategic metals.

13. *Can we evaluate the present and potential role of wetlands and other natural filters in environmental self-regulation?* The discovery that many wetlands capture and denature toxic substances raises the possibility that safety for human society can be achieved with little cost by letting nature's ecosystems work for society as they have worked for the biogeosphere in the past. Chapters 11 and 12 evaluate examples. Also see Ton et al. (1998).

14. *How can the useful properties of peat for maintaining a healthy biosphere be evaluated when economic profit evaluations appear to justify destructive uses?* Lynne (1984) had difficulty finding any method of economic analysis that would prevent the nonrenewable use of Florida peats for agriculture. Odum (1996) found the Santa Fe Swamp in North Florida was worth 1.6 billion emdollars to the public, much more than if the peat were sold at market price. This evaluation used emergy procedures, which are based on the work contributed by nature. See Chapter 4.

TOXICITY

15. *Can measurement of heavy metals in the pore waters of sediments represent the tendency for metals to be toxic to organisms or to be released to open waters?* Pore water analysis is a recent approach to determining the hazard of sediments containing heavy metals (Campbell and Tessier, 1996). For example, they report an experiment by Ankley et al. (1991) that correlated mortality of benthic amphipods to the ratio of heavy metal to AVS (acid volatile sulfides). The idea is that more metal relative to potentially binding sulfide means more metal free to affect the animals. Yet there are many binding mechanisms in sediments and wetlands (particle adsorption, base exchange, complexing with humic acids, precipitation). Are binding mechanisms complementary so that there are usually backup mechanisms to bind heavy metals not bound by the first process?

16. *How should acute toxicity tests (mortality in a short time — <96 h) and chronic toxicity tests (mortality over a long period — >96 h) be used to manage lead in the environment?* Whereas acute tests can quickly determine dangerous levels, should they ever be used for Environmental Quality standards (EQ)? The time scale of human lives is much longer, and the time for accumulating impact is greater than for short-lived test animals. Being on a larger scale, humans are more sensitive to some stresses, such as ionizing radiations, than other species of plants, animals, and microbes.

17. *Can emergy per unit mass and/or transformity (the emergy per unit energy) be used to predict toxicity of substances?* Energy hierarchy theory finds the effect of something increasing

with its transformity (Chapter 4). Emergy per unit increases with scarcity, possibly because scarcity is an evidence of more embodied work (and thus higher emergy per unit). An old pharmacological principle states that drugs that are most beneficial are also the most toxic at the wrong dose. In a sense, environmental pollution is a state of materials in the environment in the wrong dose. Does greater potential impact (for good or harm) occur when transformity of substance and interacting receiver differ by one or two orders of magnitude? The smaller impact of toxic substances on small microcosms compared to large ones may be an example.

18. *Can present concepts of heavy metal risk be put on a simpler and more general basis using emergy?* The complex concepts of risk now in use to judge heavy metals in the environment evaluate "probabilities" of metal concentration, probabilities of sensitive organisms being present, and probabilities of the interests of society being impacted (Suter, 1993). Perhaps emergy and emdollars can evaluate a potential impact and thus rank priorities and reorganize the use systems to convert impacts into benefits.

19. *What index of overall ecosystem condition is the best indicator of stress due to heavy metals?* Indices of diversity of plants and animals have been consistent in indicating pollution. See, for example, diversity of diatoms growing on slides in streams (Patrick, 1949), Bender (1973) using graphs of species vs. individuals, and a review by Cairns (1977), who finds biodiversity a measure of ecological integrity. Many papers find lower species diversity with increasing heavy metals. There is toxic action, but also an effect of helping those species that are adapted to the heavy metals to prevail and dominate. For example, a heavy monoculture of *Scenedesmus* grew on a copper plate in Silver Springs, FL, rather than the high diversity periphyton growing on glass slides. See microcosm examples reviewed elsewhere (Beyers and Odum, 1993).

INITIATIVES IN INDUSTRIAL ECOLOGY

Industrial ecology is a recently popular word for initiatives from private industry to develop a stable and efficient system of material use and recycle. Socolow et al. (1994) say the goal of industrial ecology is "evolution of the world's industrial activity into suitable and environmentally benign systems."

20. *Can industries such as battery makers take over the whole cycle of processing, use, reuse, and control of environmental recycle?* In our time many people seek to privatize many public needs. Can the private industries that now produce heavy metal products take over the whole cycle of a heavy metal including recycle, reuse, and safe control of environmental releases? The products are at the high value end of the cycle, and incentives are provided by market prices and demand for use. Recycled remnants and wastes are at the more dispersed part of the cycle where there is little market demand or source of money. To manage privatizing for the public good may require emergy-based economic subsidies or tax incentives for the dilute part of cycles.

21. *How will changes in automobile transportation affect heavy metal use and environmental impact?* Urban growth in the 1990s meant more cars, more air pollution, and more consideration of alternatives to the heavy lead batteries. The Paul Scherrer Institute (1998) compared the electrical energy per unit mass (watt hours per kilogram) from new, lighter-weight batteries. Older lead batteries delivered about 35, nickel–cadmium batteries 55, nickel–hydride 76, and zinc–air batteries 100. A net emergy evaluation (which includes changed monetary costs) is needed to see if the lighter batteries can be a net benefit. With an industrial ecology perspective Graedel and Allenby (1998) document the central role of automobiles in our civilization. They found the lead batteries still with the least cost ($70 to 100 /kWh). A switch in metals would change the kind of heavy metals impacting the environment, but those most considered are all filtered by wetlands.

Will electric cars increase heavy metal risks and costs? Is this transportation sustainable? A Carnegie-Mellon University study by Lave et al. (1995) on "environmental implications of electric cars" stated that electric car mandates like one proposed in California may expose thousands of people to toxic doses of lead while unnecessarily accelerating the demise of an important non-renewable resource (Berriman, 1995). Lead released with an electric car system was stated to be larger than the former releases by leaded gasoline. (See rebuttal letters in *Science*, 1995, 269: 741–742.) Previous evaluations of electric cars indicated they were not a net benefit and thus not likely to prevail (McGrane, 1993).

22. *What is the pattern of heavy metal use and processing appropriate and economical for a time of less fuel availability and contracting economy?* For the future, what is the long-range scenario for use of metals? Will there be a climax accumulation of people and machines on the earth, followed by decreases as the fuel energy basis for the high concentrations decreases? What are the heavy metal *needs*?

When the shortages of nonrenewable resources force the global economy to a lower level, will the uses of heavy metals be less? Will there be more conservation and reuse, with diminishing environmental toxicities? An evaluation of global fuel reserves by Campbell (1997) estimates fuel prices generally rising after the year 2009.

Initiated by Georgescu-Roegen (1977) is the idea that civilization could become limited more by the critical materials, such as heavy metals, than by the fuels. However, ecosystems concentrate critical materials several orders of magnitude as needed so long as they have their energy budget. In our society more and more fuel energy is being used to recycle, reconcentrate, and reuse the heavy metals as the concentrations available for mining diminish. Emergy evaluation allows energy reserves and reserves of critical materials to be compared on an equivalent basis. Both types of nonrenewable resources are supporting the ultimate pulse of this civilization now under way.

With the questions of this chapter in mind, consider next some of the published literature on heavy metals and environment in Chapter 3.

Background of Published Studies on Lead and Wetlands

CONTENTS

In this chapter we review some of what is known about lead and its relation to wetlands. Headings are arranged somewhat chronologically. The *mining and use of lead* began in early society and later developed into the lead industry. A scientific understanding of the *chemistry of lead* developed with research in the 19th and 20th centuries. Industrial technology developed for concentrating lead. Early recognition of the *effects on health* was proven with medical studies. After greatly increased *uses of lead* developed with civilization, chemical analytic methods made possible analysis of the concentrations and movements of *lead in the environment*. Data on lead concentrations and toxicity identified places of human and environmental health risk. Then measures were sought in technology and wetlands for removing lead from the environment of humans. Finally ecological economic methods were developed to *evaluate alternative choices* in use and dispersal of lead.

Biogeochemistry, biology, and ecological cycles of lead were reviewed in detail by Nriagu and associates (1978a), including summaries of numerical data. Here we update these summaries.

MINING AND USE OF LEAD

Metallic lead has been used by humans for about 4000 years. Easily crafted and combined with other metals in alloys (such as pewter), lead was used for food and water containers and for water pipes in ancient Rome and elsewhere before its toxicity was understood (Nriagu, 1983). In their Figure 55, graphing the production of lead starting 5000 years BC, Salomons and Förstner (1984) show 10,000 tons/year used in Roman times. By 1968 world production was 3 million tons/year (Minerals Yearbook, 1968).

Now lead and its compounds are used for ammunition, solder, batteries, paints, and pigments. Nearly 80% of lead consumed in the U.S. in 1989 was destined for use in storage batteries (Gruber, 1991). The rate of recycle of lead from car batteries for reuse has varied between 60 and 96% over the past 30 years (Putnam Hayes and Bartlett Inc., 1987), giving lead one of the highest recycle rates of any domestic commodity (Gruber, 1991). Up until the late 1970s, most batteries collected for recycle were shipped first to a "battery breaker" or "battery cracker," who sawed or crushed the battery casings, drained the acid, and extracted the lead plates, which they sold to a secondary smelter (Gruber, 1991). Behmanesh et al. (1992) found 80% of the lead going to hazardous waste incinerators in the U.S. came from two secondary smelters.

Through tougher environmental regulations, most of the rather crude battery-breaking operations closed during the late 1970s, and secondary lead smelters took over the battery-breaking process. Secondary smelters generate three main waste streams: battery casings, process wastewater, and lead slag. Plastic battery casings can be washed and recycled (Neil Oakes, personal communication). Older rubber battery casings can be used as feedstock for the smelter furnace; otherwise they must be shipped to a hazardous waste landfill (Gruber, 1991). Battery acid is impure and is typically not recycled. Process wastewater is therefore very acidic and contains dissolved and particulate lead (Watts, 1984). Neutralization, precipitation, and filtration processes are used for treatment (Gruber, 1991). Lead slag fails certain tests mandated by the Resource Conservation and Recovery Act (RCRA), so it must be disposed of as a hazardous waste.

Whereas present automobiles are fuel driven, using batteries only for starting and stabilizing the car's electric functions, electric cars run on battery electricity and require many more batteries for each car. However, there is doubt that electric cars can replace fossil fuel-powered cars except where electric power is in excess from nuclear or hydroelectric sources. Converting fuel to electricity and then to car operation is not efficient compared to running cars on fuel directly. The future use of lead batteries may depend on how widespread will be the use of other kinds of batteries, such as the nickel–metal–hydride battery, or innovations based on fuel cell technologies. Lead ores are a nonrenewable resource, and future uses of lead have to be increasingly based on recycling and reprocessing.

RELEVANT CHEMISTRY OF LEAD

In the earth's crust lead is widely distributed as a trace element (16 ppm [parts per milliom, milligrams per kilogram], according to Goldschmidt cited by Kuroda, 1982). Trace lead substitutes for ions of similar size in mineral crystals, potassium in feldspars, and calcium in basic rocks. Where lead concentrations are higher, often in reduced conditions, the mineral galena (lead sulfide) develops, and this is the main commercial source of lead.

In the laboratory or in the environment, lead in solution often reacts with sulfide, carbonate, or phosphate that may be present and precipitates as a solid, depending on the acidity (measured as pH) and oxidation-reduction potential (measured with electrodes as volts) (Garlaschi et al., 1985; Harper, 1985; Lion et al., 1982; Rea et al., 1991; Rudd et al., 1988; Salomons and Förstner, 1984; Sheppard and Thibault, 1992).

Huang et al. (1977) list 12 chemical equations and their equilibrium constants commonly involved with lead in the environment, including reactions with hydroxides, oxides, sulfides, sulfates, and carbonates. Their graphs show the attachment of lead to negatively charged solid surfaces increases sharply above pH 5, but is decreased somewhat by competition from binding by soluble organic substances and metal chelates.

Partition of a heavy metal among its chemical species depends on oxidation-reduction potential and on sediment texture and mineralogy (Gambrell et al., 1980). The valence state of lead (+2) is not changed by the range of redox potentials in most environments. However, higher oxidation potential may increase lead mobility by oxidizing insoluble sulfides, a process which also lowers pH (Gambrell, 1994). Where oxidation potential and pH are high, lead may deposit along with iron and manganese in the hydroxide form.

Moore and Ramamoorthy (1984) summarize the chemistry of lead, some of the compounds and valences ("species" of lead) found in the environment. $Pb(OH)$ is found in the sea, soluble between pH 6 and 10. $PbCO_3$ was found in river sediment as particles. Lead is methylated by microbes.

Harrison (1989) describes the widespread circulation of alkyl lead compounds in the biosphere with some industrial, automotive, and environmental processes of methylation, converting inorganic lead (divalent lead) into dialkyl lead, trialkyl lead, and tetraalkyl lead. Other processes degrade the methyl lead compounds back into inorganic lead. Some industrial processes release tetravalent lead (+4).

Patterson and Passino (1989) edited a summary of the speciation of metals. Mathews (1990) showed that high temperature incinerators vaporize lead, and if chloride is present, lead chlorides form, limiting solid formation, releasing lead as $PbCl_2$ to the atmosphere.

Fergusson (1990) summarizes forms that lead takes in the environment as a function of pH and Eh (oxidation potential). $PbCl_2$ is insoluble and $PbCO_3$ and PbS almost insoluble. Valence is greater at higher pH. In air, water, and sediment, organic-lead complexes change from tetravalence to lesser valences to inorganic lead. The lead/calcium ratio declines in the food chain (from rocks to sedge to animal). He quotes Nriagu (1978) that weathering of granite produces a profile of 200 ppm lead. A diagram summarized the global flows and pools of lead.

Senesi (1992), with spectroscopic methods, found lead and zinc competing for hard ligands.

Properties of heavy metals were compared (Tessier and Turner, 1995). Residence time is proportional to assimilation efficiency, with lead having a low efficiency and low residence time. The coefficient of variation is 16 for lead in the clam, *Scrobicularia plana*.

In solids, trace metals with similar sized atoms tend to be found together.

The ionic radius of lead is 0.099 nm and calcium 0.12 nm.

Holm et al. (1995) provided a method for separating species of zinc in low concentrations.

Lead and Humic Substance

Lead becomes attached to humic substances. One third of trees consists of lignin, that holds fibers together. When trees decompose, brown humic material from the breakdown of the lignins is released into soils, peats, and waters (example: black water streams). Humic substances are a mixture with a wide range of molecular size and properties, classified into three groups: fulvic acids, humic acids, and humin. These groups are defined according to their response to pH (acid–base scale), which affects their molecular structure, causing precipitation. Humin is insoluble when extracted in a basic solution, as well as in acid solution, while fulvic and humic acids stay in solution. With more acid added, humic acids precipitate, and fulvic acids remain in solution (Stevenson, 1982). Humic acids have a molecular weight ranging from 50,000 to 100,000 AMU (atomic mass units), with some having molecular weights over 250,000. Fulvic acids, on the other hand, have weights between 500 and 2000 AMU (Stevenson, 1982).

Vedagiri and Ehrenfeld (1991) studied lead binding in humic waters from Atlantic White Cedar Swamps with sphagnum mosses in New Jersey pinelands and determined chemical fractions. They recognized soluble lead if particles were less than 0.45 µm (10^{-6} m) and filterable lead if particles were greater than 0.45 µm and caught by a membrane filter. The soluble portion was then subdivided into: (1) labile soluble lead (here labile means that the lead is loosely bound to soluble molecules); (2) nonlabile humic soluble lead (lead tightly bound to photooxidation-sensitive small humic and fulvic molecules); (3) nonlabile soluble lead (lead tightly bound to soluble inorganic and organic compounds). The concentration of free divalent soluble lead in water was significantly greater at lower pH. The quantity of larger molecules associated with lead increased with pH, and with increased dissolved organics. Lead adsorption on clays increased with pH above 6.0, where there is less competition from hydrogen ions for negatively charged binding locations. For this experiment the authors found most of the insoluble lead was sensitive to photooxidation by the sun.

Leaching Procedure for Testing Toxicity

A procedure named TCLP (Toxicity Characteristic Leaching Procedure) has been required by federal agencies for classifying certain solid and liquid wastes as hazardous. Sediment or waters leached at pH 4.93 and 2.88 are designated hazardous if lead concentrations exceed drinking water standards by a factor of 10. This index overestimates toxicity where the environmental conditions are at high pH and oxidation potential as in some marine sediments (Isphording et al., 1992).

LEAD TOXICITY AND HEALTH

Posner et al. (1978), Rosen and Sorell (1978), Chang et al. (1984), and Moriarty (1988) reviewed lead uptake and effects on people. High concentrations of lead that are toxic sometimes come from naturally occurring processes around ore bodies, sometimes from human activity such as mining and smelting, from lead pipes and plumbing adhesives, from utensils made of pewter (lead alloy), lead solder, lead-glazed pottery, and stained glass windows, from dumps containing products made with lead, from decomposing lead-based paints, and places where there are automobiles using gasolines with lead additives (tetra-ethyl lead). The National Lead Information Center can be contacted at 800-LEAD-FYI.

Lead is a physiological and neurological toxin to humans. Acute lead poisoning results in dysfunction in the kidneys, reproductive system, liver, brain, and central nervous system, resulting in sickness or death (Manahan, 1984). Environmental exposure to lead is thought to cause mental retardation in children (Jaworski et al., 1987). It can particularly affect children in the 2- to 3-year-old range. Other chronic effects include anemia, fatigue, gastrointestinal problems, and anorexia (Fergusson, 1990). Lead causes difficulties in pregnancy, high blood pressure, and muscle and joint pain. Drinking water quality standards for lead in most developed countries and for the World Health Organization are a maximum of 0.05 mg/l (van der Leeden et al., 1990) and are likely to be reduced to lower levels.

Forbes and Sanderson (1978) reviewed lead toxicity in domestic animals and wildlife; Wong, et al. (1978) summarized lead in aquatic life.

Toxicity to waterfowl from lead shot has been extensively studied (see review by Eisler, 1988), but toxicity from other forms of lead contamination is less well known. Birds fed diets of up to 100 mg of lead per kilogram of diet (dry weight basis) showed elevated lead body burdens but apparently no symptoms of toxicity.

At moderate concentrations (1.0 to 2.0 mg/l) lead was found to increase the growth of water fern (*Azolla pinnata*) and duckweed (*Lemna minor*), but phytotoxicity was found at higher concentrations (4.0 to 8.0 mg/l) (Jain et al., 1990). Lead removal was noted for both species, and saturation effects were observed.

Ruby et al. (1992) found that the form of lead in soils made a large difference in the lead absorbed from the acid stomach as soils were ingested and passed through. Lead in urban soils was more available and toxic than that from soils around mines in Butte, MN.

Ruby et al. (1992) found human toxicity to lead affected by solubility of lead ingested into the intestinal tract. Uptake from complexes of lead in mined soil including the mineral anglesite ($PbSO_4$) and galena (PbS) was slower than in experiments that used pure crystalline lead sulfate.

SOURCES OF LEAD

Lead is widely distributed in air, waters, and land as a trace element. As summarized by Kesler (1978), lead ores form from hot solutions around sulfur-rich magma, deep sedimentary rocks under pressure, and replaced limestones. Galena (lead sulfide) is the dominant mineral in lead ores where lead may be 7%. Known reserves are about 140 million tons. High concentrations of lead are found in and around these ore bodies, veins, and associated waters such as hot springs (20 to 1800 µg/l).

Ward et al. (1977) found lead in the vicinity of a New Zealand battery factory lead smelter to be much greater than lead from motor vehicle exhaust.

Chow (1978) cited examples of mining and industrial wastes with 500 to 140,000 µg of lead per liter, with various treatment processes removing 99%.

Summarizing many papers Nriagu (1978) found 100 to 67,800 ppm lead in street dusts. Stormwater runoff contained 100 to 12,000 µg/l. Lead in sewage varies from 0.010 to 0.5 ppm/l or more in industrial areas.

Stephenson (1987) details sources of lead in wastewaters. The U.S. EPA Toxics Release Inventory (1989) summarized industry-reported lead releases and transfers in 1987, including both routine and accidental releases. The total reported lead released directly to air, surface water, and sewage treatment plants was 1.5 million kg. Aquatic lead pollution is often associated with acid pollution as in acid mine discharge. Also, acid electrolytes used in battery production are a problem in reclamation.

Mathews (1990) describes volatile lead losses from high temperature hazardous waste smelters which then condense on fly ash, on slag, and elsewhere in the environment. With chloride present, lead chlorides form before solid lead.

Callander and Van Metre (1997) summarize the dramatic decrease of 98% in lead emissions in the U.S. as lead additives to gasoline were phased out. In 1970, 182 kilotons of lead per year were released to the atmosphere. By 1992, emissions were 2 kilotons/year from vehicles and 3 kilotons/year from industrial sources.

LEAD DISTRIBUTION IN THE ENVIRONMENT

Lead released from economic activity is found in air, water, and the land. Many studies show surface horizons of high lead concentration in soils, sediments, glaciers, and stratified waters throughout the world, recording the maximum surge of lead emissions from cars and industry earlier in this century. Farmer (1987) provided an annotated bibliography of lead from motor vehicles.

Lead in the Atmosphere

Nriagu (1978) reviews data on lead in the atmosphere, the balance between emissions and fallout, with a turnover time of 2 to 10 days. Auto emissions, especially from cars with leaded gasolines (prior to the phase-out of leaded gasoline), contributed to atmospheric lead pollution, which then went to waters and lands. Friedlander et al. (1972) found 75% of lead in gasoline emitted to the atmosphere. By the 1990s, however, leaded gasoline was little used in the U.S. There was an estimated 1333 billion g annual lead production with 1.8% released to the environment and emissions to air as 0.063%. Emission from cars was given as 22 mg of lead per kilometer of road.

Lead in Waters and Sediments

Earlier work on the fate of heavy metals in aquatic systems was on the chemical reactions involved (Huang et al., 1977; Vuceta and Morgan, 1978; Brown and Allison, 1987). Although the fate of chemicals is dependent on chemical equilibria, mass balance, and microbial transformations on a time scale of days and years, the rate-limiting processes are more likely to be the larger-scale compartment storages and cycling processes rather than the chemical reactions per se (Nriagu, 1978). See models and evaluated diagrams in Chapter 4. Moore and Ramamoorthy (1984) reviewed papers on heavy metals in waters with a chapter on lead. Lead concentrations in freshwater sediments ranged from 20 ppm in natural arctic lakes to 3700 ppm in lakes near metal mining and 11,400 ppm in a Norwegian fjord receiving wastes. Furness and Rainbow (1990) review heavy metals in the sea, its algae, and animals, toxicity, and human exposure.

Förstner and Wittmann (1979), quoting Schaule and Patterson (1979), show distribution of dissolved lead to be 5 to 15 ng/kg in upper waters in the Northeast Pacific Ocean, decreasing with depth, a result of recent introductions from the air.

Förstner and Wittmann (1979) quoted Koppe that 95% of the lead in released salts was taken up and immobilized from waters flowing 70 km in the Ruhr catchment.

Nriagu et al. (1981) found concentrations of five heavy metals in particles to be equal to their concentration in the water within a factor of 2.

Förstner and Wittmann (1983) and Chow (1978) reviewed information on the distribution and geochemical cycle of lead in waters and sediments. There were large increases in the lead in recent snows on glaciers (increase from 0.01 to 20 μg/kg), in lakes, in surface waters of the sea, and in recent sediments derived from these waters. Chow found the lowest lead concentrations in seawater determined by the lead in suspended mineral particles such as manganese oxides where lead substitutes for manganese. Depending on pH, dissolved and colloidal lead may be present combined with chlorides, sulfates, and hydroxides. Below 1000 m the ocean's lead was about 0.2 ppm. Estimates of the lead cycle are in Chapter 4.

Simpson et al. (1983) found lead in runoff waters was taken up by soils of tidal wetlands in Delaware (236 to 300 μg/g), with higher values near storm drains (400 to 2260 μg of lead per gram).

Ten papers by Nriagu (1984) on the Sudbury Ontario smelter area were included. In the Sudbury lakes, Yan and Miller reported a lower diversity of aquatic plants.

Rygg (1985) found diversity of benthic fauna increasing with heavy metals in marine sediments of fjords with heavy metals.

Purchase and Fergusson (1986) found lead runoff from a battery factory and street dust in Christchurch, New Zealand was captured by river sediments (90 to 80,000 μg/g), not much reaching the estuary (2.7 to 26 μg lead per gram). Most of the lead was in the form of lead carbonate, sulfate, and sulfide mineral crystals.

Windom et al. (1988) found 50 to 350 pmol/kg transported in an estuary in Thailand. After removal from waters, metals were regenerated from organic matter.

Mobile Bay, in Alabama, is an example of the high lead concentrations in many river mouths and estuaries (Isphording, 1991). The lead flux to oceans from rivers has more than doubled as a result of human activities (Fergusson, 1990; Garrels et al., 1975). This increase is small compared with the increase due to direct atmospheric deposition on the oceans, but the contribution from rivers will become more important as atmospheric lead pollution is more closely controlled. In the range of pH 5 to 6.5, Gambrell (1994) found that oxidation reduction potential made little difference in exchange of lead with bottom sediments of Mobile Bay. Already a downward trend in lead concentration in rivers of the U.S. has been correlated with the reduction in lead additives in gasoline (Smith et al., 1987).

Borg (1995) describes the two orders of magnitude lower values of lead in natural waters compared to analyses 10 years ago which were often contaminated by collecting and processing methods. In Swedish lakes, 1 ppb (part per billion) lead (0.1 to 2.7 ppb) was often in the organic complex, whereas zinc was in soluble form (0.5 to 25 ppb).

Jenne (1995) found zinc that is absorbed by marine sediments reduced by half with a dose of penicillin to inhibit microorganisms.

De Gregori et al. (1996) found unsafe levels of lead, zinc, and copper in filter feeding marine mussels and sediments in estuaries of Chile.

Beyer et al. (1998) found 880 ppm in feces of swans feeding in the lead-rich mining areas of the Coeur d'Aleve River in Canada compared to 2.1 ppm in reference areas.

Lead on Land

In their review of geochemistry Rankama and Sahama (1950) noted similarities in the ionic radius of calcium, lead, and strontium to account for 33 ppm lead in American limestones and dolomites, and 20 ppm lead in calcareous coral reefs, which also concentrate strontium. Evaporite deposits contain 1 ppm associated with calcium sulfate. Basic igneous rocks contain 5 to 9 ppm with 9 to 30 ppm in granites.

Lead in the land reflects the geological history of the base rock, higher in ores, developed in association with mountain building and volcanism. Lead distribution in the earth's crust before industrial development was summarized by Nriagu (1978). Smaller concentrations of lead in ultramafic and basaltic rocks (2 to 18 ppm) increase with feldspars to more alkaline rocks (31 to 495 ppm). Lead is concentrated in the weathering process. Lead concentrations (1 to 400 ppm) are found in shales and other sedimentary rocks. Coals contain 5 to 99 ppm and oil 0.04 ppm. Mine tailings and battery processing contribute lead to the surface landscape. However, Allen (1995) quotes a 1995 EPA report that all primary lead production in the U.S. is now 99% efficient or better (1% or less left in the environment).

Palm and Ostlund (1996) estimate pools of storage and the budget of flows of lead and zinc into and out of the city including the sewage system of Stockholm, Sweden.

Lead in Soils

Jennett and Linnemann (1977) found lead absorbed at the top of soil columns in laboratory and in kaolinitic soils in the field around lead smelters in Missouri (1307 µg/g). Lead absorbed approached 100% of the cation exchange capacity. Little lead was leached or transported by distilled or rainwaters, but some lead was desorbed by humic solutions with chelating capacity.

Stevenson (1986) found zinc 2 to 50 ppm in soil, with some samples to 200 ppm and more from limestone.

LaBauve et al. (1988) found little lead leaching from soils and lake sediments by percolating a synthetic landfill leachate.

Harrison (1989) found that lead emission from an English highway was 281 g/m of highway per year, of which 14 g was in drainage waters.

Kuiters and Mulder (1990) describe leachates from forest soils starting as polysaccharides and polyphenols, which form metal complexes and then are changed into fulvic acids. Organic lead concentrations are correlated with ionic strength, with metal-complexing capacity, but inversely correlated with pH.

Herrick and Friedland (1990) found 106 ppm lead and 18 ppm zinc in forest soils in the Green Mountains of New England, less than in analyses made earlier.

Sheppard and Thibault (1992) found desorption of 70% of lead in sandy soil by EDTA chelating agent, but retention of lead in organic soils of reed-sedge peat. Since residual lead fractions are tightly bound, complete lead removal was considered costly.

Krosshavn et al. (1993) compared heavy metals in podsoils formed from different ecosystems, where 99% of lead remained bound at the natural acid pH, and where 97% was bound when soil suspensions were adjusted to pH 4 and 95% at pH 3. Binding of lead was similar in soils from spruce, pines, and oak forests, but 60 to 72% in peats from wetlands (fens and bogs).

Miller and Friedland (1994) considered the decrease of lead in northern forest soils following the decline of atmospheric rain-out of lead since the leaded gasoline maximum in 1980. They calculated lead removal response times (turnover times) as 17 years for northern hardwood forests and 77 years for subalpine spruce–fir forests.

Gambrell (1994) found more lead available to plants and to leaching in acid, oxidized upland soils.

To determine the differences in natural fractionation and polluted fractionation of lead in soils (vicinity of lead smelters), Asami et al. (1995) compared 38 samples from 11 different soil profiles in Japan. Of these profiles 8 were from wetland paddy fields. Lead in topsoil and subsoil of unpolluted soils was 30 and 22 ppm, respectively, and in polluted soil 237 and 130 ppm, respectively. Less than 10% of the lead was soluble. In both polluted and unpolluted soils, relatively high portions of the lead were bound by organic sites (70% of lead in the polluted soil). Polluted soils had a significantly higher percentage of lead bound to inorganic sites.

Dong (1996) reports that colloidal particles containing lead can migrate through soils depending on organic and iron content.

Lead in Plants

Reddy and Patrick (1977) found water-soluble lead and its uptake by rice plants decreasing when pH and oxidation potential were experimentally increased.

Chumbley and Unwin (1982) found only small uptake of lead by 11 vegetable crops (means: 0.1 to 2.9 ppm of lead) from land containing sewage sludge (means: 97 to 214 ppm).

Whitton et al. (1982) found lead uptake and concentration by the aquatic liverwort *Scapania* useful as an environmental monitor. Lead increased in plants from 100 to 50,000 µg/g as a function of the concentration in water increasing from 0.003 to 1.0 µg/g.

Lead uptake by sea grasses was positively correlated with temperature and inversely correlated with salinity (Bond et al., 1988). Higher temperatures and distilled water increased the accumulated lead, and there were slight variations among different species (*Zostera, Halophila, Heterozostera, Lepilaena*).

In a study of estuarine eel grass from Denmark, Lyngby and Brix (1989) found highest lead concentrations in the oldest root structures. Above ground the oldest leaves contained the highest levels of lead, similar to that in dead attached leaves. They described lead binding to the outer surface of the root in a crystalline form, as well as being sequestered in the cell walls. Concentrations of lead increased with age of the plants and during decomposition, some lead being absorbed from the water. Where there was 41 ppm in roots, leaves were 2.9 to 13 ppm.

Pahlsson (1989) reviewed the literature on lead in plants. Apparently low concentrations of lead stimulate plant growth, although lead is not essential to function in plants. Roots accumulate large quantities of lead, but little is translocated to aerial shoots. Lead is bound at root surfaces and cell walls. Lead toxicity to various plant species varies over a wide range of concentration (100 to 1000 µg/l in solution; 5 to 100 µg/g soil; 19 to 35 µg/g plant). Toxicity is reduced by phosphate; 21 to 600 µg/l interferes with mitotic processes and cell divisions. Seed germination is little affected by relatively high lead concentrations (20,000 µg/l). High levels decreased activity of the enzyme d-aminolevulinic acid dehydratase by interacting with SH groups. Organic lead compounds (tetra-ethyl lead) are toxic to forests. Mycorrhizal plants are more resistant.

Kuyucak and Volesky (1990) reviewed concentrating bioabsorption of lead and zinc by many kinds of algae, and its toxicity to the cells. Zinc is a necessary trace element at low concentration and toxic at higher levels.

Using red maple and cranberry seedlings, Vedagiri and Ehrenfeld (1991) tested the bioavailability of lead and zinc in microcosms. They concluded that the plant community as well as the soil and water characteristics play a role in the uptake of metals. Lead was "strongly immobilized" in plant cell walls. In the case of maple the presence of *Sphagnum* decreased the uptake and concentration of metals in tissues of the seedlings. The opposite effects were observed for the cranberry seedlings.

Gupta (1995) compared heavy metal accumulation in three species of mosses in India where leaded gasoline was in use, finding an urban–suburban gradient (66.4, 52.3 µg lead per gram in *Plagiothecium*; 40.7, 35.1 in *Bryum*; 28.4 in *Sphagnum*).

Eklund (1995) found lead in the wood of oak tree rings near a lead reprocessing plant in southern Sweden to be a good indicator of the local environmental history of lead. Concentration in trees near the plant reached 3.5 ppm of lead. In distant trees lead ranged from 0.02 to 0.2 ppm during the time of maximum lead-fall from the atmosphere.

King et al. (1984) added lead minerals (cerrusite and anglesite) to soils growing pine, spruce, and fir, causing more lead in plants (50 to >5000 ppm in ash with the ash 2 to 6% of dry weight).

Diaz et al. (1996), studying *Glomus* mycorrhizae from pine forest, applied lead and zinc treatment (0, 100, 1000 ppm), examining the resulting growth of leguminous trees. At high dose, plant growth was less, and there was less lead, zinc, and phosphorous uptake into plants.

Lead Uptake by Other Organisms

With summary tables, Eisler (1988) reviewed 300 papers on lead uptake in fish and wildlife. Values ranged from 1 to 3000 ppm dry weight depending on proximity to lead sources.

Microorganisms and algae may accumulate lead from the water column (Jaworski et al., 1987). Kelly (1988) reported enrichment ratios for algal uptake of lead from 1000 to 20,000. This may be due to the relatively large surface area of these tiny organisms. Lead adsorbed on low molecular weight particles may be taken up by animals, especially filter feeders (Jaworski et al., 1987). Thus, lead can enter biomass as ions, organo-lead molecules and complexes, or with ingested particulate matter (Rickard and Nriagu, 1978).

Luoma and Brown (1978), cited by Moriarty (1988), found lead in marine mollusks increasing with that of the sediments of their environment. The correlation was improved by using lead/iron ratios. Since lead in *Fucus* algae, which take up soluble lead, was not correlated, the clams may have been getting lead from ingested sediment particles.

Beyer et al. (1982) found earthworms from soils with sewage sludge application had only 1.2 times more lead (10 to 23 ppm of dry weight) than in control sites. Lead in shell was 13 to 27 µg/g.

Bourgoin et al. (1989) found lead uptake (150 to 332 µg lead per gram) by marine mussels (*Mytilus*) in three stations in a harbor in Nova Scotia to be inversely correlated with the industrial phosphorous waste releases there.

Siegel et al. (1990) found fungi taking up lead: 40 µmol/kg by *Penicillium* and 160 µmol/kg by *Cladosporium*.

In mushrooms near mercury and copper smelters, Kalac et al. (1996) found 26.4 ppm lead in *Lepiota procera* and 15.3 ppm in *Lepiota nuda*.

Garcia et al. (1998) reported lead ranging from 10 ppm in the mushroom *Coprinas comatus* near city center ranging down to 2 and 1 ppm in pasture and forest. Concentrations were higher in saprophyte mushrooms than in mycorrhizal fungi.

Absence of Lead Concentration by the Food Chain

Jaworski et al. (1987) and Förstner and Wittmann (1983) did not find concentration of lead in the food chain (biomagnification). There was less lead concentration at the top in marine food webs (Jaworski et al. 1987 quoting Patterson, 1980), in aquatic grazing and detrital food webs (Eisler, 1988), and in terrestrial grazing and detrital food webs (Grodzinska et al., 1987). Some larger animals at higher trophic levels with lead concentrations may have accumulated concentrations over their longer life span.

Simkiss and Taylor (1995), studying the clam *Scrobicularia*, found lead with short residence time and, accordingly, a low accumulation efficiency. Coefficient of variation was 16 for lead, with different values for other heavy metals.

Lead with Wastewater Irrigation

Sidle et al. (1977) analyzed the heavy metals taken up by clay loam soils when canary grass and corn crops were irrigated with wastewaters. Waters contained 140 µg/l and applied 36 to 41 lb/acre of lead over a 3-year period; soils contained 3.1 to 6.1 µg/g of soil without much difference with depth.

In irrigation canals supplying waters to rice, Chen (1992) found 2.1 to 2.4 ppm lead in Japan and 0.12 to 3.6 ppm in Taiwan.

Lead with Sewage Sludge Application

As reviewed by Nriagu (1978), sewage sludge was found with an average of 100 ppm lead, and 4 to 1015 ppm lead in topsoils receiving sewage sludge. Weathering of rocks generates soils with 20 to 200 ppm lead. Solution of limestones may concentrate lead.

Overcash and Pall (1979) found 2 to 20 ppm lead in coal, but 720 to 1630 ppm lead and 2170 to 3380 ppm zinc in sewage sludge. The EPA recommends limits depending on the cation exchange capacity of clays, allowing more lead where there is more exchange capacity of clays. Above pH 7 almost 100% of lead was bound on clay minerals (kaolinite) in competition with various valences of lead hydroxide.

Chumbley and Unwin (1982) studied the lead uptake by vegetable crops grown on soils (97 to 496 ppm of lead) with history of sewage sludge application. Lead in 11 crops was 0.1 to 3.7 ppm not correlated with soil lead.

Chang et al. (1984) studied heavy metals on soils growing barley plants before and after adding sewage sludge from Los Angeles. About 82% of the soil lead was extractable with EDTA and inferred to be in carbonate form.

Levine et al. (1989) studied heavy metals accumulating in old field succession where commercial, heat-treated sewage sludge (milorganite) was added for 10 years. Lead was not concentrated in the leafy parts of plants, but lead and zinc were concentrated many times in earthworms.

Juste and Mench (1992) found heavy metals accumulating with sludge applications to agricultural soils but remaining in the upper 15 cm.

McBride (1995) reviews research on heavy metal availability and toxicity to agricultural plants on land receiving sewage sludges, and questions safety of practices and regulations on soil loading which permit 300 ppm of lead. Milligrams per kilogram were converted to kilograms per hectare using a factor of 2. Although lead uptake in corn leaves was small, McBride found regulations for lead levels in soils receiving sewage sludge set too high for safe agriculture because older soils release lead initially bound.

Luo and Christie (1997) found that lime stabilized heavy metals in sewage sludge down to 100 ppm, with little effect at 33 ppm. EDTA-extracted lead was proportional to the soil lead (250 ppm range). Liming reduced zinc uptake.

Berti and Jacobs (1998) studied the lead budget in Michigan soils fertilized with sewage sludge recovering 45 to 155% of that added. Lead remained in the upper 15 to 30 cm, with very little removed by plant uptake, soil movement in tillage, deep leaching, or wind erosion.

Release of Lead from Sediments into Waters

If oxidizing conditions are experienced, degradation of organic materials may result in release of some lead, which may diffuse through pore water to the sediment surface, where it may be caught by hydrous oxides (Nixon and Lee, 1986) or suspended particles. Although lead flow is largely unidirectional into sediments (Rickard and Nriagu, 1978), the released lead may precipitate back slowly and be subject to export.

Mass lead movement has been associated with turbulent transport (Everard and Denny, 1985 in Jaworski et al., 1987). Rickard and Nriagu (1978) also report desorption of trace metals due to dilution where lead-rich surface waters are continuously flushed by incoming low-lead water.

Windom et al. (1988) evaluated dry season and wet season behavior of trace metals in Bang Pakong estuary in Thailand. During high runoff, lead was 0.3 to 67 mol/kg; at low discharge, lead was 0.1 to 12 mol/kg. Most of the lead was removed with the organic matter.

Borg and Johansson (1989), studying heavy metals, found 4 to 10% of the lead in rainfall on forests was transported into Swedish lakes mainly with humic substances.

Paulson et al. (1989) estimated the budget of lead flowing in and out of waters and sediments of a section of Puget Sound, Washington, from natural and anthropogenic sources during 1980–1983. Of the 109 metric tons of lead, 45% was from municipal and industrial sources, 21% from the atmosphere, and 16% from rivers. Most of the lead (72%) went to the sediments, whereas 28% was passed down the estuary. Manganese was correlated with lead; manganese precipitating at the sediment surface may have helped capture lead.

Clevenger and Rao (1996) conducted experiments to represent field conditions where solid wastes were on top of old dolomitic mine tailings in Missouri (810 to 1280 ppm lead, pH 8.1). Leachates from solid wastes moved 440 ppm lead in an hour.

LEAD IN WETLANDS

Because their soils are anaerobic when wet and many substances are in states of low oxidation potential, heavy metal behavior differs from that in ordinary soils (previous sections). As reviewed by Gambrell (1994), previous studies show more efficient uptake and binding of lead in wetlands than uplands. Literature summaries and discussion of mechanisms of lead binding follow.

Valiela et al. (1974) found *Spartina* in salt marshes taking up heavy metals in the growing season, returning them to organic sediments during the winter.

Hirao and Patterson (1974, in Thibodeau and Ostro, 1981) reported that "wetlands in the High Sierras retain 98% of the 9 grams per hectare per year (0.008 pounds/acre/year) of lead aerosol which reaches them from west coast air pollution."

Kelly et al. (1975), cited by Nriagu (1978), found 120 kg of lead per hectare in the upper soil layer of marsh and floodplains in the lower end of Lake Michigan after years of industrial emissions. Lee and Tallus (1973) found over 500 ppm lead in the top levels of peat.

Banus et al. (1975) found that most of the lead added to a salt marsh ecosystem was captured by the top sediments with a little in the grass.

Cassagrande and Erchull (1976, 1977) found uniformly low concentrations of heavy metals in the Okefenokee Swamp.

Gardner et al. (1978) found 25 to 42 ppm lead and 80 to 162 ppm zinc in wetland sediments. With depth there was a maximum at about 40 cm decreasing to a minimum of 25 ppm lead and 100 ppm zinc in older sediments.

Mudroch and Capbianco (1979) found low concentrations where sewage waters had passed through wetlands for 45 years. While studying sediments of the natural marsh receiving waters with heavy metals, they found duckweed and *Myriophyllum* accumulated more lead, zinc, chromium, and cadmium than in control areas. The plants were *Glycera*, where *Typha* had also been observed earlier.

Simpson et al. (1983) found lead imported into tidal wetlands from nonpoint sources at the time of dieback of aquatic macrophytes. However, more lead was found near outflow of storm drains.

In 25 wetland soils in an area of heavy metal processing in Sudbury Ontario, Taylor and Crowder (1983) found little copper, zinc, nickel, or other heavy metals in leaves, although heavy metals were in the roots.

Turner et al. (1985) found 98% of incoming lead was retained by the muck soil and vegetation.

Crist et al. (1985) found leaf decomposition in wetland microcosms was not affected by pH 3 to 5 and lead in the range 0 to 1000 ppm.

Darby et al. (1986) describe element mobilization in man-made estuarine marsh.

Nixon and Lee (1986) write that the loss of metals in decomposition is not large — that in fact during decomposition the absolute mass of metals actually increases rather than decreases, due to continuous uptake from surface water. Ecological engineering practices for retaining lead should keep the sediment covered with water and prevent oxidation (especially by burning). Maintaining the water level for these goals may also ensure continuous accretion of organic matter (Nixon and Lee, 1986; Giblin, 1985), which will both continue to sorb lead from the surface waters and will eventually form a natural "cap" as lead-contaminated sediments are progressively buried. Maintenance of vegetation resists water flow and decreases wind-driven turbulence, decreasing downstream transport of particulate-adsorbed lead. What soluble lead does escape may be at or near environmental background levels and pose no unreasonable toxic hazard. In this way wetlands may act as a buffer for high concentrations of toxic metals as well as a filter.

Glooschenko (1986), for bogs in northern Europe, finds 16 to 68 ppm zinc and 3.8 to 32 ppm lead. Maximum of 60 ppm was found at 30 cm. Sphagnum was suspended in bags to evaluate lead emissions. Accumulations were 2 to 63 mg/m^2/year. A diagram showed exchanges in water and sediment among organic lead species. Zinc decreased from 140 to 20 ppm 70 km from a smelter in Quebec.

Lead in peatlands of the Pungo River in North Carolina was mostly in immobile bound humic fraction with about 12.8 μg of lead per gram at the surface (recent deposition), 2.7 μg/g at 10 m, and 3.6 μg/g at 1 m (Pace and Di Giulio, 1987). *Rangia* clams in brackish waters draining these peatlands had little lead.

Kufel (1991) studied the seasonal uptake, decomposition, and release of lead in littoral plants in Lake Gardynskie, Poland. Where sediments contained 5.0 g/m^3, cattails (*Typha*) accumulated 116 μg/m^2 in underground roots and stems, 182 μg/m^2 in shoots, and 11.7 μg/m^2 in derived detritus, leaving 286 μg in standing litter at the end of the season. Reeds (*Phragmites*) growing where sediments contained 1.28 g/m^3 lead bound 5.3 mg/m^2 in underground organs, 3.6 mg/m^2 in shoots, and 1.5 mg/m^2 in derived detritus, leaving 0.28 mg/m^2 in standing litter.

Stockdale (1991) reviewed nutrient uptake of wetlands including heavy metals.

Oberts and Osgood (1991) found a pond and wetland series in Minnesota removed 74% of the lead from storm runoffs with 10 to 87 μg of lead per liter going into sediments which contained 14 to 138 ppm.

Vedagiri and Ehrenfeld (1991) studied lead and zinc uptake in wetland microcosms containing sphagnum moss, peat, and seedlings of red maple and cranberry. The plant species had opposite responses in their heavy metal binding with the addition of sphagnum waters.

In subtropical China, a wetland with cattails removed 95% of lead in waters (1.6 mg/l) draining a lead–zinc mine (Lan et al., 1992). They found 4942 ppm lead in sediments, 350 ppm in cattails, and 850 ppm lead in *Paspalum*.

In Louisiana coastal wetlands, Pardue et al. (1992) found the ratio of lead to aluminum useful in separating sediments with lead pollution from normal wetlands.

Dobrovolsky (1994) relates the lead and strontium in mangroves to that in the substrate, finding more strontium and lead in coral islands than silicate islands. Both strontium and lead have ionic radii that tend to substitute for calcium.

Kittle et al. (1995) described the effects of acid on plant litter decomposing in bags in wetlands, finding considerable species differences. Woolgrass and rushes decomposed more slowly than cutgrass and *Calamus*. Decomposition and nutrient release from cattails was slower at low pH; and 50 to 80% of organic matter remained in bags below pH 7 compared to 40% in controls.

Shotyk et al. (1998) used the immobile property of lead in sediment layers of bogs of Europe to trace the accelerating lead emissions of civilization starting 3000 years ago, reaching a maximum of 1570 times background in 1979, given as 0.01 mg/m^2 deposited before 5320 years ago.

Bindler et al. (1999) used analyses of lead and stable isotope ratios (206 Pb/207 Pb) in bores from peat bogs of Sweden to determine the lead through historic times of metal production to the pristine condition of 0.1 µg/g 5500 years ago.

Physical Filtration

Wetland plants slow the flow of water through wetland systems, allowing fine particles to settle out. Emergent plants may trap and hold sediments, preventing turbulent resuspension. Wixson (1978) reported that particles of lead-rich rock flour and minerals were trapped by mats of algae in Missouri streams. In the Biala River wetland in Poland large particles of lead were found trapped by emergent vegetation (Wójcik and Wójcik, 1989; this book Chapter 9).

Wolverton and McDonald (1975) found lead removed by alligator weed (*Alternanthera philoxeroides*) in microcosm experiments. With water hyacinths (*Eichhornia crassipes*) there was 65% lead removal (10 mg/l solution) within 1 h and 96% removal in 96 h (Wolverton and McDonald, 1978). No significant difference was found when lead was in combination with mercury and cadmium. In hyacinth systems in Texas concentrations of heavy metals in the bottom sediments exceeded that in the living plants above by a factor of 10 or more (Reed et al., 1988, p. 135).

Absorption on the Negative Charges of Organic Matter and Clays

Adsorption (and precipitation) on dead organic matter is important in metals removal. Organic soils typically have a high cation exchange capacity, which may range from 300 to 400 µeq/100 g (Manahan, 1984). As the proportion of organic matter in a wetland soil increases, the cation exchange capacity increases, and the proportion of cation exchange capacity saturated by metal cations decreases (Mitsch and Gosselink, 1986). With an inverse correlation between pH and organic matter content in wetlands, the chemical exchange capacity (CEC) tends to be saturated with hydrogen ions at high organic matter levels. However, despite the prevalence of hydrogen ions on organic matter at low pH, there is a high rate of adsorption of heavy metals in wetlands (Giblin, 1985; Baudo, 1987).

Wieder (1990) reported 1320 µeq/g cation exchange capacity in *Sphagnum* peat. Drever (1988) noted porosity of ocean sediment 0.4 at 600 m. Lead absorption by hallocysite, a silicate, was greater than equlibrium concentration with lead carbonate. Although lead was not included, organic fractions in sphagnum peat and sawdust were found similar in binding of heavy metals when equilibrium constants were evaluated with the Langmuir equation.

Precipitation as Insoluble Lead Sulfide Where Oxygen Is Low

Anaerobic conditions (low oxygen, low redox potential) prevail in continuously flooded soil because microbes use up the small amount of dissolved oxygen in soil waters. If sulfates are present and oxygen is not, sulfate-reducing bacteria reduce sulfate to sulfide which then reacts with lead and some other metals to precipitate as microscopic solid particles. Lead sulfide is not soluble in wetland conditions (only soluble in oxidizing acids such as nitric acid [Förstner and Wittmann, 1983]). Forming sulfides removes acid, and pH may rise from 2.7 to 6.5. Bicarbonate is a by-product (Nixon and Lee, 1986; Giblin, 1985). Reducing conditions (low oxygen) also liberate lead ions from oxidized lead compounds (example: ferromanganese oxyhydroxides), which then react with sulfide to precipitate lead sulfide (Rickard and Nriagu, 1978). Sediments under conditions of acid mine drainage, or other lead and sulfuric acid pollution, are rarely deficient in sulfide. In other words, wetlands promote the formation of immobile lead sulfide.

At the Colorado School of Mines, Wildeman et al. (1996, 1997) constructed wetland mesocosms as pilot plants for treatment of acid mine drainage. Later plants were omitted and an artificial reducing sediment was prepared with mixture of cattle manure, soil, and limestone sands. Mine waters were processed through the system at a rate that would generate enough hydrogen sulfide from sulfate reduction to precipitate and hold the heavy metals while also raising the pH. Lead levels were reduced by 94% or more, changing pH from 2.9 to 6.5. In one system lead initially 0.4 to 0.6 ppm and zinc 0.18 ppm were reduced below detectable limits (0.02 ppm lead and 0.008 ppm zinc), holding pH 8.

The distribution and flow of lead in wetland ecosystems depend on the whole system of water flow, biomass, microbes, recycling processes, and sedimentary storages. Whether wetland sediments can form a permanent sink for lead is a matter for further study.

Combination with Peat and Humic Substances by Complexation

Singer (1973), Rubin (1974), and Reuter and Perdue (1977) review the chemical complexes formed between heavy metals and humic substances, thus reducing their toxicity, solubility, and ability to be leached.

Vedagiri and Ehrenfeld (1992) found acidic pineland wetlands with urban runoff to be less acidic, to contain less dissolved organic carbon, and with less labile lead (64 to 71% compared to 95%). Fractionation of lead was different (hydroxy links) in urban runoff from that passing through wetlands (humic binding).

Krosshavn et al. (1993) studied binding of heavy metals with humus from spruce, pine, oak forests, and wetland mires at pH 3, 4, and the natural level; 56 to 99% of the lead was bound.

Heavy Metals in Florida Wetlands

Cypress swamps in Florida were found to filter nutrients including silver from secondarily treated sewage (Odum et al., 1977; Tuschall, 1981; Ewel and Odum, 1984). Best et al. (1982) reported studies of uptake of cadmium, copper, zinc, and manganese in a cypress strand swamp near Waldo, FL. Fiberglass barriers were used to bound two parts of the swamp with enclosures 40 × 10 m. The heavy metal solutions were mixed into partially treated sewage flow through one of the enclosures. The other received only the sewage waters. Most of the heavy metals were taken up by the time the water had passed 40 m.

Pat Brezonik and students (Thompson, 1981; Tuschall and Brezonik, 1983) found the ten common heavy metals all being bound in organic matter, peats, lake sediments, and wetlands as fast as they were introduced by air pollution and local wastewater runoffs and field experiments in North Florida.

By analyzing sediments dated with the lead isotope method in two lakes, Thompson (1981) found rate of input of lead to be 320 and 1020 g/ha/year. The top sediments deposited in recent years contained 2 times more lead and 1.6 to 24 times more zinc than in earlier times.

Klein (1976), Carriker (1977), and Boyt et al. (1977) found wastewaters with low concentrations of heavy metals reduced further in passing through Florida wetlands. Klein (1976), Tuschall (1981), and Best et al. (1982) found low levels of heavy metals including lead (<1 to 2 ppb) and zinc (<10 to 40 ppb) in Florida cypress wetlands at Waldo, Jasper, and Gainesville, FL, and in sewage effluents applied to these areas for tertiary treatment. In a field experiment where swamp waters were flowing slowly in a cypress strand at Waldo, a bounded strip was enriched at its upstream edge with cadmium, copper, zinc, and manganese; 80 to 90% was taken up by the wetland in 40 m. The immobilization rates are in grams per hectare per day: cadmium 7.2, copper 36, manganese 72, and zinc 72. In a still microcosm containing the cypress swamp soil, 18 to 67% of the enriched metals was absorbed.

In the Cypress dome wetland-wastewater project in Gainesville (1973–1979), Carriker (1977) found lead in the trailer park package plant effluent was 44 (6 to 690) ppb, in an intermediate pond 4.6 (0 to 20) ppb, in the swamps receiving 2.5 in./week wastewaters 8.1 (0 to 39) ppb, in a swamp receiving groundwater 7.6 (0 to 30) ppb, and in a control swamp in the university experimental forest 9.2 (1 to 36) ppb. Lead in these sediments ranged from <0.1 to 0.75 ppm. Zinc in the effluent was 73 (15 to 311) ppb, in the intermediate pond 20 (4 to 40) ppb, in the wastewater swamps 18 (3 to 60) ppb, in the groundwater swamp 21 (3 to 80) ppb, and 70 (31 to 101) ppb in control swamp. Zinc in these sediments ranged from <0.5 to >5.0 ppm. Heavy metals were immobilized in the upper few centimeters, with concentrations decreasing with depth into the sediments. Floating duckweed plants immobilized heavy metals, competing with the binding of humic substances. There was 6 to 10 ppm lead in plant matter and 71 to 205 ppm zinc. When the plants decomposed only 9 to 12% of the lead was released and 13 to 18% of the zinc.

A detention basin–wetland mix of cypress and understory vegetation of hyacinths, cattails, and small trees near Orlando, FL removed 83% of inflowing lead from urban runoff (62 μg/l) (Martin, 1988).

METHODS OF HEAVY METALS REMOVAL

The background on heavy metals removal may be a useful perspective on filtration by wetlands and technological alternatives. Harrison and Laxen (1981) reviewed ways of controlling lead. Lead and zinc, as well as other heavy metals, exist in wastewaters and environmental waters in many forms, including soluble, insoluble, inorganic, metal organic, reduced, oxidized, free metal, precipitated, adsorbed, and complexed. For removal of metals, lead must be converted to a suitable form compatible with removal by precipitation, or coagulation, or otherwise attaching to an insoluble form. Processes include precipitation, coagulation, adsorption, ion exchange, filtration, and ultra-filtration. Conventional biological treatment is efficient in removal of heavy metals. However, high concentrations of these metals may negatively affect biological treatment processes. Most lead is removed by solid contact water softening treatment.

Bioremediation

Jerger and Exner (1994) classified bioremediation approaches as including: bioaugmentation (example: adding bacterial cultures), biofiltration (example: microbial stripping columns), biostimulation (example: augmenting indigenous microbe), bioreactors (example: stirring liquids), bioventing (example: drawing in oxygen), composting (example: adding "bulking"), land farming (example: processing on soil particles).

Isphording et al. (1992) found algae-scavenging heavy metals in the Upper Bear Creek Reservoir.

Johnson et al. (1994) suggested three alternative approaches to revegetation of abandoned mine lands in Britain where lead was 48 to 76,500 ppm, zinc 26 to 42,000 ppm, and copper 30 to 72,600 ppm. The ameliorative approach fixes the land first so that ordinary vegetation can prevail. The adaptive approach utilizes whatever species are adapted to the special conditions, often by selecting naturally colonizing species that have already appeared on site. The third is to develop agriculture and/or forestry. They suggested spiny shrubs to keep people off toxic areas. Pathogenic soil fungus (*Phytophplethora cinnammemi*) was described causing vegetation dieback in Australian sites.

Some species of plants and animals concentrate heavy metals without being harmed by the high concentrations. Cultivating these species to remove heavy metals is called phytoremediation, reviewed by Brown (1995).

Dushenkov et al. (1995) call heavy metal uptake by plant roots, rhizofiltration. They reported experiments with Indian mustard (*Brassica juncea*), which accumulates 3.5% of its weight in lead, concentrating lead 131 to 563 times. The uptake immobilizes the lead, and not much of the lead is translocated from the roots. The rate of removal was proportional to the mass of live roots with little uptake by dead roots. The more lead in solution, the longer time was required to absorb half (40 h when lead was 500 mg/l). Exudates helped precipitate some lead phosphate in the container. In a study where terrestrial hydroponic plant roots removed lead from water, roots also mediated precipitation of insoluble lead phosphate and extrusion from the root. Chlorosis was evidence of toxicity when solutions used were higher than 300 mg/l.

Vymazul (1995) published a summary table of concentrations in algae from 1 to 70,000 ppm in the lead belt of Missouri. Lead inhibits chlorophyll and photosynthesis in light, not in dark; inhibited by 8.5 ppm lead in diatom tissue; concentration factor 91 to 37,500 in the brown algae *Fucus*.

Fifteen papers on hyperaccumulation are included in the review edited by Brooks (1998). Many plants were identified that can accumulate heavy metals without toxic effect, some through microbinding and isolation of the heavy metal among the cells. These plants compete well on areas with high metal content. Some species concentrate metals (nickel and cobalt) in quantities that might be commercial (called phytomining). Examples are plants growing on serpentine, other ultramafic rock areas, or areas enriched by industry, where nickel is 750 ppm or more, and the metal in plant biomass may be 0.5% or more.

Plants concentrate lead less in dry lands because of soil binding. Also, lead in soil binds phosphates. However, lead was dramatically concentrated and removed by wetland species including water hyacinths, algal mats, willows, and poplars (rhizofiltration). Lead and zinc were concentrated by factors greater than 2000 times.

Precipitation and Coagulation

Precipitation and coagulation are most often used for removal of both ionic and non-ionic forms of heavy metals. Stages are

1. Adding chemicals with turbulent fast mixing, 1 to 2 min
2. Slow mixing to form flocs
3. Sedimentation of flocs
4. Filtration of the remaining flocs if necessary

Flocculant chemicals are added to aid coagulation. The chemicals include lime, ferric chloride, sodium aluminate, ferrous sulfate, aluminum sulfate, activated silica sol, bentonite or other clays, calcium carbonate, sodium xantogenian, and polyelectrolytes such as polymeric amines or polycationic polymers.

The choices of chemicals, necessary doses, coagulation conditions, and pH are based on laboratory studies and/or pilot-plant experiments. Dosage is somewhat critical, since relatively small excesses over the optimum may seriously impair the coagulation or even give increased dispersion. Usual doses range from 0.1 to 5 mg/l for polyelectrolytes or from 10 to 300 mg/l for other chemicals. Economical analysis and access to local resources of the chemicals are considered.

In the 'slow mixing' stage where flocs form, peripheral stirring velocity is 0.1 to 0.5 ft/s, and retention time in the flocculation chamber 10 to 30 min. Flocculation is aided by the recirculation of preformed sludge (2% by volume is usually sufficient) to the inlet of the flocculation tank.

Settling and sedimentation may be carried out in tanks operated either batchwise or continuously. The tanks operated continuously are either of the vertical-flow or horizontal-flow types. In the continuously operated tanks, the clarified water is withdrawn from overflow, while the sludge, which collects at the bottom of the tank, is removed either continuously or intermittently.

An average rating for a horizontal-flow sedimentation tank is from 15 to 30 gal/ft²/h. The retention time is 2 to 6 h. For vertical-flow tanks of sludge-blanket type the retention time is usually only 1 to 2 h with throughputs of 25 to 100 gal/ft²/h.

The maximum upward flow rate is that at which the sludge blanket begins to rise. With aluminum coagulants it is about 12 ft/h; with the larger and heavier floc, such as that produced by activated silica or chalk, flow rates can be 50% greater.

Filtration

Filtration methods are

1. Without preliminary coagulation; only the coarse material is removed and most of the colloidal particles pass through
2. With preliminary coagulation and settling
3. With preliminary coagulation, but without settling

Where coagulation is required to remove colloidal forms, settling precedes filtration. With the third procedure settling tanks are avoided, but filters must be backwashed more frequently, and the backwashing wastewater has to be treated.

With upward flow filtration, chemical is injected just before the filters, so that formation and removal of the flocs occur in the filtration bed. Sand, anthracite, diatomaceous earth, activated carbon, zeolite, crushed dolomite, or limestone is used for the filter beds. Up to 200 gal/ft²/h is commonly used in rapid gravity filters, and filter beds vary from 1.5 to 3 ft.

Adsorption

Most of the chemicals used in precipitation can be used for adsorption. Biosorbents may be used including dead bacteria or fungi, which adsorb the ionic and colloidal forms of heavy metals. Some adsorb selectively enough to be used for recovery of metals.

Ion exchange, extraction, ultrafiltration, and reverse osmosis are used mostly for high quality water treatment after pretreatment for removal of the suspended and colloidal particles and organic substances that could disturb the process.

Conventional methods for removing heavy metals from wastewater require large facilities (tanks, pumping stations, and mixers, for example) and a large quantity of chemicals, energy, and trained people. These methods produce a large volume of sludge to be processed.

Activated Sludge

Activated sludge from wastewater treatment plants has great capacity for absorption of heavy metals on the web of polymer fibrils that enmesh discrete cells. Experimentally calibrating a kinetic model, Neufeld et al. (1977) found the uptake of metal increased until an equilibrium was reached between the absorption and solution. More than other heavy metals, lead uptake reached equilibrium with a 1:1 ratio of lead to sludge by dry weight.

Reprocessing of Lead Wastes through Smelters

Earlier, "suck, muck, and truck" methods were used to move toxic wastes to be buried again in another place. This was done with the Sapp Superfund site upstream from the Swamp studied in this book. Now, however, Staley (1996) reports the successful processing of Superfund wastes with 3 to 57% lead through secondary lead smelters like those used to reprocess and recover battery lead. Costs ranged between $35 and $375 per ton (see reprocessing industry evaluated in Appendix A11b).

EVALUATION OF ALTERNATIVES FOR LEAD PROCESSING

Many published papers deal with wetland valuation for a range of products and services. Reviews are by Bell (1989), Douglas (1989), Leitch and Ekstrom (1989), and Scodari (1990). Gosselink et al. (1974) used the replacement value concept in a widely publicized study to calculate the value of tidal marshes for various services and products, including waste treatment. They estimated the replacement value of an acre of marsh/estuary to be between $10,000 and $280,000 (the annual benefits estimated at between $480 and $14,000/year).

Thibodeau and Ostro (1981) calculated the capitalized value of Charles River Wetlands for pollution reduction of BOD and nutrients at $16,960/acre (at a discount rate of about 8%, 1977 dollars). Each acre substituted for $85 of plant cost and annual operation and maintenance of $1475. They recognized the value of wetland peat for adsorption of heavy metals and organic pesticides, but did not calculate the value because of uncertainty about long-term retention.

Folke (1991) calculated the value of wetland "life support" for a progressively degraded 34-km^2 Swedish wetland system. He found that it had actually cost at least 32,000 SEK (1989 Swedish Crowns; about $5300) to partially replace the water quality functions of the wetland with a water purification plant. They estimated the fuel equivalent embodied energy to be between 180 and 230 billion fuel joules or 9.7 to 12.4 quadrillion solar emjoules (about 6000 U.S. emdollars).

Baker et al. (1991) used a simulation model to examine the effects of different loading rates of iron on treatment efficiencies and the economic costs comparing constructed wetland with conventional treatment of acid mine drainage. They found that at low loading rates and treatment efficiencies of less than 85%, constructed wetlands were less costly than conventional treatment systems. The values of wetlands under this approach are the savings obtained by using wetlands instead of conventional treatment. The values depend on the loading rate.

Chereminisoff (1993) provided remediation technology, standards, and guidelines for lead workers including costs of replacing painted surfaces.

SIMULATION MODELS OF HEAVY METALS

Understanding of heavy metals in the systems of environment and human civilization has matured in the last three decades as part of the progress and application of computer simulation methods. Dynamic perspective was provided on heavy metal flows and concentrations on many scales from the small, fast biochemical and microbial processes to the global flows of the biosphere. Some examples are cited chronologically.

In a review Chadwick (1973) generalized the relationship of material cycle with a three-block minimodel. Block A was the reservoir of less active storage of the material; Block B was the main center of processing of the cycle; and Block C drawing on Block B is the high quality use by plants and animals, in a lesser cycle.

Rolfe and Haney (1975) provide a budget of lead flows for a watershed in Illinois, estimating flows of lead from automobiles and other sources and the movements in the landscape, with small amounts passing out of the area in stream discharge. Most remained in soils.

Neufeld et al. (1977) modeled heavy metal uptake rate by activated sludge as proportional to the concentration in the waters decreased as the bound quantity reaches the capacity limit in the sludge. These limits were calibrated in batch experiments evaluating equilibrium thermal equation. Order of affinity of heavy metals was lead > cadmium > mercury > chromium > zinc > nickel.

Finn (1978) evaluated a cycling index for a nine-compartment model of lead flows in a tropical moist forest. Matrix algebra was used to calculate the ratio of recycled lead to throughflow lead, which was 0.155, a lower index than for some nutrients.

Jorgensen (1979) summarizes data from the literature for modeling heavy metals including uptake by organisms, exchanges of toxicants with sediments, exchanges with suspended particles, and exchanges with organic substances. Jorgensen (1979, 1984) used graphs of generation time and decreasing unit metabolic rates with increasing size of organisms to calibrate rates of biological uptake and release of heavy metals.

Seip (1979) simulated uptake of zinc by the benthic alga *Ascophyllum* in waters of different zinc concentration. Zinc content increases with age in the benthic alga *Ascophyllum*, with 2500 ppm in algae absorbed when water contained 150 ppb. A model of logistic growth and mortalities of algae by age classes included an age inhibition factor. Zinc uptake was proportional to the biomass and to the zinc concentrations in water, minus zinc secreted.

Harrison and Laxen (1981), summarizing lead in environment and humans, showed lead accumulating with age in humans. They include a human lead flow and pool network diagram. Included from Webb (1978) is a bar graph of the percent lead in stream bed sediments with decreasing percent of samples with increasing concentration to 320 ppm or more.

Nyholm et al. (1984) simulated the distribution of lead, zinc, and cadmium being released from mining operations in a Greenland fjord using compartments for water areas and levels. Concentrations in water relative to sediments were determined from Langmuir curves.

Thoman (1984) listed the features needed in environmental models of hazardous substances giving equations for the mechanisms:

1. Sorption–desorption mechanisms between water and sediments or particles
2. Losses of toxicant through biodegradation, volatilization, chemical reactions, and photolysis
3. Advection transport or dispersion of toxicant
4. Settling mechanisms
5. External inputs
6. Sorption by organisms
7. Feeding intake by organisms
8. Assimilation into growth of organisms
9. Prey–predator transfers
10. Depuration or excretion by organisms

The system model was a combination of the separate equations, many of which assumed equilibria.

Jorgensen (1993) uses classification of six kinds of models for ecotoxicology: food chain, static model of mean flows, dynamic models of toxic substance, ecotoxicological models in population dynamics, ecotoxicological models with effect components, and fate models with or without a risk assessment. Main factors are concentration, adsorption, solubility, excretion, and biodegradation. Fugacity models are where the escaping tendency at an interface $f = c/z$, where c is concentration and z the fugacity capacity.

Mitsch et al. (1993) simulated the uptake of metals in constructed wetlands receiving acid mine drainage. In this model flows of mine drainage contributed metals to water column exchanging with pore space waters in sediments, and mobilized into substrate by seasonal growth of cattails, with rates controlled by pH as controlled by acid inflow.

Wixson and Davies (1993), while providing guidelines for reducing lead toxicity in people, provide a simulation model in BASIC that calculates the blood lead in humans based on soil and other environmental data.

Pierzynski et al. (1994) in Hester and Harrison (1994) provided a diagram and equations of a general model for heavy metals in mined sites. Inputs and output flows of water and heavy metals were connected to a land compartment with removal pathways for transpiration, adsorption to plant roots, bonding to exchange sites, and adsorption to soil material.

Weinstein and Buk (1994) plot net production as a function of frequency of impact, finding an optimum pulse frequency with maximum production. Toxic impact pulls down biomass and diversity, which springs back as toxic substance disperses.

Jorgensen (1993, 1995) summarizes chemical and biological processes and evaluations to include in heavy metal models: adsorption, precipitation, acid–base and hydrolysis, oxidation–reduction, complex formation, uptake and release by organisms including biomagnification and biodegradation, air–water exchanges, and water–sediment exchanges. He describes models for heavy metals in food web simulation, steady-state mass balances, trophic level aggregates, ecotoxicology in population dynamics, ecotoxicology including toxic effects, and heavy metal fate with risk evaluation, where risk evaluation may include perception of hazard by people.

On a larger scale, Jorgensen (1986, p. 326) provides a quantitative budget model diagram of lead flows and pools in Denmark in 1969. Global models of the cycles of heavy elements are considered in Chapter 4.

Biogeochemical Cycle of Lead and the Energy Hierarchy

Howard T. Odum

CONTENTS

Chemical elements such as lead circulate in the biogeosphere and through the economy of civilization. It is customary to overview chemical cycles by making simplified diagrams of principal components, pathways, and places of storage. Such simplifications are called systems models. On some diagrams symbols are used to show causal relationships. On other diagrams numerical values are placed on the pathways to show at a glance which flows and storages are more important. This chapter uses systems models to overview the principles of heavy metal distribution using the cycle of lead.

New perspectives come from relating the elemental cycles to the natural energy hierarchy by which the earth is organized. When people in an organization converge their work to fewer supervisors, and these in turn send fewer inputs to even fewer people at the top of the organization, we call it a hierarchy. In turn, those at the top spread their influence among those back at the lower levels.

The biogeosphere processes energy through series of units, including the atmosphere, oceans, continents, living organisms, industrial processes, human beings, etc. Each unit transforms input energy into a small amount of higher quality output energy that goes to the next higher level. A

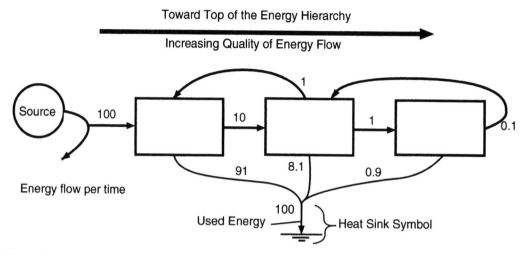

Figure 4.1 Diagram of a three-unit system arranged from left to right according to its energy hierarchy. In the series of energy transformations available, energy flow decreases but energy quality increases.

series of energy transformations is an energy hierarchy because abundant energy at the base of the organization is transformed and converged into smaller but higher quality energy and units at the top of the chain. The top units send small, controlling energy flows back to the lower levels.

In our diagramming of systems the energy hierarchy is arranged from abundant low quality energy on the left to high quality energy on the right. For example, Figure 4.1 shows a series of three units with available energy flow being transformed into an output of higher quality but less energy. Notice also the feedback from right to left of small energy flows (1 and 0.1) dispersing influence to the lower levels. System symbols are given in Appendix A1.

In the process most of the energy is degraded, losing its ability to do work. As required by the law of conservation of energy, the energy that inflows and is not stored inside has to flow out. An energy diagram has to include pathways for energy to the outside. Energy that can no longer do work is indicated as used energy by showing it flowing down and out through a degraded energy symbol (heat sink symbol). This used energy disperses as heat, eventually leaving the earth. Figure 4.1 has an energy source, energy pathways, and a heat sink.

Whereas available energy flows in, causes transformations, and is dispersed, materials circulate in cycles. When the systems of the earth organize, they recycle material elements in loops. In Figure 4.2 elemental materials are shown circulating in an ecological system aggregated into two units. Dilute nutrient elements on the left are concentrated and passed to the right. The materials recycle back to the left (called feedback), becoming dispersed in the process. Energy is used to converge materials from dispersed, dilute distribution to centers where the material is concentrated (on the right). The cycle is completed when the concentrations of materials are dispersed outward to the larger area again. The converging and concentrating of elemental materials followed by dispersal are a part of the natural hierarchy of environmental organization.

McNeil (1989) showed a three-dimensional picture of converging and diverging of materials in circulation. He gave the example of the tree (Figure 4.2b), which draws chemical elements into roots that converge to the hierarchical center, the trunk, then diverge again into the leaves. Chemicals drip from the leaves and fall when the leaves fall. After leaf decomposition the chemical materials are released into the soil to make the cycle again. He proposed the geometrical toroid form (Figure 4.2c) as a general systems concept for circulation. Some heavy metals follow this pattern.

The universe has many levels of hierarchy, with materials converging to small centers, and these in turn converging to larger centers. Familiar examples are the villages, towns, and cities of the human-populated landscape. Figure 4.3 shows circulation with three levels of hierarchy. The

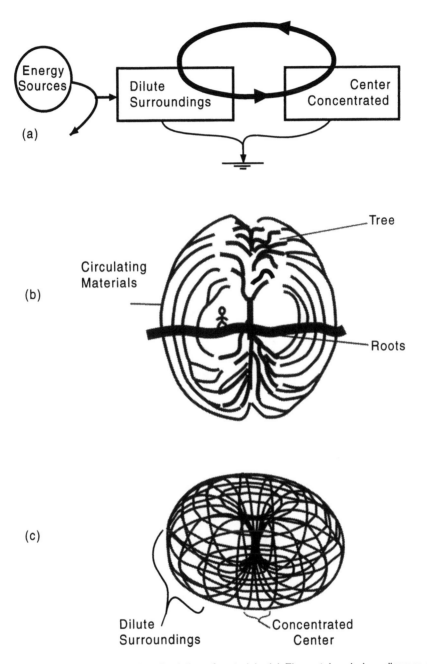

Figure 4.2 Two-unit system showing the circulation of materials. (a) Elemental cycle in a diagram of energy flow; (b) convergence and divergence of nutrient elements circulating between a tree and its environment (McNeil, 1989); (c) three-dimensional circulation represented as a toroid (McNeil, 1989).

material circulation is above (Figure 4.3a), a systems diagram of these materials circulating is in the middle (Figure 4.3b), and energy sources and sinks are included in Figure 4.3c.

MATERIAL CYCLES IN THE HIERARCHY OF THE EARTH

After millions of years the self-organizing processes of the earth developed a hierarchy of energy processing including the atmosphere, the ocean, the lands, and the mountains. Figure 4.4a

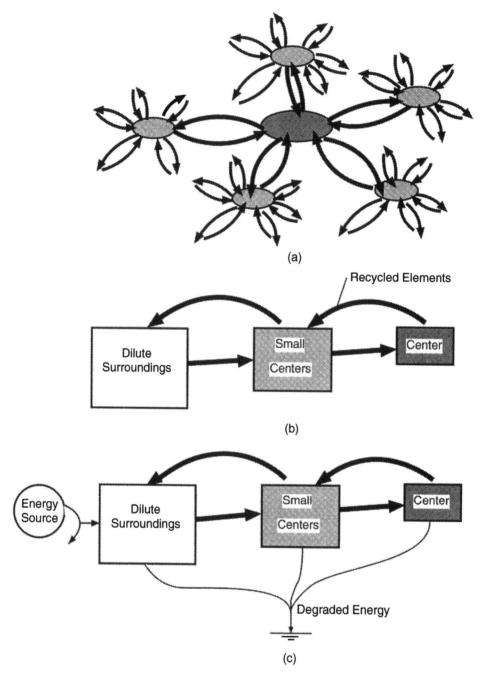

Figure 4.3 Convergence and divergence of materials circulating in a three-level hierarchy. (a) Spatial pattern; (b) systems diagram with circulation of elements; (c) systems diagram with energy source and sink added.

shows a simplified model of the main units of the earth, with the ocean and atmosphere on the left and land formation and mountain building centers on the right. The circulation of matter is shown with thick pathways. Processes on the left are relatively fast, requiring only days or years to cycle, whereas those on the right take millions of years.

Many kinds of material circulate between the units of the biogeosphere (Figure 4.4a). Some material cycles such as water are concentrated at the left end of the chain of units (Figure 4.4b). Water vapor from the ocean becomes atmospheric storms and rain. The rains on land and mountains support

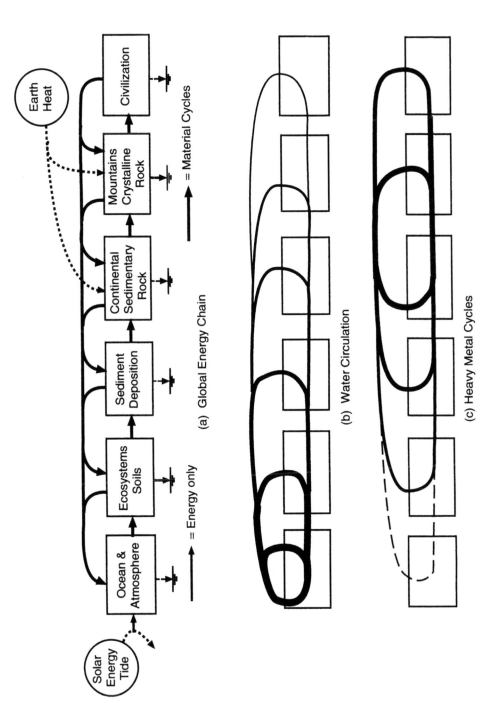

Figure 4.4 Main features of the global geobiosphere arranged from left to right in the order of the hierarchical organization of energy. (a) Chain of main components with thick pathways representing the circulation of materials; (b) water circulation concentrated at the lower energy part of the earth chain; (c) heavy metal circulation concentrated at the higher energy part of the chain.

the ecological systems, with water being transpired back to the air as water vapor. Runoff water carries sediments back to the sea where they are deposited, becoming sedimentary rock and land again.

Other materials, such as the heavy metals, circulate primarily among units at the higher levels of the hierarchy (Figure 4.4c). For example, before the recent additions of air pollution there was little lead in the ocean, but more lead in rocks of the land. The process of forming crystalline rocks concentrates heavy metals into ore bodies. With the development of civilization, the heavy metal ores, such as lead, were mined as an important part of technology. Lead was important in the Roman civilization and even more important in modern technology because of extensive use of batteries. In Figure 4.4a the urban centers of the human civilization are on the right, a place of concentrating materials such as lead for high technology purposes. Even in a biological food chain, there is a tendency for some heavy metals to go to the top of the chain, to the right in systems diagrams.

Yet other materials, such as quartz sand, circulate in the center of the hierarchy, being uplifted as sand dunes or cemented as sandstone in land formation. After weathering processes, sands wash back to the sea to become coastal sediments again.

Many of the material cycles are controlled by water as it carries sediments and deposits them in wetlands and river deltas (sediment deposition unit in the center of Figure 4.4a). Wetland ecosystems are a prominent part of the sediment depositing system located between the mountains and the sea. Freshwater wetlands are along the rivers and saltwater wetlands in the estuaries. As we read in Chapter 1, wetlands filter heavy metals from air and waters, returning them to the geological cycle in formation of sediments and coal.

INCLUDING MECHANISMS IN SYSTEMS DIAGRAMS

We can improve the diagram of the main units of the biogeosphere (Figure 4.5a) by showing some of the main operating mechanisms. Figure 4.5b shows the main pathways of interaction between units, the circulation of lead, and its connections to the main flows of energy. Two more symbols are used. The hexagon-shaped symbol is for units that have storages that feed action back to the left to augment inflow. Feedbacks that reinforce their own intakes are called autocatalytic processes. An interaction symbol is shown where two different inputs join in a production process.

The diagramming shows all the processes and cycles coupled together. To be coupled is to be joined to the action of energy sources. The diagram shows solar energy interacting with seawater to make water vapor, clouds, storms, ocean currents, and waves. These generate rain that combines with land to form ecosystems, soils, and glaciers. The runoff waters carry sediments down rivers to the deltas and wetlands where the sediment and lead are captured, ultimately to be recombined as land. Lead that escapes to the open ocean deposits with offshore sediments.

There are heavy metals such as lead in all the phases of the earth and flowing between the main components of the earth's surface. There are heavy metal elements circulating in all the shaded pathways in Figure 4.3a along with the water and sediments. Widely distributed in very dilute concentrations in oceans and air, the element converges to become more concentrated in centers of geobiospheric action of land formation and mountain building. The unit labeled economy (our modern civilization) uses rich deposits of fuels as energy for development of the assets of civilization that also require mined materials.

BIOGEOCHEMICAL BUDGETS

Previous authors have summarized data on the distribution of elements by putting estimates of average flow rates and storage quantities on simplified diagrams of the main features of the geobiosphere. Just as we call the average values of money stored and flowing each month in our

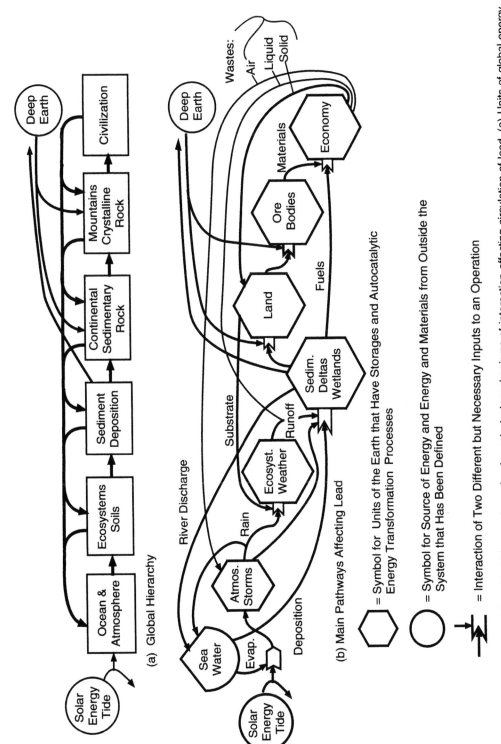

Figure 4.5 Main features of the global biogeosphere showing principal mechanisms of interaction affecting circulation of lead. (a) Units of global energy hierarchy from Figure 4.4; (b) main pathways affecting lead (Appendix A4).

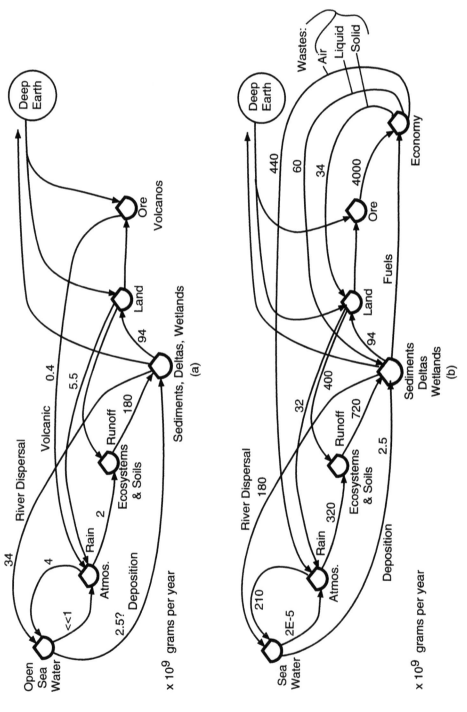

Figure 4.6a-b Main flows of lead in the geobiosphere (Appendix Table A4.2). (a) Lead circulation before civilization; (b) modern circulation of lead.

family accounts a budget, we can refer to the summary diagram and numerical values of a chemical material as a biogeochemical budget.

Garrels et al. (1975) assembled data for the quantity of lead in different phases of the earth and estimated the flows of lead along the pathways from one part to another. Nriagu (1978b) evaluated the main pathways of flow of lead in its global cycle. Pritchard (1992) summarized these flows with a complex energy systems diagram.

In Figure 4.6 we overview the global lead cycle by including only the most important pathways (from Figure 4.5), thus showing how flows are processed through the main units of each level of the biogeosphere's hierarchy. After assembling data from literature (Appendix Table A4.1), the flows of lead in billion grams per year (109 g/year) were written on the pathways.

Salomons and Förstner (1984) assembled graphs by Whitfield and associates (Whitfield and Turner, 1982) that explain the concentrations of heavy metals in the sea in terms of element flux as part of the global sedimentary cycle evaluated as in approximate steady state. Depending on the elements, positive charged atoms are bound to negatively oxidized charged sediment particles that wash to the sea, settling to the sediments, which are eventually uplifted in the earth cycle. The more tightly they are bound (greater electronegativity function), the less they exchange with waters (partition coefficient). The more tightly they are bound, the less time they remain in river and seawaters (smaller residence time). The shorter the residence time the lower the concentrations in the seawaters.

The concentrations of lead in the sea were kept very small by several biogeochemical mechanisms. Goldberg and Arrhenius (1958) found lead ions in aquatic chloro-complexes becoming bound in deep sea manganite 20 to 200 ppm in sediment and 2000 ppm in manganese nodules. Chow and Patterson (1962) found 21 ppm lead in deep sea ooze, 38 to 84 ppm in clays. They

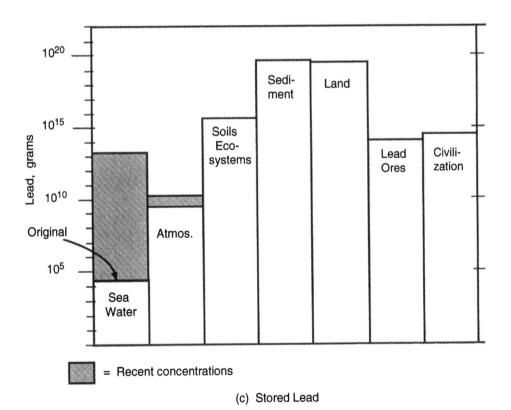

Figure 4.6c   Lead storages in the biogeosphere (Table A4.1).

estimated mechanical deposition rate 2×10^{-6} g/cm^2/1000 years and chemical rate 4.7 in these units. Pelagic lead was 2/3 precipitated and 1/3 as particles. Tatsumoto and Patterson (1963) found 0.002 to 0.20 ppb lead in Atlantic and Mediterranean seawaters, and in the Mediterranean and Pacific up to 0.38 ppb in surface waters, diminishing to 0.01 ppb below 1000 m.

Figure 4.6a has estimates of the flows of lead cycle before civilization. The circulation of lead was relatively small. This diagram has no civilization-economy unit on the far right. By 1971 Bertine and Goldberg recognized that the fluxes of heavy metals due to civilization were approaching those of the natural cycle of land uplift and weathering. The lead emission soon exceeded the natural lead cycle (Volesky, 1990).

Figure 4.6b has estimates of lead flows in our current condition. Adding civilization to the biogeosphere added higher levels to the energy hierarchy, and the result was a further concentrating of heavy metals. From cars and industry on the right, the high values of lead recycle as air, liquid, and solid wastes dispersed to waters and land to the left. The actions of humans in using and dispersing lead increased the lead circulation ten times (Figure 4.6b).

Lantzy and Mackenzie (1979) compare the emissions from the human civilization to the regular biogeochemical cycle of the elements. Heavy metals in soils were in proportion to the levels in shales from which soil was derived. They defined an interference factor as the ratio of anthropogenic to natural fluxes of an element. For lead the factor was 34,583. Lead in the rainout was 21% higher than in the stream load. As stimulated by human use and releases, lead was atmophilic. Lead cycle was given as 5×10^8 g/year in its continental part and 8.7 in its volcanic part, 0.012 in volcanic gas, and 0.016 in fumaroles and hot springs. The industrial part was $16,000 \times 10^8$ and 4300×10^8 g/year from fossil fuel use.

Förstner and Whittmann (1979) provided an environmental index of relative pollution potential equal to the metal concentration divided by the average metal content. The ratio for lead was 35. Another index, the Technophility, was defined as the ratio of annual output of lead to the mean concentration in the earth's crust (sometimes called a Clarke in honor of a pioneer in evaluating geochemical cycles).

EMERGY OF MATERIALS IN A BIOGEOCHEMICAL CYCLE

There is a natural tendency for concentrated things to disperse. This tendency is the second energy law. It takes work to concentrate things and keep them concentrated against the natural dispersal tendency. As we explained in Chapter 1, various kinds of work can be put on a common basis as emergy. Emergy is defined as the memory of available energy of one kind previously used up directly and indirectly to make a product. Its unit is the emjoule. In this book we use solar emergy (solar emjoules, abbreviated sej).

Since work is required to concentrate materials, higher concentrations of material require more emergy per mass. In other words, emergy is required to concentrate materials and keep them concentrated. The ratio of emergy to mass of materials is a useful measure of work that has been applied to materials.

Thus, emergy can be related to the hierarchical position of elements circulating as part of systems. Emergy is added to the material cycle as it is converged to a hierarchical center where it is more concentrated. For example, in the simplified model of a tree in Figure 4.2, elements become more concentrated in producing the organic matter of the trunk. The organic product carries the emergy of the inputs that went into that development. When the product is decomposed, the elements that are released carry the emergy of the product. Emergy per mass decreases when a material disperses as it recycles outward, becoming less concentrated (passing to the left in Figures 4.2 and 4.3).

The lowest emergy per mass is zero. A chemical substance which is at the lowest background concentration of the biogeosphere has no available energy and thus has no emergy. It cannot disperse or depreciate any further by diffusion, being already at the lowest concentration.

EMERGY PER MASS OF LEAD

In the hierarchy of units of the biogeosphere (Figures 4.4 to 4.6), higher concentrations of lead are on the right (Table 4.1) where more emergy has been processed to sustain them. In the biogeosphere the processes of the earth add emergy as they converge and concentrate heavy metals in making the ore bodies that develop in and around high temperature mountain building (Figure 4.5b). Lead ore (in the form of crystals of lead minerals dispersed in rocks) is associated with centers of mountain building to which the earth cycles converge.

Table 4.1 Values of Lead Circulation

Note	Item	Lead Flow (g/yr)	Emergy/ Mass (sej/g)	Emergy/ Year (sej/yr)	Value (E9 EM$/yr)
1	Land cycle	9.36 E10	1 E9	9.36 E19	0.062
2	Economic use	4.0 E12	4.5 E9	1.8 E22	12
3	Dilute wastes	5.34 E11	2 E8	1.06 E20	0.071

Abbreviations: E9 = × 10⁹; Em$ = emdollars.

Note: Emergy divided by global emergy/money ratio for 1995: 1.5 E12 sej/$ (Brown and Ulgiati, 1999).

1   Land cycle: (2.4 cm/1000 years)(1.5 E14 m²)(1 E4 cm²/m²)(2.6 g/cm³) = 9.36 E15 g/year; lead fraction 10⁻⁵; (9.36 E15 g/year)(1 E-5) = 9.36 E10 g/year.

   Emergy/mass that of the land cycle: (9.44 E24 sej/year)/(9.36 E15 g/year) = 1 E9 sej/g.

2   Mine production (Nriagu, 1978).

   Emergy/mass from Appendix Table A4.3.

3   Lead flows as dilute wastes from Figure 4.6: air, 4.4 E11 g/year; liquid, 6.0 E10 g/year; solid, 3.4 E10 g/year; total, 53.4 E10 g/year. Emergy/gram for dilute concentrations assumed from Figure 4.7.

The human society adds emergy from fuels, machinery, and people when it mines minerals and refines elements further into technological form such as electrical storage batteries or gasoline additives (now discontinued in the U.S.). The more work goes into concentrating the lead, the higher the emergy per mass. Spatially, refined lead is concentrated in cities and transportation corridors.

Where lead is processed as a trace element within the cycle of the earth matter, we can assign a small part of the annual emergy budget that drives global land cycle according to the lead proportion (1×10^{-5} g lead per gram of land). The global emergy budget of 9.44×10^{24} sej/year divided by the global land cycle 9.36×10^{15} g/year equals 1.0×10^{9} sej/g. The share of emergy budget for lead within the land cycle is

$$(1 \times 10^{-5} \text{ g/g})(9.36 \times 10^{15} \text{ g/year})(1 \times 10^{9} \text{ sej/g})$$
$$= 9.36 \times 10^{19} \text{ sej/year}$$

At the hierarchical center of the cycle, the material may be at its highest emergy content per gram because much work was exerted in developing the concentration, first by the earth and then by the human economic system. Pritchard, in Appendix A11, Figure A11.7 and Table A11.6, evaluates the emergy of lead processing, obtaining an emergy per mass of refined lead as 7.34 E10 sej/g.

For lead, values for different degrees of concentration were plotted in Figure 4.7 as a function of emergy per mass of lead expressed as emjoules per gram (data from Appendix Table A4.3). The resulting graph shows higher emergy/mass for higher concentrations of lead consistent with the ideas about materials and energy hierarchy. Graphs of this type may be useful for estimating transformities from observed concentrations.

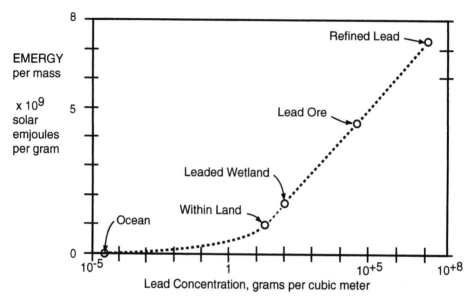

Figure 4.7 Graph of emergy per unit mass of lead for different concentrations of lead in the earth system. Values are explained in the text with calculations in Appendix Table A4.3.

Another way to evaluate a lower concentration is to evaluate how much additional emergy would be required to concentrate the lead further to the refined state. Then this amount of emergy can be subtracted from the emergy per gram of the refined state to get a value for the lesser state.

TRANSFORMITY, THE EMERGY PER UNIT ENERGY

If the available energy in a flow is known, transformity, defined as the emergy per unit energy, can be calculated. Transformity increases along the energy hierarchy. For example, Figure 4.1 shows decreasing energy flows for the same emergy flows, which means the transformity increases. If the energy source is solar energy, then the solar transformity of the inflow to the first unit is 100 sej/100 J = 1 sej/J; the flow to the second unit is 100 sej/10 J = 10 sej/J; the flow to the third unit is 100 sej/1 J = 100 sej/J; and the output of the third unit is 100 sej/0.1 J = 1000 sej/J. We can use the transformity to mark position in the energy hierarchy, high values to the right.

Genoni and Montague (1995) calculated transformities for heavy metals and compared these with transformities of items in the food chains. Higher transformity substances were found higher in the food chains with high transformity species. This was evidence that products that took more emergy to make are used higher in the energy hierarchy where their effects are greater.

THE WETLAND AS A HEAVY METAL FILTER

As more wetlands are studied it is becoming apparent that wetlands self-organize in great variety, adapting to various kinds of inflows of water, organic matter, sediments, and various chemicals, including the heavy metals. Many materials including heavy metals are captured and recycled largely within the wetland ecosystem.

In the diagram in Figure 4.8 a wetland is aggregated to show the main source of emergy and the recycle of lead. Emergy per mass in dilute recycling lead was estimated by evaluating annual emergy flow maintaining the lead-containing wetland ecosystem in Florida reported in Part II. The

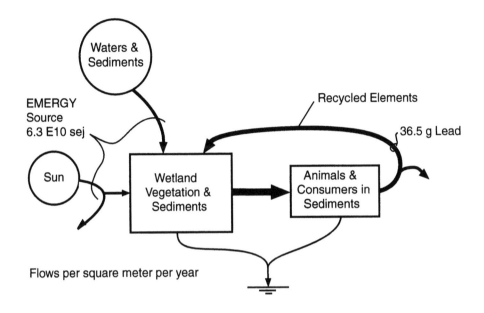

$$\text{Emergy per Lead Mass} = \frac{6.3\,E10\;\text{sej/m}^2\text{/year}}{36.5\;\text{g/m}^2\text{/year}} = 1.73\,E9\;\text{solar emjoules/gram}$$

Figure 4.8 Aggregated diagram of a wetland ecosystem and evaluation of the emergy per mass of its recirculating lead (see Note 3 in Appendix Table A4.3).

inflowing waters carry emergy of the flooding physical energy developed by geopotential work upstream and that in the chemical potential energy of the water used in the transpiration that makes the vegetation productive. To obtain the emergy per mass of the circulating lead, the annual emergy driving the recycling loop was divided by the annual lead circulating in the ecosystem, being concentrated by plants and sediments and released again by consumers = 1.73 sej/g lead (see Appendix Table A4.1, Note 3).

ECOSYSTEMS DIAGRAM SHOWING MECHANISMS

More of the details found in many wetlands are shown with their relationships in Figure 4.9. The interaction symbols (pointed blocks) show the action of one input on another and vice versa. There is physical absorption of particles by plant biomass and uptake of dissolved substances by the plant roots, facilitated by the uptake and transpiration of water by the plants. Heavy metals are bound to the humic substances of the peaty organic sediments formed from plant decomposition (lignin binding, Lb).

Consumers, including other microorganisms, small animals (microzoa), and larger wildlife (hexagon symbol defined in Figure 4.5), release and recycle some heavy metals as they carry out their metabolism. Very little heavy metal flows out with overflowing waters. The organic sediments hold heavy metals by several mechanisms. Some metals are precipitated as insoluble sulfides, where metabolism without oxygen (anaerobic) forms sulfide gas (H_2S) from sulfates. Where sulfates in fresh waters are abnormally high, there is too much sulfide gas and trees are stressed (Richardson

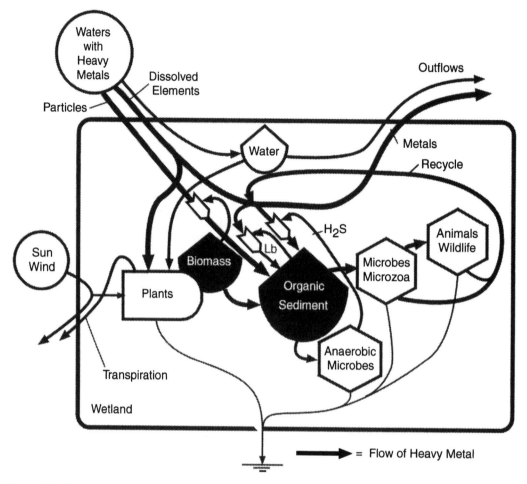

Figure 4.9 Diagram of a wetland system showing the way heavy metals are filtered and stored. See explanation in text. Lb = binding by lignin.

et al., 1983). However, where sulfates are normally high as in salt waters, the salt marsh and mangrove plants that prevail are adapted to function well.

Mathematical relationships are indicated by the configurations of symbols and pathways in Figure 4.9. Equations for computer computation that are implied by the symbols and connections in the diagram can be used to simulate the behavior of the system. For example, Chapter 8 contains a simulation of a model of lead uptake by a lead-filled Florida cypress swamp.

Wójcik (1993) calculated emergy requirements for uptake of lead and zinc by a Polish wetland by summing the input pathways shown in Figure 4.10. Whereas Figure 4.9 has detailed interactions of energy and materials, the diagram in Figure 4.10 shows only the pathways of emergy input that were evaluated. Wójcik found economic costs and emergy of purchased inputs less for the wetland compared to technological treatment. See Chapter 12.

SPATIAL PATTERN OF DISPERSAL

We showed in Figures 4.2 and 4.3 the way a circulating material recycles out from its most concentrated and valuable state in a hierarchical center. Before civilization, lead was most concentrated in ore bodies and dispersed outward when these were recycled by earth processes, as in volcanic emissions and erosion. Spatially, the centers of concentration in ore bodies and

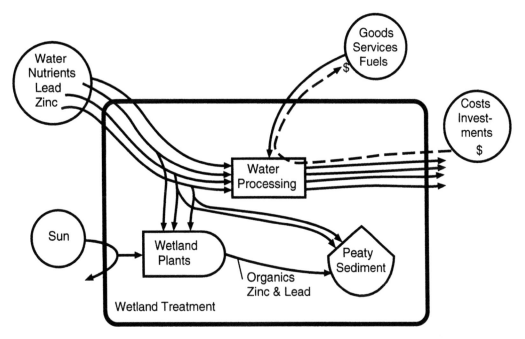

Figure 4.10   Diagram of the sources contributing emergy to Polish wetland treatment of lead and zinc wastewaters as evaluated by Wójcik (1993).

the lead concentrations in derived soils scattered across a landscape are not unlike a pattern of scattered villages. See, for example, the map of lead in England (from J.S. Webb cited by Nriagu, 1978).

With the further concentrations in civilization, first by Romans, and even more by modern economy, the highest concentrations and values are in the centers of the energy hierarchy of civilization, the industries and cities. Nriagu (1978) documented in great detail the high concentrations of lead as they flow out from cities in air, water, and solid waste disposal. For example, Figure 4.11 shows the concentrations in air to be highest around the areas of lead smelting and most urbanized use. Note the high atmospheric concentrations in Poland, where the wetlands provided catchment of water and air wastes, as described in Part III.

These centers of concentrations are located on the surface of the land, and going away lead concentrations decrease. The lead in the air is greatest near the ground with lesser concentrations higher in the atmosphere (Figure 4.12). Within the lower atmospheric system considerable emergy is processed to develop air, water vapor, and heavy metals in the upper atmosphere. When some lead is pumped into the upper air, its emergy per mass is increased. Perhaps trace elements that reach the tops of high mountains may have interactions with vegetation and land commensurate with the higher emergy concentrations there.

Lead in the atmosphere in particles and aerosols has a turnover time of 2 to 10 days before falling on the land or the sea (Nriagu, 1978). Near the urban centers the content is maintained at about 500 to 5000 ng/m^3. (A nanogram = 10^{-9} g.) Away from cities and developed countries the air content is about 0.1 to 10 ng/m^3.

The fallout of anthropogenic lead from the atmosphere over the ocean created high concentrations (0.5 to 3.5 μg/kg), mainly in the upper 1000 m (Chow, 1978). Apparently much of this lead is captured by the processes of coastal sedimentation and wetlands before it can disperse into deeper waters. Since the oceanic systems may not be adapted to benefit from high lead concentrations, short circuiting atmospheric dispersal appears to be a better global design for use of its emergy value.

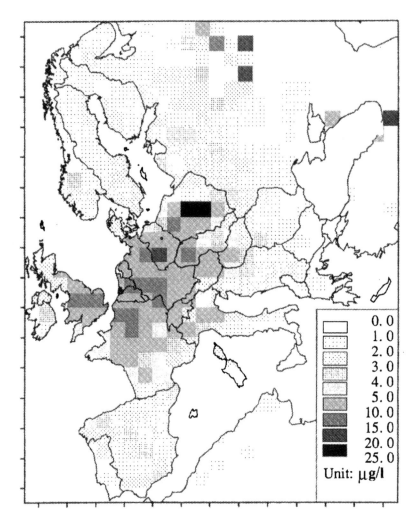

Figure 4.11   Concentration of lead in European rainfall in the mid-1980s. (From Alcamo, J., 1991. *Options*,
September, International Institute of Applied Systems Analysis, Laxenburg, Austria. With permission.)

On land, lead recirculation flows out from cities along transportation corridors, highways, and
railroads in lead gasoline additives, fuel combustion, in lead batteries, automobile dumps, and solid
waste deposits (Nriagu, 1978). Increasingly now, the concentrated lead of batteries is recycled to
battery recovery plants that reuse the lead. However, when they are too dilute for economic recovery,
the dilute wastes require wetland recovery (see Chapter 11).

Many papers have reported the decreasing concentrations of lead out from emission centers.
With a dense pattern of sampling and analysis, for example, Simpson (1985) used a statistical
method (Kriging) to locate isopleths of lead concentration in soils at distances of 0 to 12,500 ft
from a lead smelter. The dispersal of heavy metals from a hierarchical energy processing center
provides a neat example of how matching of outward dispersal feedbacks of high impact can result
from the inward converging that concentrates value and transformity (Figure 4.3a). Impact on
environment decreases with lead concentration, and dispersal decreases concentration. The impact
of the dispersing emissions is commensurate with the emergy concentrated by the earth and/or
humans toward the center. The impact forms patchy circles of concentration measured by transformity of the dispersal and the empower density of the landscape work. Hierarchical element distribution is readily understood in situations with anthropogenic point source pollution, but hierarchical
elemental distributions are usual in nature and in civilized landscapes where mechanisms are many.

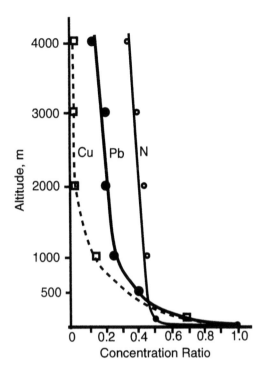

Figure 4.12   Distribution of atmospheric heavy metals with altitude: fraction of those at 50 m. N = condensation nuclei. (Zhigalovskaya et al. given by Dobrovolsky, 1994.)

FREQUENCY DISTRIBUTIONS

Frequency distributions are bar graphs that show many areas with small concentrations and few areas with high concentrations. These graphs have a typical hollow shape representing the common and the rare. See, for example, in Figure 4.13, the distribution of lead in granite rocks (Ahrens, 1954). Many theoretical papers consider the uneven distribution of geochemical elements, often represented with log-normal plots. See, for example, Skinner (1986), Miller and Goldberg (1955), Middleton (1970), and Roberts et al. (1998). Graphs of decreasing frequency with increasing concentration result from the spatial pattern that goes with the energy hierarchy (Figure 4.3). The details and mechanisms may vary, but the most general explanation seems to be that materials are distributed according to the universal energy hierarchy, possibly in all systems.

Figure 4.13a is the distribution of lead in crustal rocks, which Ahrens (1954) showed was a close fit to a lognormal distribution. On any scale it takes available energy appropriate for that scale to concentrate materials. For example, Genoni (1998) relates the Gibbs free energy used up in concentrating chemical substances to higher specific Gibbs free energy; or when generalized to available energy of all kinds, the principle is expressed in emergy terms. Emergy has to be used to concentrate materials, as already explained with Figure 4.7. Depending on its nature, each kind of material has a range of emergy per gram that determines its place of cycling in the universal energy hierarchy. From a dispersed state over large areas it is coupled to the self-organizational concentrating and diluting circulation (Figure 4.3). The emergy available to a landscape (solar energy and energy from geologic processes below) is proportional to the area, but at each hierarchical step transformation, the emergy is concentrated at a hierarchical center and the materials with it. Thus, the skewed pattern of chemical distributions in the environment may be

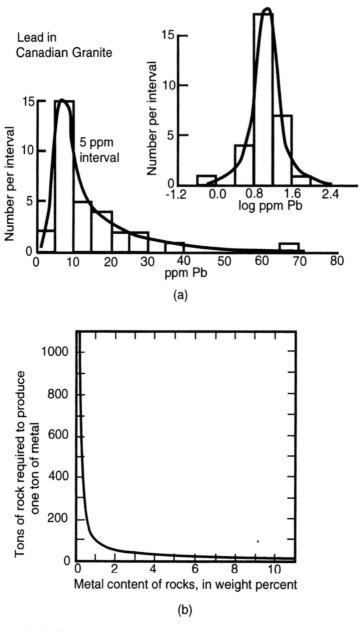

Figure 4.13   Metal distributions and energy hierarchy. (a) Lognormal distribution of lead in granite rocks. (From Ahrens, L.H., 1954. *Geochimica et Cosmochimica Acta*, 5:49–73; [Part 2] 6:121–131. With permission from Elsevier Science.) (b) Rock required to concentrate metals (Page and Creasy, 1975).

explained by the coupling of materials to the emergy concentrating, transformity increasing pattern of universal energy hierarchy. The steep left side of the distribution connects the general background concentration (the peak of the distribution dependent on the crustal abundance) to the minimum concentration, something greater than zero, the realm of small-scale processes.

In human economic self-organizing, as with geobiologic self-organizing, available energy concentrates and transports materials to centers. Huge fuel and electrical energies are used by industry to mine and process the heavy metals from landscape to city use. Page and Creasy (1975) published steep hollow curves for resources required to concentrate ores of different concentration (Figure 4.13b). Expressing those results in another way, for the same fuel, the higher the levels of concentration the less quantity is transformed. In other words, distribution of metal concentration within civilization is like that in nature for a similar reason, the coupling of materials to the energy hierarchy.

The spatial organization of cities also has the pattern in Figure 4.3, with empower density (solar emjoules per area per time), transformity, and money circulation increasing toward the center (Odum, 1996). Rolfe et al. (1972) and Rolfe and Haney (1975) mapped the way heavy metals circulated with society and environment in Urbana, IL. There were annual pulses, with most lead immobilized in soils and stream sediments without biological magnification. Palm and Ostlund (1996) measured lead flows in Stockholm, Sweden. Heavy metal concentrations and their high transformities have parallel distributions increasing toward the city center. The emergy measures show where circulating materials tend to concentrate during the self-organization of the economy and environment. According to the theory, materials will tend to interact and impact (for benefit or disruption) with items in the landscape with transformities within one or two orders of magnitude.

HUMAN INTERACTIONS WITH LEAD

Several authors have summarized the distribution and flows of lead in normal human beings with diagrams of the daily budget of lead (Patterson, 1973; Rabinowitch et al., 1976; Fergusson, 1990). Values from the later budget have been applied to an energy systems diagram in Figure 4.14, which has the parts of the human arranged according to the energy hierarchy with the brain to the right, the most controlling and valuable component.

According to energy hierarchy concepts, self-organization reinforces the interaction of items which can amplify the productive output of their mutual participation. Typically, a small flow of higher transformity can have most effect by interacting with a matching flow with more energy but somewhat lower transformity flow. Heavy metals with high transformities thus have amplifier, controlling effects by interacting with parts of the biogeosphere with lesser positions in the energy hierarchy. Such interactions are system reinforcing where the interaction produces a useful output.

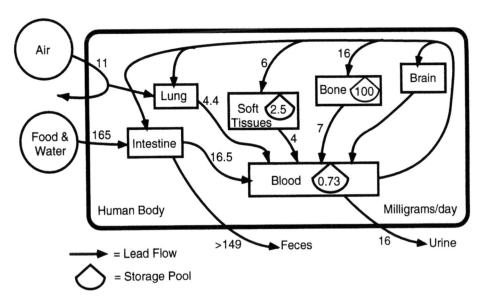

Figure 4.14   Energy systems diagram of the flows and storages of lead in a human being using values from Fergusson (1990).

However, heavy metals can combine and divert many aspects of the physiological system of life (enzymes, chlorophyll operations). Toxicity develops when high transformity substances are not appropriately organized with beneficial interactions.

Human beings, their brains, and their information processing have higher emergy per gram and transformity than heavy metals. In a functioning organization, the humans, their brains, and information interact with, but control, the lead processes and cycles. The system is functionally disturbed when the reverse happens and the heavy metals affect the humans by interfering with the living physiology. Förstner and Wittmann (1979) reviewed damage of lead poisoning to brains and kidneys, often causing early deaths.

Thus, a good system channels high transformity flows to reinforce the larger system functions, organizing to insulate their capabilities of disruption. The long-evolving biogeosphere does this by providing humic substances in all its ecosystems, especially in wetlands where peat deposits helped regulate the earth's heavy metal cycles. Humans are learning to isolate pathways by controlling the use of lead in paints, in gasoline, and batteries that can impact humans or the environment.

EVALUATION PERSPECTIVES

Nriagu (1994), in a later summary of lead in the environment, finds soils as lead sinks, aquatic environment as most vulnerable, less lead in the atmosphere than earlier, but metal pollution still increasing. With the U.S. using 22%, the world from 1901 to 1990 received 2 million tonnes of lead and 13 million tonnes of zinc from industrial emissions. Lead emission from energy processing was estimated as 13, mining 2.6, smelting and refining 23, waste incineration 2.4, leaded gas 250, and total industrial 330. Total natural emission was 12 million kg/year.

Skinner (1986) ranked heavy metals by the concentration over the general background level of the earth's crust necessary to be commercial. The more abundant the element, the less cost in concentrating and the higher the percent required for mining. Whereas commercial iron and aluminum deposits need to be 25 and 30% (5 times background), zinc needs to be 2.5% (300 times background) and lead 4.0% (4000 times background). The higher the crustal abundance, the larger are the sizes of the largest deposits discovered. For many elements there is a thousand times greater energy requirement for retrieving metals from crustal rocks compared to mining sulfide ores. The sulfide ore bodies are hierarchical centers of geologic work (example: volcanoes) with high transformity. Their concentration was made possible by large earth empower processing. The concluding implication was that ore body metals are economical, but those dispersed in the crustal rocks are usually not.

Emergy per mass suggests the appropriate policy regarding recycle. Wastes with concentrations above about 10,000 g/m^3 (Figure 4.7) have enough value to justify economic reuse. Wastes with uneconomical concentrations may still have enough emergy per mass to make a contribution to the natural recycle of land and water.

As explained in Chapter 1 (Figure 1.8), emergy flows have emdollar equivalents for evaluating processes in terms of gross economic product. For perspective on the importance of lead, global flows are given their emdollar values in Table 4.1. The value of lead ores used by the economy is 12 billion emdollars per year, 200 times more contribution than in the lead dispersed in earth of the main land cycle. Lead dispersed in wastes is about 71 million emdollars, similar in magnitude to the lead in the natural cycle.

Lead in a Cypress-Gum Swamp, Jackson County, Florida

Part II has four chapters summarizing the studies of a lead-filled swamp in Florida. Chapter 5 contains the ecological studies by Lowell Pritchard, Jr. Chapter 6 contains the chemical studies by Shanshin Ton and Joseph J. Delfino. Chapter 7 has the studies of leaded wetland microcosms by Shanshin Ton. Finally, Chapter 8 reports the computer simulation of a model of lead in the wetland by Shanshin Ton and Howard T. Odum.

Ecological Assessment of the Steele City Swamps*

Lowell Pritchard, Jr.

CONTENTS

At the time of this study the Steele City Swamps showed varying degrees of impact, caused less by the large quantities of lead that was filtered than by the high acidity of the battery plant water. Figure 5.1 shows the distribution of lead in sediments, with larger values upstream and along the pathway of water flow. At Station A (Figure 5.2) there were a few dead trees, and the wood (in cross section) was strangely red colored. Much of the pond at B was devoid of trees but covered in part with floating plants (*Nymphaea*). The waters were turbid with underwater objects visible only a few inches below the surface. Gum trees were scattered at Stations C and F, but these had just a few leaves each. In backwater areas (Stations D and H) more trees were found including cypress. In other words, trees were absent in the areas where the most lead had been absorbed (Figure 5.2). Station RF (reference forest) was in a pond which had not received lead wastes and had a normal concentration of cypress and swamp black gum trees.

A limited ecological assessment was made by testing seedling survival and growth in the field, by measuring the area of green leaves, by estimating plant productivity with two methods, and by sampling enough underwater invertebrates to calculate indices of biodiversity. Details on the methods are given in Appendix A5A.

---

* Condensed by the Editor.

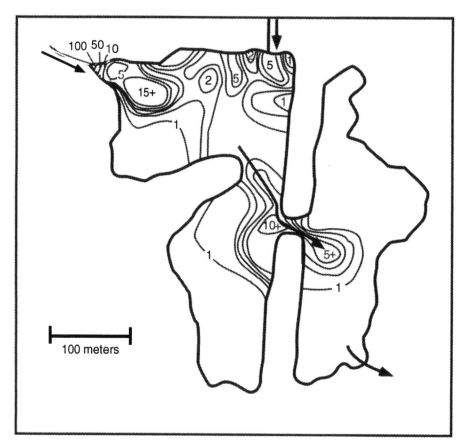

Figure 5.1 Sediment lead concentrations in Steele City Bay, in hundreds of parts per million sediment by weight. Isoconcentration lines were calculated with data from Mundrink (1989).

TOXICITY ASSESSMENT WITH PLANTED TREE SEEDLINGS

Tree seedlings (pond cypress, bald cypress, and swamp black gum) were planted where bare bottoms had become exposed, measured September 24, 1990 and monitored again June 6, 1991. In the interim there were very high water levels because of heavy rains. Since wetland tree seedlings die if covered with water, mortality was large (Table 5.1). Surviving seedlings showed growths 10 to 30 cm. Growth and survival at station F may indicate that the sediments, although still containing lead (Chapter 6), were not toxic to new seedlings.

BASAL AREA OF TREE TRUNKS

By measuring the diameter of live trees, the area of the trunk cross sections may be calculated, which is called the forest basal area. Table 5.2 shows zero values in the upper stations without trees, increasing downstream to 57 cm²/m² of land, a bit less than that in the reference forest.

GREEN LEAF AREA

In a normal wetland forest there are several layers of leaves on top of each other. The area of leaves per area of ground below is called the leaf area index. Leaf area of the sparse tree areas was

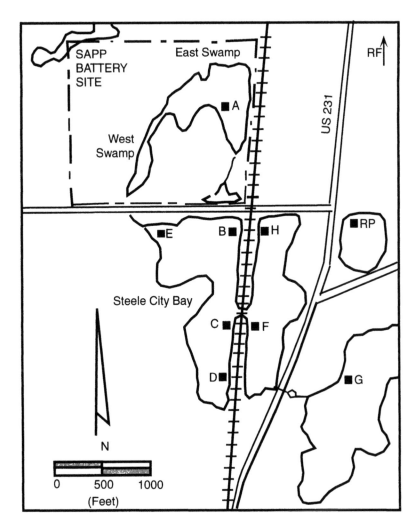

Figure 5.2 Site map of Sapp Battery, Steele City Bay, and other downstream wetlands. Locations of samples and measurements are shown. Overall drainage pattern in this area is from northwest to southeast.

Table 5.1 Tree Seedling Survival

Location	Species	Number Planted	Number Surviving	Average Growth of Survivors (cm)
F	TAAS	21	10	20
	TADI	21	12	10
	NYSY	21	5	29
G	TAAS	20	0	—
	TADI	20	0	—
	NYSY	20	0	—
H	TAAS	14	0	—
	TADI	13	13	29
Reference forest	TAAS	20	0	—
	TADI	20	19	10
	NYSY	20	0	—

Note: TAAS = *Taxodium ascendens*; TADI = *T. distichum*; NYSY = *Nyssa sylvatica* var. *biflora*.

Table 5.2 Summary Statistics for Water Lilies and Trees

Location	Water Lilies LAI (mean ± S.D.)	Trees Basal Area (cm²/m²)	Trees LAI (mean ± S.D.)	Trees Leaf Area/Basal Area (m²/m²)
A	0.0 ± 0.0	0.0	0.00 ± 0.000	—
B	1.2 ± 0.3	0.0	0.00 ± 0.000	—
C	1.3 ± 0.5	0.0	0.00 ± 0.000	—
D	1.1 ± 0.1	0.0	0.00 ± 0.000	—
F	1.1 ± 0.2	10.1	0.04 ± 0.007	42
G	1.1 ± 0.3	26.8	0.25 ± 0.004	95
H	0.0 ± 0.0	57.0	2.10 ± 1.527	368
RefFor	0.0 ± 0.0	77.9	2.85 ± 1.807	366
RefPond	0.7 ± 0.2	0.0	0.00 ± 0.000	—

Note: LAI = leaf area index; RefFor = reference forest; RefPond = reference pond; S.D. = standard deviation.

measured by the annual litterfall into baskets set above the water (Table 5.2). In a fully developed forest there may be 4 to 7 m² of leaves per square meter of ground, but much less (0 to 2.9) in the impacted wetland with few healthy trees remaining (Table 5.2). The index was about 3 in the reference wetland without lead (Figure 5.3, Table 5.3).

Floating lilies (*Nymphaea odorata*) were present at Stations B, C, D, F, and G and the index on July 7, 1991 ranged from 0.70 to 1.25 (Table 5.2). A statistical analysis of variance comparing values from the different stations did not find any difference between stations that was greater than the general variation among data samples.

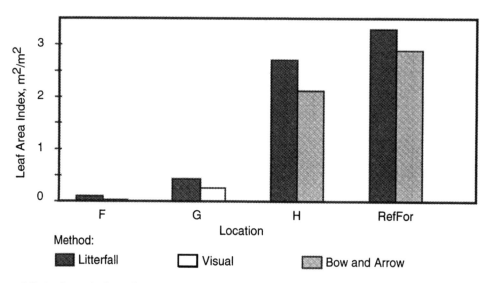

Figure 5.3 Leaf area indices of trees using several estimation methods.

Table 5.3 Litterfall Trap Data and Calculations

Location	Mean ± S.E. Litterfall (g/m²)	Fraction Canopied	Corrected Litterfall (g/m²)	Leaf Area/Mass Ratio (m²/g)	Corrected LAI (m²/m²)
F	90.3 ± 23.9	0.079	7.2	0.00985	0.068
G	288.6 ± 93.0	0.139	40.1	0.00985	0.378
H	355.5 ± 29.6	0.800	284.4	0.00985	2.682
RefFor	348.2 ± 67.0	1.000	348.2	0.00985	3.284

Note: S.E. = standard error of the mean; LAI = leaf area index; RefFor = reference forest.

PRODUCTIVITY OF EMERGENT PLANTS

Since cypress and black gum drop their leaves in winter, and water lilies in this area die back in winter, the leaf area at the end of summer is a measure of a year's leaf production. However, the total organic matter made by plant photosynthesis (gross production) is much greater than this estimate of net storage in the leaves, since much of the organic matter that was made during the year went to support necessary respiration of leaves, limbs, trunks, roots, and insects. The ratio of gross plant production to net production from studies of similar wetlands was used to estimate gross production (Table 5.4).

METABOLISM OF THE UNDERWATER ECOSYSTEM

The oxygen generated by the photosynthesis of underwater plants (algae and macrophytes) goes into the water as dissolved oxygen, increasing during hours of sunlight. At night plants, animals, and microbes use this oxygen to operate their normal metabolism, and the dissolved oxygen goes down (community respiration). Oxygen also diffuses into waters from the atmosphere until the molecules going into water equal those diffusing out (equilibrium). If the dissolved oxygen in the water gets higher than the atmosphere's equilibrium level, it diffuses out. If the dissolved oxygen is less than the equilibrium level, then oxygen diffuses in. Summarizing these processes, the dissolved oxygen is the balance between gross photosynthesis and diffusion in, minus community respiration and diffusion out.

By measuring the dissolved oxygen every few hours for 24 h or more and plotting a graph, a curve of dissolved oxygen is usually observed going from a minimum at sunrise after a night of respiration to a maximum near sunset after a day of photosynthesis. With methods given in Appendix A5A, one may subtract out the diffusion so that respiration can be calculated from the oxygen

Table 5.4 Ecosystem Productivity in Sampled Locations

| | Gross Primary Productivity (E6 J/m²/year) | | | | Empower[d] (E10 |
| | Herbaceous | | | | |
Location	Canopy[a]	b	Aquatic[c]	Total	sej/m²/year)
A	0	0	7	7	0.9
B	0	34	3	38	5.0
C	0	37	11	48	6.3
D	0	32	16	48	6.3
F	3	33	8	44	5.8
G	16	34	9	59	7.8
H	134	0	0	134	17.6
RefFor	185	0	0	185	24.3
RefPond	0	21	15	35	4.7

Note: RefFor = reference forest; RefPond = reference pond; GPP = gross primary production; LAI = leaf area index.

[a] GPP was calculated for the reference forest using an LAI/GPP regression from data on wetland forests in Brown et al. (1984: p. 317). GPP for trees in other locations is a fraction of reference forest GPP based on the ratios of their LAI.

[b] GPP was calculated for *Nymphaea* by setting the highest LAI equal to two times a conservative estimate of freshwater marsh net primary production (1000 g dry wt/m²; Mitsch and Gosselink 1986: p. 274) and assigning productivities based on the ratios of LAI.

[c] Aquatic GPP is the average of summer and winter values from Table 5.5. GPP in units of grams O_2 m²year^{-1} was multiplied by 3.5 kcal/g O_2 (Cole 1975: p. 179), 4186 J/kcal, and 365.25 d/year.

[d] Using the emergy/hectare calculated for Northwest Florida wetlands in Pritchard (1992, Appendix F), a transformity for gross primary productivity (GPP) of the reference forest was calculated (1317 sej/J). Empower of run-in and rain was divided by the energy flow. This transformity was used to convert the GPP at other locations to emergy.

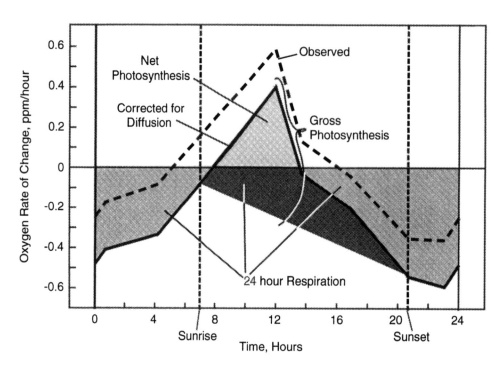

Figure 5.4 Example graph for calculating aquatic metabolism from the rate of change of dissolved oxygen, in parts per million per hour.

decrease at night. The oxygen increase in daytime is the gross photosynthesis minus the concurrent respiration. You can add the night respiration to the daytime oxygen increase to get an estimate of gross photosynthesis.

It helps to plot the hourly changes in dissolved oxygen on a "rate of change" graph after corrections for the diffusion (Figure 5.4). Points above the horizontal line are increases (positive), and points below the line are decreases (negative). The shaded area above the line is net photosynthesis, and the shaded area below the line is night respiration. To get the gross photosynthesis, an estimate is made of the daytime respiration using some assumption about the way it varies. In Figure 5.4 the daytime respiration (darkly shaded) is made to increase during the day, based on the idea that the more sugar made by the plant the more respiration there is. See Appendix A5A for more details.

In Figure 5.4 the respiration (below the line) is greater than the daytime net photosynthesis (above the line). In other words, in the course of a day and night, more oxygen is used than is produced. This means that the swamp water remains below equilibrium with the atmosphere most of the time, a condition resulting from the quantities of decomposing organic matter from litterfall stream transport, tree roots, and animals.

The results of diurnal curve measurements of oxygen and the resulting calculations of metabolism are summarized for several stations in Table 5.5 and Figure 5.5. As expected, respiration was consistently higher than underwater photosynthesis; rates were higher in summer, and slightly higher downstream.

The gross primary production of the below-water ecosystem and the above-water canopy estimates were combined in Figure 5.6. Because so many trees were missing, the upper stations had lower totals.

INVERTEBRATE ANIMALS

With procedures detailed in Appendix A5A, small animals, mostly aquatic insects, were sampled with a cylindrical collector from underwater and the taxonomic family identified. (Lists of types, dates, and stations are in Appendix A5B, Tables A5B.2 and A5B.3.) The numbers of animals per square meter (density) increased a little downstream away from the lead source (Figure 5.7, Table 5.6).

Table 5.5 Total Aquatic Metabolism for Locations in and around Steele City Bay

Location	Distance to Site (m)	Depth (m) S	Depth (m) W	P_g (g O_2/m³/d) S	P_g (g O_2/m³/d) W	R_{24} (g O_2/m³/d) S	R_{24} (g O_2/m³/d) W	P_g (g O_2/m²/d) S	P_g (g O_2/m²/d) W	P_g (g O_2/m²/d) AVG	R_{24} (g O_2/m²/d) S	R_{24} (g O_2/m²/d) W	R_{24} (g O_2/m²/d) AVG
A	0	—	0.6	—	1.2	—	1.7	—	0.6	1.3	—	0.9	2.4
B	40	0.8	0.9	1.2	0.3	3.7	0.8	0.9	0.3	0.6	3.0	0.7	1.9
C	244	0.7	0.9	3.8	1.3	6.1	2.0	2.9	1.1	2.0	4.6	1.7	3.2
F	259	0.4	0.7	3.7	2.8	8.4	4.6	1.3	1.8	1.6	2.9	3.0	3.0
D	387	0.9	1.0	5.5	1.0	6.2	1.4	4.9	1.0	3.0	5.6	1.4	3.5
G	700	0.5	0.7	5.8	0.8	9.7	2.0	2.9	0.6	1.7	4.9	1.4	3.1
RefPond	800	—	0.7	—	1.3	—	1.9	—	0.9	2.7	—	1.3	3.0

Note: P_g = gross primary productivity; R_{24} = 24-h community respiration; S = summer measurement, August 21, 1990; W = winter measurement, February 1, 1991; AVG = average; g = grams; m = meters; d = days; O_2 = dissolved oxygen.

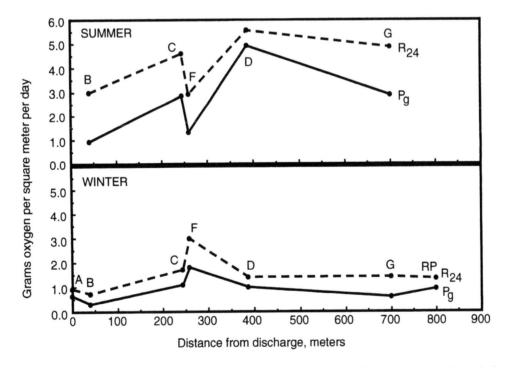

Figure 5.5 Aquatic metabolism on area basis. P_g = gross primary production; R_{24} = 24-h community respiration.

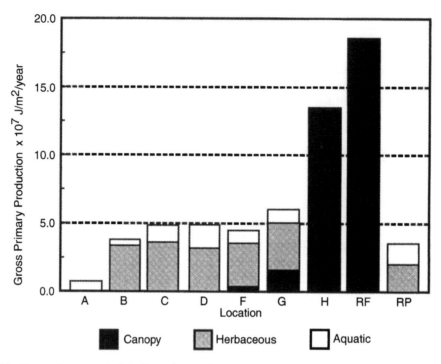

Figure 5.6 Gross primary productivity for each sample location estimated from diurnal oxygen curves, litterfall, and/or leaf area indices.

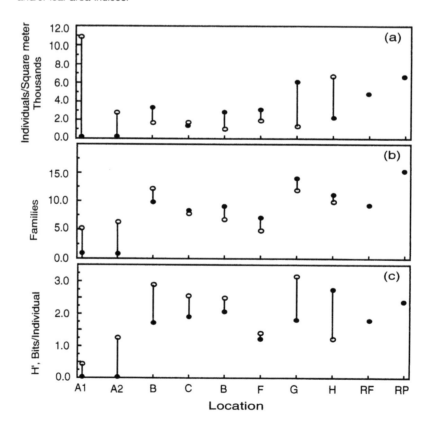

Figure 5.7 Indices of macroinvertebrate community structure for sample locations: (a) density, (b) family richness, (c) Shannon diversity. Sample dates: 08/21/90 (●) and 02/03/91 (○).

INDICES OF BIODIVERSITY

Often pollution causes the biodiversity of ecosystems to diminish. There may be large numbers of one species (example: Station A1 in Table 5.6). In this study four indices of diversity were calculated from a series of winter and summer data on invertebrates (Table 5.6). Details on these indices are included in Appendix A5A:

S is the number of kinds (families) represented (richness)
H is the "information theory content in bits per individual (Shannon)"
D is the Simpson index
M the number of species in a set of individuals counted (Margalef)

In Figure 5.8 three indices were adjusted to a common scale showing similarity among indices. The lowest diversity values were in areas formerly most stressed with lead-acid (Stations A and F), but for most of the stations the results were highly variable, not consistent indicators of lead content.

OVERVIEW CONCLUSION

The measurements in this chapter show that, in the absence of a tree canopy, the processes of the aquatic ecosystem are dominant and returning to normal, with restoration little affected by residual lead. A much longer time may be required for restoration of the wetland forest, a delay inherent in the slower turnover time and reseeding of trees.

Table 5.6   Diversity Indices for Macroinvertebrates

Location	Distance (m)	S	N	Density (N/m²)	Shannon H'	Shannon s²	Simpson's D$_s$	Simpson's S²	M$_a$
				August 21, 1990					
A1	0	5	251	10780	0.42	0.01	0.115	0.001	0.7
A2	0	6	63	2706	1.23	0.04	0.436	0.004	1.2
B	40	12	41	1761	2.87	0.05	0.806	0.003	3.0
C	244	8	38	1632	2.52	0.03	0.801	0.002	1.9
D	259	7	26	1117	2.46	0.04	0.806	0.003	1.8
F	387	5	46	1976	1.36	0.04	0.504	0.005	1.0
G	700	12	31	1331	3.12	0.05	0.871	0.002	3.2
H	500	10	154	6614	1.24	0.02	0.347	0.002	1.8
				February 3, 1991					
A1	0	1	3	129	0.00	0.00	0.000	0.000	0.0
A2	0	1	2	86	0.00	0.00	0.000	0.000	0.0
B	40	10	77	3307	1.72	0.05	0.511	0.004	2.1
C	244	8	35	1503	1.91	0.07	0.617	0.007	2.0
D	259	9	65	2792	2.08	0.04	0.671	0.003	1.9
F	387	7	72	3092	1.23	0.04	0.414	0.004	1.4
G	700	14	141	6056	1.84	0.03	0.534	0.002	2.6
H	500	11	54	2319	2.73	0.04	0.797	0.002	2.5
RefFor	NA	9	11	4767	1.80	0.03	0.558	0.003	1.7
RefPond	NA	16	153	6571	2.38	0.02	0.691	0.001	3.0

Notes: Distance is distance from site; S is number of families; N is number of individuals; H' is Shannon diversity in bits/individual; D$_s$ is Simpson diversity; and M$_a$ is Margalef diversity. Reference sites were added for the winter sampling event.

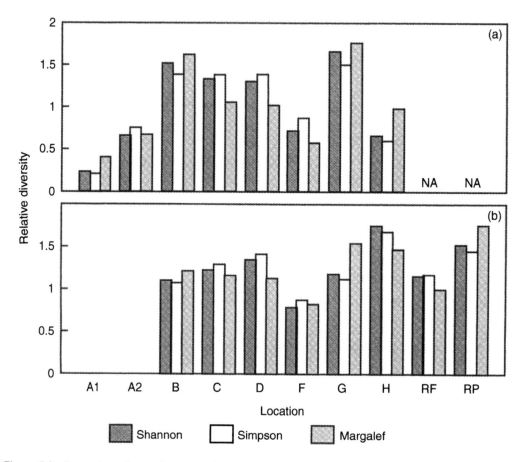

Figure 5.8 Comparison of normalized macroinvertebrate diversity indices for locations A to H, reference forest (RF), and reference pond (RP). (a) 08/21/1990, and (b) 02/03/1991.

CHAPTER 6

Lead Distribution in Steele City Swamps*

Shanshin Ton and Joseph J. Delfino

CONTENTS

In 1970, Sapp Battery Service, Inc. initiated its operations to process lead recovery from used automobile batteries. The company gradually expanded its operation to process approximately 50,000 used batteries per week in 1978. Wastes from operations were dumped outside the plant and allowed to run through adjacent wetlands, finally being discharged to Steele City Bay (Figure 6.1).

After 7 years of operation, in 1977, the first complaint about damage to cypress trees in adjoining wetlands was reported to the Florida Department of Environmental Regulation (FDER). FDER closed the site in January 1980.

After the site was abandoned, EPA undertook emergency cleanup actions under provisions of the Clean Water Act, Section 311. The Sapp Battery site was included on the final National Priorities List in August 1982.

After that date, EPA cooperated with FDER to conduct the Remedial Investigation/Feasibility Study (RI/FS). In this study, the on-site soils, groundwater, surface water, and sediments were examined.

In 1985, Ecology & Environment, Inc. (E&E) became involved in field investigations to further delineate the extent of the contamination. Another draft feasibility study was finished by E&E in January 1987. See Appendix Table A6B.2.

The analyses of lead were made from April 1989 to September 1992. Samples first collected at stations A through G (Figure 6.1) later extended further downstream in a series of sites: A, B, C, F, OF1, G, and OF2 (Figure 1.3). Descriptions of the sampling sites are given by Ton (1993). The chemical methods used are given in Appendix A6A, and Appendix A6B has a tabular listing of data on lead in waters, sediments, vegetations, and related limnological data for April 1989 to May 1992.

---

* Condensed by the Editor.

1-56670-401-4/00/$0.00+$.50
© 2000 by CRC Press LLC

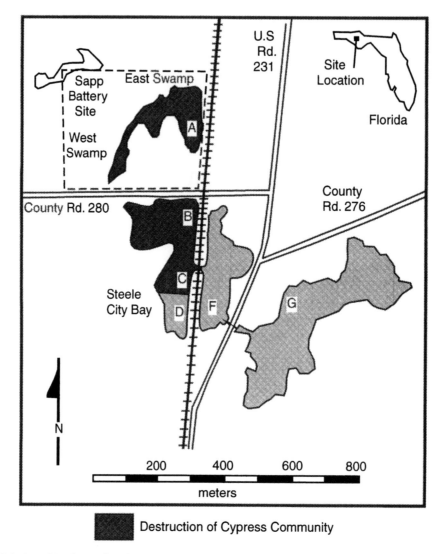

Destruction of Cypress Community

Figure 6.1 Location of sampling sites, Jackson County, Florida (Ton, 1990).

LEAD IN SURFACE WATERS

Lead concentrations in surface waters decreased downstream (Figure 6.2). Concentrations in most samples were less than those reported in earlier years (Appendix Tables A6B.2 and A6B.3), and most concentrations were less than the 0.03 mg/l regarded as safe for recreation, fish, and wildlife.

LEAD IN SEDIMENTS

Concentrations of lead in sediments also decreased downstream (Figure 6.3). For the most part, lead concentrations were highest in the surface sediments with lower concentrations 15 to 45 cm below (Figure 6.4). These results indicated that the distribution of lead in sediments corresponds to the surface water drainage pattern. Relatively low lead concentrations outside the boundary of the study area suggest that the wetland acts as a filter to retain lead (Appendix Table A6B.5).

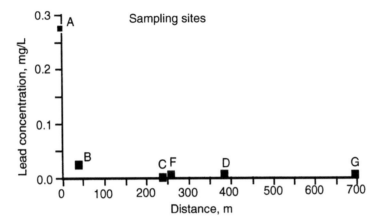

Figure 6.2 Concentration of lead in surface waters as a function of distance from original discharge (Station A in Figure 6.1) (Ton, 1990).

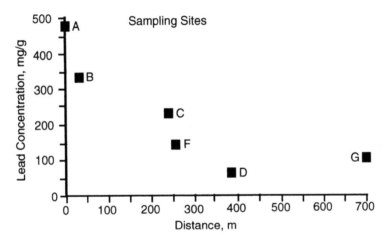

Figure 6.3 Concentration of lead in upper sediments (0 to 15 cm) as a function of distance from original discharge (Station A in Figure 6.1) (Ton, 1990).

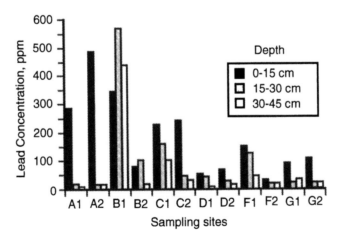

Figure 6.4 Lead concentrations in sediment profiles at Stations A through G in Figure 6.1 (Ton, 1990).

LEAD IN VEGETATION

Water lily (*Nymphaea odorata*) was the most popular species for the entire wetland, except site A. Leaves, stems, and roots of water lilies were separated for lead analysis. Generally, concentrations of lead in leaves and stems were slightly higher than those in roots. However, high concentrations of lead accumulated in roots were found commonly in other species (Appendix Table A6B.8).

CHEMICAL PROPERTIES OF LEAD IN SEDIMENTS

Sequential chemical extraction (Appendix A6A, Table A6A.1) was used to separate six components of lead in sediments, each followed by lead determinations. Exchangeable lead was extracted with potassium nitrate solution; adsorbed lead was removed with potassium fluoride; organically bound lead with sodium phosphate; inorganic precipitated lead with EDTA; sulfide lead with nitric acid; and residual lead. Results are given in Appendix A6B, Table A6B.9 and Figure 6.5.

BINDING OF LEAD TO HUMIC SUBSTANCES

As already known from the literature and confirmed with the study of lead fractions (Figure 6.5), much of the lead combines with humic substances. To measure the binding to the humic substances in the waters of the study area, 40 gal of surface water was collected from the control pond near the Sapp swamp in June and July 1991. A dialysis apparatus was set up (Figure 6.6) so that lead in solution on one side of a membrane could diffuse through the tiny pores, some becoming

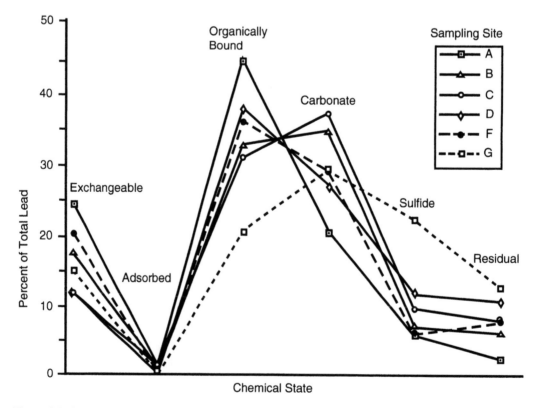

Figure 6.5   Percentages of total sedimentary lead in each of six fractions at six stations in Figure 6.1 (Ton, 1990).

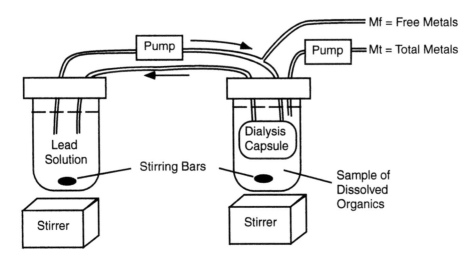

Figure 6.6 Dialysis apparatus used for measuring the binding of lead by humic substances.

bound by the humic substances in solution on the other side. One side of the membrane had lead solution only, whereas the other side had lead solution plus that bound in the organic matter. After suitable calculations were made using Scatchard graphs (Appendix Figure A6A.3), the ratio of lead bound to humic matter was found to be about 1.5 g lead per kilogram organic matter, slightly more at higher pH, and about half the binding by a sample of humic materials (Aldrich Chemical Co., purified and freeze dried by Davis [1993]).

A small amount (2.9 to 6.2%) of the organic humic molecules diffused in the other direction across the membrane, as measured with an instrument measuring the absorption of ultraviolet light. The lower the pH the more diffused, suggesting an effect of pH in making the molecules more compact (already known from past publications). The humic substances from the bay diffused more than the sample of standard humic substance (less than 1%). In other words, the humic molecules from the control swamp were smaller than those in the standard humic material.

CHAPTER 7

Experiments with Lead and Acid in Wetland Microcosms*

Shanshin Ton

In order to study the toxicity of lead in wetland conditions like those of the Steele City Bay, 129 microcosms were made, each with a seedling of cypress or swamp black gum (Figure 7.1; Appendix Table A7.1). A microcosm is a small ecosystem developing in a container. In this experiment peaty materials and a seeding of life from the outdoor swamp were placed in each container and allowed to develop for 2 months. Each microcosm had a teaspoon of fertilizer nutrients placed under the root zone. Seedlings were added, and the microcosms were arranged at random in a greenhouse. Arrangements were made for chemical solutions containing lead and acid in different concentrations to drip into these containers (Figure 7.1). In this way, some conditions in the Sapp swamp were duplicated. All microcosms received similar quantities of water so that the effect of water on growth was made uniform. Microcosms were not crowded together so that there was abundant natural light. Height of seedlings was measured and related to the treatments.

Lead concentrations applied were 0.5 and 1.0 µg/ml (a microgram per milliliter [µg/ml] is the same as a gram per cubic meter [g/m^3]). Solutions with zero lead were also applied. A comparison without special treatment is called a control. Because some growths developed in the tubes delivering the waters with lead, the concentrations actually reaching the microcosms, when analyzed, were found to be somewhat less than the intended concentration. (See Appendix Table A7.2.)

Acidity like that in the Steele City Bay was duplicated in microcosm by adding sulfuric acid to the dripping waters. pH is the scale of acidity (concentration of hydrogen ions on a logarithmic scale; the lower the value the more acid; pH 4 is ten times more acid than pH 5). Treatment waters dripping into microcosms for 8 months included pH 2, 4, and the neutral pH 7. Neutral pH is neither acid nor basic and was included for comparison (control). Cypress and gum trees are found naturally in waters between pH 4 (normal acidity of swamps) and pH 8 (the slightly basic conditions of most spring waters in Florida).

Microcosms were started in June 1991 and a period of 2 months was designated for the stabilization of seedlings and the development of a small ecosystem in the container. Figure 7.2 shows the average addition to height of seedlings over the 16 months of growth starting in August 1991. Both cypress and gum are deciduous, dropping leaves in November and replacing them in April. Thus, there was about 2 months growth before the leaves fell. Growth stopped for about 4 months, resuming after the leaves returned.

The rate of growth (centimeters increase in seedling height per month) for this period can be observed as the changes between monthly observations in Figure 7.2. Notice that growth was

---

* Condensed by the Editor.

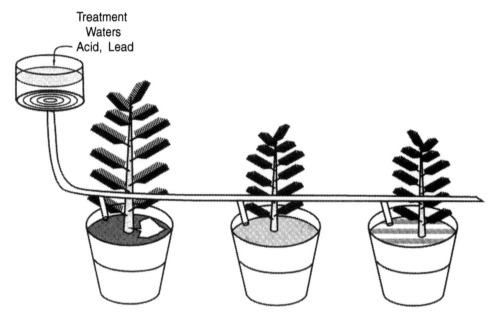

Figure 7.1 Sketch showing seedling microcosms and arrangement to supply steady flow of water, acid, and lead.

moderate at first during the autumn months when light was less, stopping during the leafless winter months, accelerating during the summer, and stopping again during the next winter. Under these conditions, the gum seedlings were slow to start growth and the cypress were slightly taller at the end. As often observed when new seedlings are planted, growth accelerates as the leaf area and root systems increase, capturing more sunlight for photosynthesis and transpiration (processing of water from roots through microscopic leaf holes into the air, keeping leaves cool and drawing up nutrients).

At the end of the summer of the second year, growth stopped 3 months before the normal time for leaf fall. This may be due to a combination of nutrient limitation and water stress which result from higher respiration rate of rapidly growing seedlings during summer months. Due to the large size of the seedlings and the hypothesis of nutrient limitation and water stress, I cut the seedlings to their original height before the next growing season. They did regrow in the third year from March to July 1993 before I stopped the experiment.

Because the experiment had many replications (six duplicates for each treatment or control), it was possible to use statistical methods to learn if the differences between treatments were greater than the variation among microcosms with the same treatment. Thus, an analysis of variance was made of the height growth data as detailed in Appendix Tables A7.3 to A7.5. None of the treatments with lead or acid or their combinations was proven to have any effect greater than the general variation among the replications. Therefore all the data were averaged in Figure 7.2. Because of the self-organizing abilities of a set of species in an ecosystem, isolated microcosms, although started similarly, normally develop different individual characteristics, adding variability to micro-cosm research.

The range of acidity and lead in the Steele City Bay swamps from 1990 to 1993 was less toxic (lower pH [1 to 2] and higher lead concentrations [3 μg/ml]) than that which occurred during the period of battery washing when the trees were killed. The range of acidity and lead treatments in the microcosm was that of the present swamps many years after the initial toxicity.

Therefore, the question addressed in the microcosm study was whether the present levels of acid and lead were toxic to regrowth of swamp trees. The results showed that the present levels

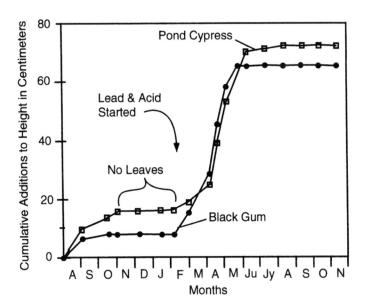

Figure 7.2 Cumulative record of average height increases in growth of seedlings in wetland microcosms 1991 to 1992. The change between points is the height increase per month (centimeters per month).

were not toxic to the two species of swamp trees predominant in this area. Cypress and gum require bare swamp soils for germination and regeneration. Because the waters had been dammed, there was no normal hydroperiod, and thus no bare swamp exposed in drought periods. The field observations showed many cypress seedlings growing back along the edges of the swamp where germination was possible. There was abundant growth of floating plants in open waters. It may be concluded that the present conditions are not toxic to regrowth.

CHAPTER 8

Simulation Model of a Lead-Containing Swamp Ecosystem

Shanshin Ton and Howard T. Odum

CONTENTS

After receiving the acid waters from years of washing lead batteries, the cypress-gum swamps downstream from the Sapp operations filtered and stored much of the lead (Chapter 6). Many trees had been killed, and the swamps were full of dead wood and detritus. Chemical studies (Chapter 6) showed lead present in tiny amounts in the waters, but larger amounts were chemically bound in the organic sediments. This chapter describes a simulation model of lead and its simplified representation of the storages and flows within the swamp ecosystems. First we describe our concept of how the swamp operates in processing lead by discussing parts and processes in a simplified model. The systems diagram of that model (Figure 8.1) helps you visualize how this model is structured. Causal relationships are listed in Table 8.1. Some of the details are in Appendix 8.

Figure 8.1 shows the main features of the model using standard symbols for representing systems (Chapter 1). The boundary of the swamp is represented by the rectangular frame with rounded corners. Lines represent flows of material and energy. Crossing the boundary into the swamp systems are the flows of sunlight, the wind, and the inflows of water carrying lead. The water inflow includes stream and the rain. Outflows shown crossing the boundary are the reflected sunlight, the winds carrying water vapor, the stream outflow carrying organic matter and a little lead, and the heat energy by-product (in delicate lines passing out the bottom of the frame [the heat sink]).

Shown within the swamp (Figure 8.1) are the plants, the organic sediments, and consumers (microbes and small animals). The plants produce the organic matter of their own biomass, some of which goes into the sediments (sometimes called organic detritus). These plants take up some of the lead from the water, including it in the biomass. The organic detritus (which includes microorganisms) also takes up lead from the water. Some of that organic matter is consumed by the combined action of small animals and microorganisms. When that organic matter is consumed some inorganic minerals that remain, such as lead, may return to the water (see pathway labeled "recycle"). Two important places where flows intersect and branch are shown, one for the water

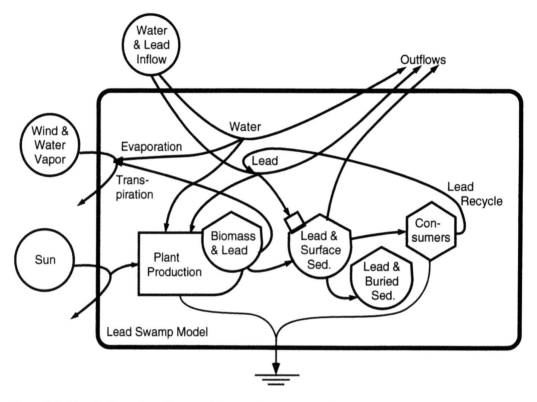

Figure 8.1 Simplified overview diagram of the simulation model of the lead-absorbing wetland ecosystem, Steele City Bay, Jackson County, Florida. Appendix Figure 8A.1 shows all of the relationships programmed into the model.

Table 8.1   Description of Relationships in Simulation Equations for the Lead Swamp Model (Figure 8.1 and Appendix Figure A8.1)

Available light is the inflowing light minus that in use.

Water flowing out is the inflowing water minus evaporation minus plant transpiration.

Lead flowing out is the lead flowing in plus the recycle minus that taken up by plants and by sediments.

Addition to organic sediments is the sum of the contribution from plants minus that used by consumers minus that in outflowing stream water.

Gross production of plants is proportional to the available light and the plant biomass, but diminished by lead in the plants.

Addition to plant biomass is the sum of the gross production minus plant respiration minus organics that fall into the sediments.

Addition to lead stored in the plants is proportional to the lead in the water and in the plant biomass and decreased by the lead in the organics that go to sediments.

Addition to lead stored in surface sediments is proportional to flow from plants (proportional to the biomass of plants and to the lead concentration in the plants) plus uptake from water (proportional to lead concentration in the water and to the quantity of organic sediments) minus the lead in the organic matter consumed minus the lead in organic matter in water flowing out minus that going into buried sediments.

Consumption is proportional to the quantity of organic sediments diminished by toxic action of lead in the sediments. Recycle of lead to the water is proportional to the consumption.

A small flow of organic sediment and lead goes into buried sediments.

Note: When something is proportional to each of two properties (A and B), it means that it is proportional to their product (A * B).

and one for the lead in the water. Part of the water that enters the swamp is evaporated and transpired by the plants and goes out as vapor in the wind. The lead flowing into the swamp receives some lead from recycle, has the two uptake pathways already mentioned, and what is not taken up flows out of the swamp with the water. The outflow stream also carries organic matter that contains some lead. The main storages which are important to lead and its processing are represented with the "tank" symbol. They are plant biomass, sediments, lead in plants, and lead in sediments.

To make the model generate patterns with time on the computer, equations are written that have causal relationships. For example, if we believe that the flow of organic matter from the plant storage (P) to the organic sediment storage (D) is in proportion to the amount of plants, the mathematical term for the flow is K_7*P (where P is the plant storage, * means to multiply, and K_7 is a coefficient that is evaluated from the numbers in Figure 8.2). The term says that some fraction of P flows per unit time along that pathway (Figure 8.1).

Because the details of the equations may be tedious for the general reader, they are placed in Appendix A8, including a more detailed version of the model's systems diagram (Figure A8.1) that has all of the relationships that are in the equations of the computer program. However, here we provide Table 8.1 to state in words most of the relationships that are contained in the equations.

CALIBRATION

Putting numerical values for storages and flows is calibration, an important step in simulation. Some values are estimated from the chemical analyses and others are estimated as what is necessary to make the whole set of flows consistent. For this model the quantities present in these storages

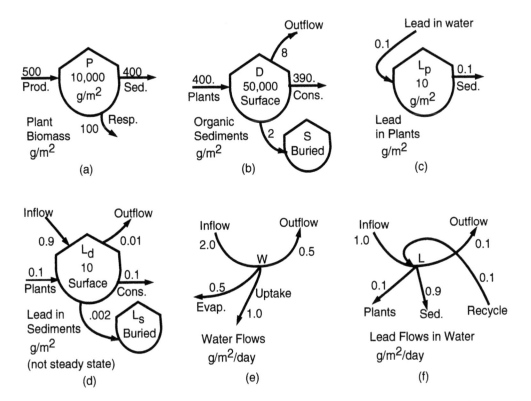

Figure 8.2 Values of storages and flows used for calibration of the lead swamp simulation model diagrammed in Figure 8.1 and Appendix A8.1.

and the flows in and out on the average are labeled for each tank and pathway junction in Figure 8.2. Appendix A8 has the spreadsheet Table A8.2 that was used for calculating the coefficients for the computer program listed in Table A8.3.

TYPICAL SIMULATION

Figure 8.3 is a typical result of simulating the calibrated model. The graph shows a run of 60 years. The first 5 years are the results before the lead inflow starts. Because the sunlight has a seasonal variation, the plant biomass goes up and down each year. As might be expected, the sediments also show an annual pulse, since they are supported by the surge of annual plant growth.

After 5 years, the program starts the inflow of toxic lead in the stream. The plant growth is impacted and is reduced to a lower level, and the organic sediments are diminished. Quantities of lead rise in water, then in plants, and finally in sediments. Surface sediments respond to changes in several years. In this model there is a slow growth of buried sediment and lead which is no longer exchanging with the surface.

Then, after 20 years, the program reduces the lead inflow (Figure 8.3). The lead in the water drops to a low concentration, followed more slowly by decreases of lead in plants and surface sediments. As the lead concentrations decrease, some plants recover and again go into their annual oscillation. The scenario observed in this simulation is consistent with the events observed over time thus far. Other "what if" experimental runs were made of the model and its variations by Ton (1990).

Usually the larger sense of understanding from a systems view results if combining known mechanisms and calibrating a model with measurements from detailed studies generate a reasonable simulation of the main events. At present, plant growth is returning, including surviving trees and floating vegetation.

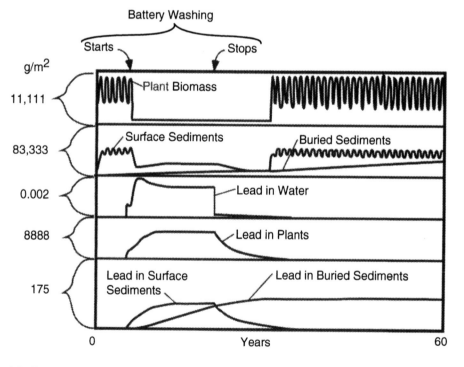

Figure 8.3 Typical graphs obtained from simulating the lead swamp model. See Appendix A8 for more details on calibration and equations.

PART III

Lead and Wetlands in Poland

Part III contains studies by the research team at Krakow, Poland on the Biala River marshes that have received lead and zinc wastewaters for 200 years. Chapter 9 contains the ecological and chemical studies by Wlodzimierz Wójcik and Malgorzata Wójcik. By quantitatively summarizing the sources, mining, manufacturing, and environmental dispersal of lead and zinc in Poland, Chapter 10 shows the need and potential for a national policy on wetlands filtration.

Lead and Zinc Retention in the Biala River Wetland of Poland

Wlodzimierz Wójcik and Malgorzata Wójcik

CONTENTS

INTRODUCTION

The Biala River Wetland in southern Poland has received mine water discharges containing high concentrations of zinc and lead for about 400 years, providing a rare opportunity to study long-term filtration of heavy metals by a wetland. Application of the wetlands for treatment of industrial wastewater and as a sink for pollutants is questioned by some professionals. Among their concerns is a problem of long-term tolerance of wetland plants to high concentrations of heavy metals in their tissues as well as in soil. This chapter presents the results of investigations of these heavy metals in the wetland and the rates of filtration.

SITE DESCRIPTION AND HISTORY

The Biala River is located 60 km west of Krakow City, Poland. The distance between the river source and its outlet is about 11 km. The wetland study site is located at the confluence of the Biala River and two discharge channels in the vicinity of Laski village (Figure 9.1). The wetland extends 3.5 km along the river course with a longitudinal slope of 0.01 to 0.6%. Between 70 and 300 m wide (most often between 100 and 150 m), the wetland–stream complex covers some 70 ha. In the upper part the Biala River meanders throughout its course and within the wetland area forms curves with radii from 10 to 200 m, branching into two or more arms which return downstream to the main channel. Starting from its central part there occur local impoundments, and gradually the water covers the entire valley in its lower part.

The Biala River had its source in an area rich in springs with a catchment basin of 53 km². As a result of the activity of the Mining and Metallurgical Works "Boleslaw," the groundwater table has been lowered to more than 100 m below ground surface. Consequently, a complete disappearance of the natural flow in the Biala River occurred in 1975. Currently, natural runoff is only possible in the case of large rainfall events or during periods of extensive thaws. Figures 9.2 and 9.3 show the recent flow of the Biala River including the wastewaters from the mines.

The story of human activity in this area is a long one. The Ponikowska Adit discharge channel was built in the 16th century to remove mine waters from ore deposits and into the Biala River. Increased flow has been as much as four times greater than the natural flow of the river. Maximum discharge was reached about 1910. This increased flow was greater than the river bed could hold and as a result the valley flooded. In the next decades discharge of the water was reduced, with periodic increases during 1961 to 1966 and 1974 to 1979. Currently flow is about 120 cm/min. The main discharge carrier to the Biala River is the Dabrawka Channel. Of the discharge 90% is mine water, 7% is effluent from a municipal wastewater treatment plant, and 3% comes from industrial wastewater treatment.

To help make proper decisions about changes likely to occur to the wetland and environment quality in the future, an extensive sampling program was required.

METHODS

Extensive vegetative sampling and physical and chemical analyses of soil and sediments were undertaken. Soil samples from various depths were collected at 49 locations within the study site (Figure 9.4a). For tissue analysis of heavy metals in wetland plant species, 85 samples were taken from 17 sites along the study area (Figure 9.4b).

Vegetation sampling was undertaken to characterize plant communities and identify plant species for tissue analysis. Cover maps were made and spatial extent was calculated for the most common plant communities. For evaluation of annual biomass growth (net production), nine randomly selected 1-m² plots were cleared of vegetation in early spring and then harvested in late autumn. This method was used to estimate annual deposition of dead organic matter on the surface of the wetland.

Water quality of mine waters flowing into the wetland was monitored from 1977 to 1990, and the results were compared with the historical records.

The ability of wetland plants to remove heavy metals from flowing water was evaluated with upstream–downstream analyses of substances in water as it passed from one end of the wetland to the other. Concentrations were monitored at ten sampling stations (Figure 9.4c), with the time of water flow taken into account in order to capture the same wave of water with each sample.

Field experiments were conducted in June 1990 studying the passage of water marked with dye passing through wetland vegetation with the control arrangements shown in Figure 9.5. Data were used to evaluate coefficients of the Manning equation. The experimental plot was 96 m long and 4 to 8.5 m wide with an average depth between 0.069 and 0.294 m.

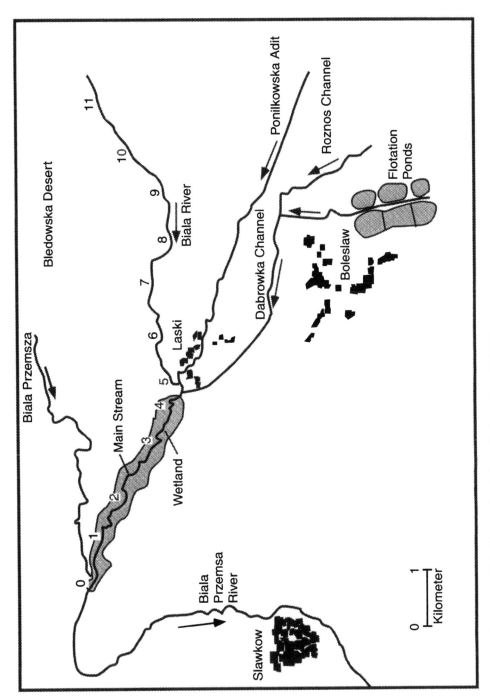

Figure 9.1 Location of Biala River wetland study site in southern Poland.

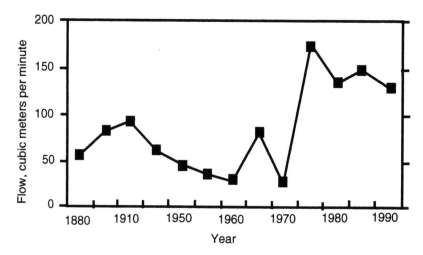

Figure 9.2 Flow of the Biala River including mine water, 1880 to 1993.

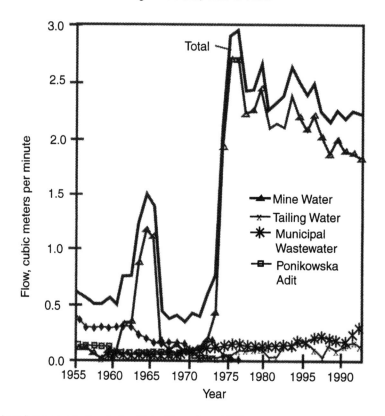

Figure 9.3 Water flows contributing to the Biala River since 1955.

RESULTS

Soils and Sediments

The results of the analyses of physical and chemical properties of soil samples collected from various depths are listed in Appendix Tables A9.9 to A9.11. The shallower layers of soil are characterized by high density (up to 3.09 g/cm³) for samples containing small concentrations of

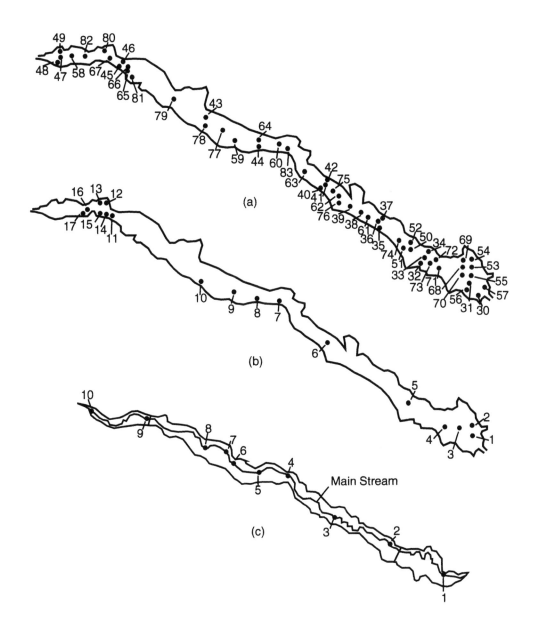

Figure 9.4 Location of sampling stations for soil, vegetation, and water analyses in the Biala River wetland. (a) Soil; (b) vegetation; (c) water.

organic matter. The size of particles was similar in all soil samples. Sediments are characterized as silty sand to silt. Dolomite and calcite keep the soil alkaline, generally above 7 for most samples, with up to 30% calcium. Currently pH of soil deposits ranges between 5 and 7.8.

The concentration of zinc (Zn), lead (Pb), and cadmium (Cd) in sediments was high down to a depth of 1.5 m. Maximum concentration of zinc was 4.46% and lead was 1.34%. These high values were most often observed in upstream reaches of the wetland, from the outlet of Dabrowka Channel to about 1.5 to 2.0 km downstream (Figure 9.6a). There was little deposit of heavy metals in sediments below the wetland, and in sites close to the valley sides. Figure 9.6b shows data near the end of the wetland.

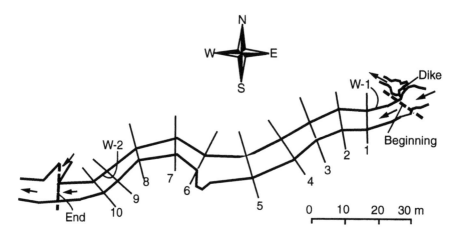

Figure 9.5 Experimental plot used for field experiments (June 6, 1990) showing the locations of cross-section calculations. W-1, W-2 are water level gauges.

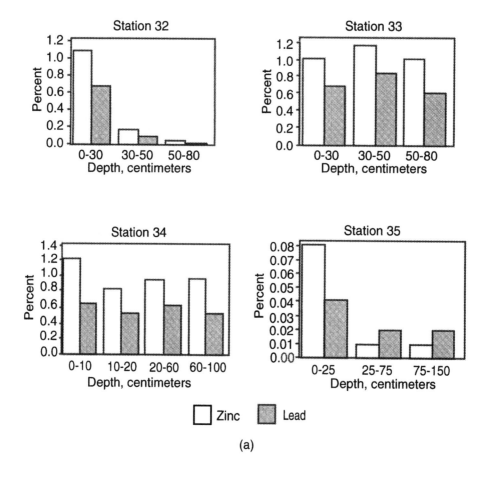

Figure 9.6a Percent concentrations of lead (Pb) and Zinc (Zn) with depth in soil in the Biala River wetland. Stations near the wastewater inflow (Stations 32, 33, 34, and 35 in Figure 9.4).

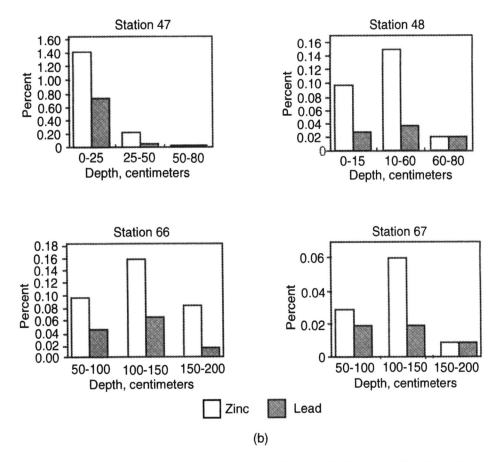

(b)

Figure 9.6b   Stations near the outflow end of the wetland (Stations 47, 48, 66, and 67 in Figure 9.4).

Zinc and lead in wetland soils were compared with soils of the adjacent region (Figure 9.7). Concentrations of metals in the vicinity of the wetland are about 200 parts per million (ppm) for zinc and 50 ppm for lead. These concentrations exceed average concentrations of zinc and lead in unpolluted Polish soils by five and three times, respectively. An estimate of total zinc accumulated in the wetland measured 3927 tons; the amount of lead was estimated at 1887 tons (see Appendix A9).

At several sampling stations, the concentrations of metals in sediments were lower at a depth up to about 10 to 15 cm. This is likely a reflection of the historic overloading of the wetland system with mine water. Prior to 1980, mine effluent contained large amounts of suspended solids after ore flotation. This resulted in sediment accumulation and simultaneous erosion of the river bottom in the main stream of the wetland. The consequence was a successive disappearance of the wetland progressing downstream from the outlet of Dabrowka Channel. As the wetland reestablished itself following reductions in mine water discharges, the soil-forming processes increased as a function of community production. This may explain why the top soil layer in many stations contains more organic matter and less heavy metals than deeper and older sediments.

Figure 9.7 Lead content in parts per million in soils of uplands surrounding the Biala River wetland.

Plant Communities and Biomass

Seventeen wetland plant communities were documented. Most plant communities cover small areas, some of them being remnants of earlier forms of land use. The most common plant communities are listed below, in order of the frequency of occurrence (Figure 9.8):

1. Sedge marsh (*Caricetum gracilis*)
2. Reed marsh (*Scirpo-Phragmitetum* with *Phragmites communis*)
3. Community with Deschampsia (*Deschampsia caespitosa*)
4. Alder swamp (*Carici elongatae Alnetum*)
5. Typical marsh (*Scirpo-Phragmitetum* with *Typha latifolia*)
6. Wet meadow (group of various communities)
7. Sedge marsh with Alder (*Caricetum gracilis* and *Carici elongatae Alnetum*)
8. Horsetail marsh (*Equisetum limosum*)
9. Fresh meadows (order Arrenatheretalia)

Missing Species

The flora of the receiving wetland shows some peculiar features. One of them is the absence of certain species, which are common on other wetlands. Examples are willow (*Salix* sp.), poplar (*Populus alba* and *P. nigra*), and some species typical for meadows, such as *Bellis perennis*.

Diversity

There was a dominance of a few species occurring in great abundance (Appendix Tables A9.12a to A9.12h). Starting with the most common, these include: reed (*Phragmites communis*), deschampsia (*Deschampsia caespitosa*), sedge (*Carex gracilis*), great stooled sedge (*C. paniculata*), water mint (*Mentha aquatica*), cattail (*Typha latifolia*), sedge (*C. rostrata*), and black alder (*Alnus glutinosa*). Species richness ranges from 4 to 33, averaging 13.3 for all samples.

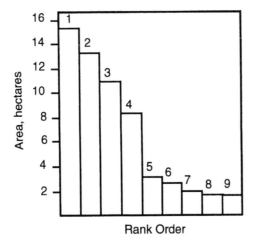

Figure 9.8 Area cover of the most common plant communities in the Biala River wetland arranged in rank order. (1) Sedge marsh (*Caricetum gracilis*); (2) reed marsh (*Scirpo-Phragmitetum* with *Phragmites communis*); (3) community with deschampsia (*Deschampsia caespitosa*); (4) alder swamp (*Carici elongatae-Alnetum*); (5) typical marsh (*Scirpo-Phragmitetum* with *Typha latifolia*); (6) wet meadow (group of various communities); (7) sedge marsh with alder (*Caricetum gracilis* and *Carici elongatae-Alnetum*); (8) horsetail marsh (*Equisetum limosum*); (9) fresh meadows (order Arrenatheretalia).

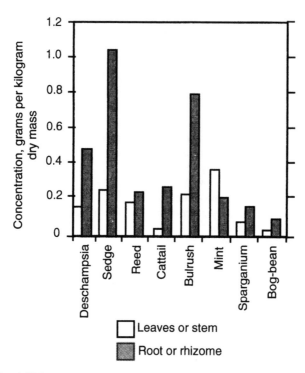

Figure 9.9 Maximum lead (Pb) concentration in plant tissues of species within the Biala River wetland.

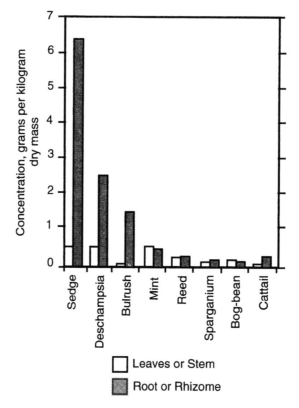

Figure 9.10   Maximum zinc (Zn) concentration in plant tissues of species within the Biala River wetland.

Average biomass regrowth on the cut plots was 846 g/m^2 dry matter (dm) ranging from a minimum of 522 $g\text{-}dm/m^2$ to a maximum of 1798 $g\text{-}dm/m^2$. Dominant plants in each plot are given in Appendix Table A9.16.

Concentration of Heavy Metals in Plant Tissues

The concentrations of metals found in plants ranged from 25 to 6500 ppm dry matter for zinc, 5 to 1050 ppm for lead, and 0.5 to 48 ppm for cadmium (Figures 9.4 and 9.5 with details of measurements given in Appendix Table A9.14). Average concentration of zinc in plant leaves ranged from 29.3 ppm in *Typha latifolia* up to 270 ppm in *Deschampsia caespitosa*. For lead, mean concentration ranged from 16.3 ppm in *T. latifolia* up to 91.6 ppm in *Mentha aquatica*. Average cadmium concentrations ranged from 1.3 ppm in *Phragmites communis* and *T. latifolia* up to 3 ppm in *Scirpus lacustris*.

In plant stems, average concentrations of zinc ranged from 106 ppm in *M. aquatica* to 126 ppm in *P. communis*. Lead in plant stems varied between 20.4 ppm in *P. communis* and 33.8 ppm in *M. aquatica*. Concentrations of cadmium in plant stems averaged 1.6 to 1.8 ppm in *P. communis* and *M. aquatica*.

Concentrations of metals in plants belowground were 2 to 20 times higher than in leaves and stems aboveground. Average concentrations of both zinc and cadmium in rhizomes of *P. communis* were 2 to 3 times greater than in leaves and stems and 3 to 7 times greater for lead. In *T. latifolia*, the mean zinc and lead concentrations in rhizomes were 5 to 8 times greater than in aboveground tissues; concentrations of cadmium in plants were similar in above- and belowground tissues. The largest differences in metal concentration between below- and aboveground parts were in *Carex* sp., 10 to 20 times greater for lead, zinc, and cadmium. The highest belowground concentrations, 6500 ppm for zinc and 1050 ppm for lead, were found in sedge (*Carex* sp.). The highest cadmium concentration, 48 ppm, was found in whole plants of *Potamogeton natans*.

Concentrations of heavy metals in the five most common plant species measured in this study were significantly higher than values published in the literature (Appendix Tables A9.14 and A9.15 and Figures 9.9 and 9.10). For example, in *Phragmites* sp., concentration of zinc was 2 times higher, lead measured 8 times higher, and cadmium was as much as 29 times higher. Lead concentrations in *Sparganium* sp. averaged 5 times higher, zinc in *Carex* sp. as much as 100 times higher, and *Mentha aquatica* 7 times greater. Figure 9.11 shows a wide range of concentrations in different parts of plants. Figure 9.12 is typical of the analyses, showing higher concentrations of zinc and lead in the sediments than in the plants growing on them.

Based on the many plant associations found, their cover, and diversity indices, long-term exposure to high concentrations of heavy metals in soils and plants do not appear to inhibit plant community growth. The vegetation has existed on the wetland many years, even under long-term exposure to heavy metals.

Physical and Chemical Analysis of Waters

Heavy metal concentrations in wastewater discharged to the wetland at different times are given in Figure 9.13. There is a correlation between concentrations of total suspended solids (TSS) and zinc and lead concentrations, suggesting that heavy metals are present mainly in non-ionic forms — an assumption supported by filtration tests.

Upstream–Downstream Measurements

Results of the upstream–downstream measurements showed that there were no significant changes in lead and zinc in the upper half of the wetland (Figures 9.14 and 9.15), but there were

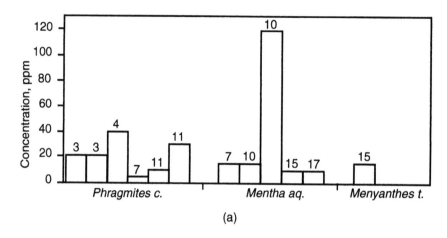

(a)

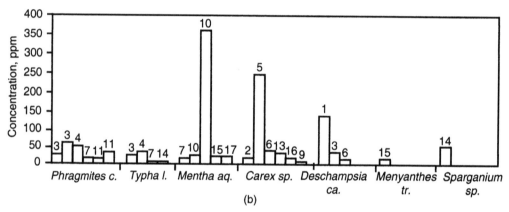

(b)

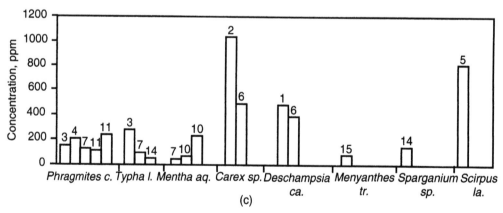

(c)

Figure 9.11   Concentrations of lead (Pb) in the parts of plants in the Biala River wetland. (a) Stems; (b) leaves;
(c) roots. Stations are indicated by number.

lead and zinc decreases in waters passing through the lower area in some runs. Reduction of
zinc and lead across the entire Biala River Wetland was much lower in the studied period 1989
to 1990 than in the year 1978. There was no difference in the reduction during summer and
winter periods. This suggests more reduction of zinc and lead due to physical and chemical
interactions with surface properties of the wetland plants, than to biological uptake by plants
(example: through osmosis).

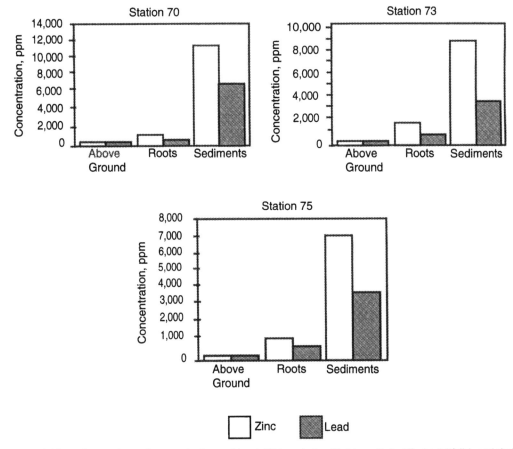

Figure 9.12   Comparison of concentrations of lead (Pb) and zinc (Zn) in cattails (*Typha latifolia*) and their sediments at three stations.

Field Experiment

The field experiment provided a quantitative estimation of the role of the plant beds in retarding flow, helping filter and sediment the particles containing lead and zinc. In the experimental plot (Figure 9.5) the average plant density was 91 plants per square meter. The effect of the plants in retarding the flow was evaluated by calculating a coefficient for the Manning equation that relates water flow in channels to the head of water. For this density of wetland plants it was 0.392, which is three times higher than the coefficients cited in the literature by Lee (1980).

PROPOSAL TO RESTORE WATER FLOW OVER THE WETLAND

Based on this research it is concluded that present water flows in channels bypass much of the wetland plants, especially in the upper area. Greater filtration can be achieved by allowing the receiving wetland to function as it did 15 years ago with waters flowing within the vegetation more. Although mining wastewaters may decline in the future, municipal wastes are expected to increase (Figure 9.17). The wetlands will be needed to filter excess nutrients like phosphorus and nitrogen.

The upper part of the receiving area could be reestablished into a wetland again. In order to accomplish this, it is proposed that a dike be built across the river valley at a distance of 400 to

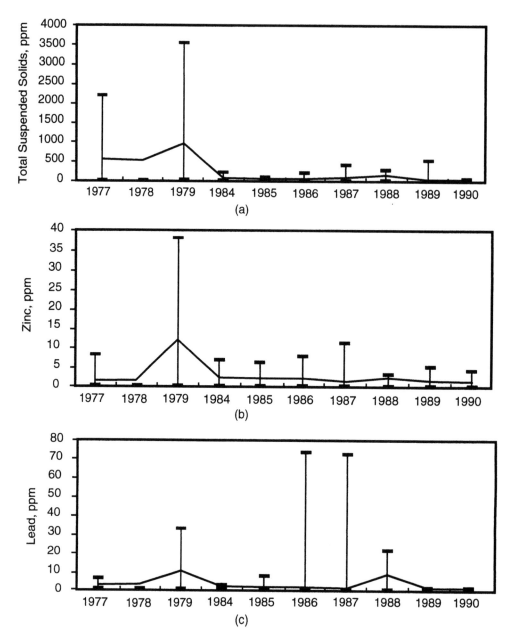

Figure 9.13   Substances in the wastewaters discharged into the Biala River wetland from 1977 to 1990. (a) Total suspended solids; (b) zinc concentrations; (c) lead concentrations.

500 m from the Dabrowka Channel outlet. The dike (Figure 9.18) would be 0.80 m high and 2.5 m wide, resulting in the formation of an intermittent impoundment with a surface area of about 7.4 ha and a volume of 2600 m³ (Figure 9.19). Longer retention periods would result in better contacts between wastewater and the wetland vegetation, which may then increase the removal of heavy metal pollutants. However, if the waters were continuously impounded, changing the hydro-period and circulation, the vegetation would be lost and the purpose defeated.

The proposed dike experiment could provide a test of the concepts. With proper ecological monitoring and measurements of efficiency of heavy metals removal in the changed condition, guidelines could be obtained for using this approach downstream and elsewhere.

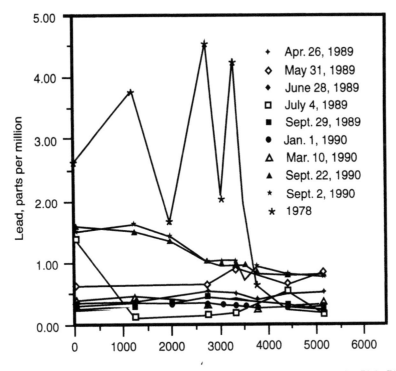

Figure 9.14   Changes observed in lead concentrations as waters passed through the Biala River wetland.

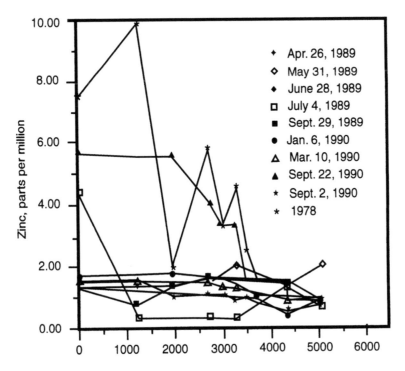

Figure 9.15   Changes observed in zinc concentrations as waters passed through the Biala River wetland.

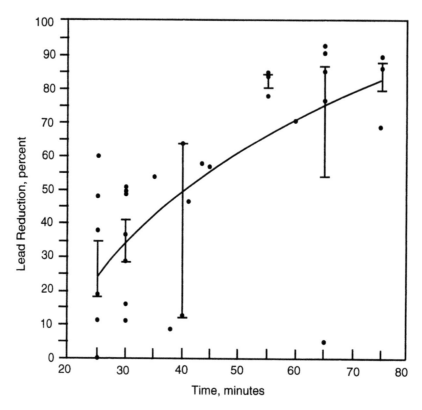

Figure 9.16   Reduction in concentrations of lead in waters during the field experiment on Plot #1. See Appendix A9 for explanation of the way the line was drawn to the data.

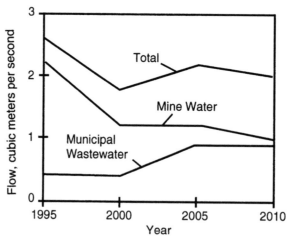

Figure 9.17   Discharges into the Biala River wetland expected in the future.

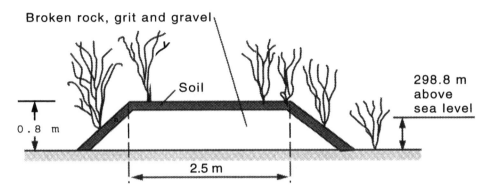

Figure 9.18   Dike proposed across the Biala River (Figure 9.19) to spread more water over the wetlands.

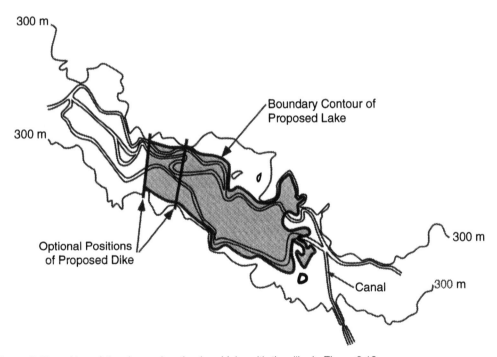

Figure 9.19   Map of the changed wetland and lake with the dike in Figure 9.18.

CHAPTER 10

Perspectives on Lead and Zinc Manufacturing, and Environment

Wlodzimierz Wójcik

CONTENTS

In Poland there are no geological ore formations with lead only, except for small resources in Jaroszowiec-Pazurek. Lead (Pb) occurs jointly with zinc (Zn) and copper (Cu). The amount of lead in zinc ores in the Silesia-Cracovian formation is estimated to be 3.69 E6 tons lead with an average lead concentration in the ores of about 1.15 to 1.8%. Lead in copper ores in the Low-Silesia formation is estimated at 7.67 E6 tons lead, with about 0.12 to 0.2% lead concentration in ores. Total lead extraction in 1993 was 107,700 tons (Table 10.1).

Some predictions suggest lead extraction in Poland through the year 2000 will be about 100 to 110 E3 tons, and after 2000 substantial decreases will be observed due to shutdowns of some zinc–lead mines.

Zinc in Poland is found mainly in the Silesia-Cracovian dolomite formation with an estimated 8.39 E6 tons. Average concentration of zinc in these ores is 3.9%. Decreases in zinc and lead extraction were observed in 1990 (Table 10.1), just after political changes which caused economic recession. Zinc production decreased more.

Zinc and lead mining in Poland is declining. The Boleslaw mine shut down in 1996, and shutdown is planned for two additional mines: the Trzebionka mine in 2005, and the Olkusz mine in 2010. The largest mine (Pomorzany mine) will be shut down in 2020.

Table 10.1   Structure of Lead and Zinc Extraction in Poland (in thousand tons)

	1989	1990	1991	1992	1993
Lead					
In Zn-Pb ores	65.8	61.3	63.6	70.1	69.4
In Cu ores	31.0	29.0	38.8	49.8	39.3
Total	96.8	80.3	102.4	119.9	107.7
World	3330.5	3352.0	3328.7	2983.9	2832.3
Zinc					
Total	203.7	177.8	175.6	186.2	182.6
World	7147.8	7364.0	7522.3	7141.2	6794.9

Compiled from Bolewski et al., 1995 and Ochrona Srodowiska, 1994.

LEAD AND ZINC ORES PROCESSING

Removal of lead in zinc–lead ores by flotation processes produces commercial concentrates. They contain about 73% lead at the Trzebionka mine and 53% at the Boleslaw mine (Table 10.2).

Flotation of copper ores does not concentrate lead. The concentrates of copper contain between 2.16% lead (at Rudna) and 1.17% lead (at Polkowice). Lead is recovered and concentrated as raw lead in the next step, metallurgical processing of copper ores.

Zinc concentrates are produced as by-products of processing of zinc–lead ores. At ZGH Boleslaw mine these concentrates contained 46% zinc. Production of zinc concentrates decreased after 1989 to about 150,000 tons zinc/year (Table 10.2).

Table 10.2   Balance of Lead and Zinc Concentrates in Poland

	1989	1990	1991	1992	1993
Lead (in thousand tons)					
Production	51.1	45.2	47.0	51.0	49.1
Import	—	—	—	1.8	—
Export	18.1	24.8	14.7	38.2	30.4
Consumption (in-country use)	33.0	20.4	32.6	14.6	18.7
Zinc (in thousand tons)					
Production	174.8	153.4	144.7	151.7	150.9
Import	1.0	—	23.6	35.6	44.2
Export	1.0	25.4	10.0	43.4	42.0
Consumption (in-country use)	174.8	128.0	158.3	143.9	153.1

Compiled from Bolewski et al., 1995 and Ochrona Srodowiska, 1994.

Lead and zinc are processed into metallic forms in the metallurgical mills. There are two basic metallic forms of lead: raw and refined. Raw lead is produced in Miasteczko Slaskie, ZGH Orzel Bialy, HM Glogow, HMN Szopienice, and ZPWMN Wtormet Bytom (Table 10.3). Concentration of lead is between 96 and 99%.

In 1989 about 55% of raw lead production was from recycled wastes. In 1993, 46% was from recycled wastes, but the total amount of raw lead decreased from 45,700 to 35,500 tons.

Table 10.3   Balance of the Raw Lead Production in Poland (in thousand tons Pb)

	1989	1990	1991	1992	1993
Primary production	37.6	24.2	20.8	22.3	30.4
HC Miasteczko Slaskie	33.6	24.2	20.8	22.3	30.4
ZGH Boleslaw	4.0	—	—	—	—
Recycling of own wastes					
HM Glogow	7.0	9.1	9.9	9.8	10.5
Recycling of wastes	38.7	37.0	23.1	23.1	22.4
ZGH Orzel Bialy	34.1	34.0	21.2	23.1	22.4
PPWMN Wtormet	1.6	—	—	1.1	0.6
HMN Szopienice	3.0	3.0	1.9	2.2	2.0
Total	83.3	70.4	3.6	58.5	65.9
World	3155.7	2939.3	2983.1		

Compiled from Bolewski et al., 1995 and Ochrona Srodowiska, 1994.

Refined lead (95 to 99.99% lead) is produced by the HC Miasteczko Slaskie and HMN Szopienice. Production of 62,300 tons of lead in 1993 was only 50% of mill capacity (Table 10.4). Production of metallic zinc in Poland is compared to that of the world in Table 10.5.

Table 10.4   Balance of Refined Lead in Poland (in thousand tons Pb)

	1989	1990	1991	1992	1993
Production	78.2	64.8	47.5	53.7	62.3
Import	—	3.5	8.3	6.0	7.3
Export	7.3	6.0	9.9	10.6	13.1
Consumption (in-country use)	70.9	62.3	45.9	49.1	56.5
World	5920.0	5668.1	5544.4	5671.3	5736.3

Compiled from Bolewski et al., 1995 and Ochrona Srodowiska, 1994.

Table 10.5   Production of Metallic Zinc (in thousand tons Zn)

	1989	1990	1991	1992	1993
Electrolytic zinc	88.4	82.9	82.1	84.3	88.5
Refined zinc	75.3	49.3	43.9	50.3	60.6
Total	163.7	132.2	126.0	134.6	149.1
World	7210.2	7053.7	7189.3	6929.0	7184.9

Compiled from Bolewski et al., 1995 and Ochrona Srodowiska, 1994.

CONSUMPTION OF LEAD AND ZINC

In Poland lead is used mainly in the production of batteries, cable protection layers, and chemicals (mostly chemical pigments). Consumption of metallic zinc decreased from 148,600 tons in 1989 to 81,100 tons in 1993 (Table 10.6). On the other hand, exports increased from 17,600 to 69,500 tons. In addition, imports contain lead. For example, in 1993, 4,769,333 m^3 of leaded gasoline was used in

Table 10.6   Balance of Metallic Zinc in Poland (in thousand tons Zn)

	1989	1990	1991	1992	1993
Production	163.7	132.2	126.0	134.6	149.1
Import	2.5	0.9	0.9	0.3	1.5
Export	17.6	31.3	26.0	58.7	69.5
Consumption	148.6	101.8	100.9	78.7	81.1

Compiled from Bolewski et al., 1995 and Ochrona Srodowiska, 1994.

Poland. This gasoline contained 0.15 g Pb per liter which released 715.4 tons of lead after combustion. Data were not available on lead in other imported goods. Zinc is used mainly in galvanization (about ~45%), production of alloys (35%), and synthesis of various chemicals (about 20%).

It is very difficult to predict future lead and zinc production and use in Poland. However, some experts predict that lead production and use will remain at 1993 levels. There may be a 30% increase in zinc with modernization of the metallurgical industry. New automobile plants in Poland may consume more zinc.

ENVIRONMENT POLLUTION FROM LEAD AND ZINC IN POLAND

Wastes from manufacturing and use of lead and zinc can cause pollution of the environment. Lead is more toxic than zinc. The most dangerous impacts may occur with emissions from production plants.

POLLUTION OF THE SOIL BY LEAD

Heavy metals are readily absorbed in soils from rainfall, fallout of dust, fertilizers, and land application of wastewater. In Poland average natural concentrations of lead in soil are estimated at 18 ppm. Recommended maximum permissable concentrations are 100 ppm (Kabala-Pendias and Pendias, 1993). Soils in industrial areas and close to lead manufacturing plants can reach concentrations as much as 1.5% lead (Swiderska-Broz, 1993). Lead concentration in soil near a battery production plant was 3800 ppm (Swiderska-Broz, 1993). Soil along highways in Poland are polluted with emissions from combustion of leaded gasoline with lead levels measured up to 2715 ppm.

Fertilizers can cause soil pollution because they contain up to 1250 ppm lead. In cases where municipal wastewater or manure is used as the fertilizer, soil contamination may occur. Municipal wastewater can contain up to 3000 ppm lead dry matter; manure can contain as much as 15 ppm lead dry matter (Kabala-Pendias and Pendias, 1993). A study of Polish soils from 1980 to 1985 showed that the highest input of heavy metals was from rainfall (Table 10.7).

Lead is common in Polish surface waters. Regulations allow concentrations up to 0.05 ppm. Concentration of lead in the mouth of the Vistula and Odra Rivers is a good estimate of water pollution levels for Poland, since these rivers drain almost the entire surface of Poland. In 1993, lead concentrations decreased along the Vistula River: Krakow City — 0.026 ppm, Warsaw City

Table 10.7   Balance of Inputs and Outputs of Some Heavy Metals in Soil, 1980–1985 (in g/ha/year)

	Cd	Pb	Zn
Input			
Mineral fertilizers	1.0	2	10
Liming	1.5	5	4
Manure	2.5	38	75
Wastewater sludge	1.5	15	30
Plant remains	3.0	20	50
Rainfall	5.0	200	5400
Total input	14.5	280	709
Output			
Harvesting	3.0	40	100
Infiltration	3.0	40	180
Total output	6.0	80	280
Balance (increase)	8.5	200	429

After Swiderska-Broz, 1993.

— 0.016 ppm, Gdansk — 0.003 ppm (Ochrona Srodowiska, 1994). In the Odra River, average concentrations were 0.022 ppm at Wroclaw City and 0.015 ppm at its mouth in Szczecin City.

In 1993, 298.2 tons of lead were discharged to the Baltic Sea from Poland. In 1993, maximum and average lead concentrations in the lake sediments were 118 ppm and 29 ppm, respectively. Monitoring was done by the Environmental Protection Inspectorate at special environmental monitoring stations. Table 10.8 contains concentrations in the rivers of seven water management regions. In some rivers close to lead manufacturing plants the concentrations reached 1200 ppm. Unfortunately, national environmental monitoring stations do not cover all the potential areas with lead pollution.

Table 10.8 Concentration of Lead in River Sediments at National Monitoring Stations in Poland during 1993 (in ppm)

Water Management Region	Max	Average
Gdansk	414	15
Katowice	414	97
Krakow	114	17
Poznan	177	19
Szczecin	97	17
Warszawa	125	10
Wroclaw	410	72

Compiled from Ochrona Srodowiska, 1994.

AIR POLLUTION

The main sources of air pollution by heavy metals are emissions from manufacturing plants, metallurgical plants, and power and heat generation plants. Heavy metals are monitored around these plants. Also, automobiles contribute lead from gasoline. Use of leaded gasoline decreased in Poland from 1985 to 1993. Emissions from this source decreased from 1.01 thousand tons in 1985 to 0.72 thousand tons in 1993.

Power and heat generation plants also emit airborne lead. Fly ash in Poland contains from 17.5 to 30 ppm lead. High concentrations of dust remove lead as fallout. Lead concentrations in dust along highways in Poland were monitored up to 3500 ppm (Kabala-Pendias and Pendias, 1993).

Average lead emission (fallout) from all sources in Poland for 1992 was 200 g/ha/year. For comparison, typical lead fallout in Sweden was 72 g/ha/year and 110 g/ha/year in Germany. In industrialized areas this emission was several times higher. For example, lead emission near our studied wetland was about 1500 g/ha/year. Studies of air pollution near the lead manufacturing plant ZGH Boleslaw in 1989 measured lead fallout up to 5000 g/ha/year.

Lead concentrations were high in large cities (Table 10.9). Highest concentration of lead in air was monitored to be in Warsaw (0.623 $\mu g/m^3$), Krakow (0.554 $\mu g/m^3$), and Bytom (0.379 $\mu g/m^3$).

Table 10.9 Concentration of Lead in Air in Selected Large Cities in Poland in 1993

City	$\mu g/m^3$
Warsaw	0.623
Bialystok	0.151
Bytom	0.379
Chorzow	0.334
Dabrowa Gornicza	0.217
Gdansk	0.151
Katowice	0.290
Krakow	0.554
Lodz	0.333
Szczecin	0.274

From Ochrona Srodowiska, 1994.

CHAPTER 11

The Ecological Economics of Natural Wetland Retention of Lead*

Lowell Pritchard, Jr.

CONTENTS

With the reorganization of the biosphere by human economic and industrial development, a new and more symbiotic pattern of environment and economics is emerging. Linked by the biogeochemistry of chemical elements, ecological systems, particularly wetlands, are becoming

* Condensed by the Editor.

recognized as part of the economy through their work in filtering toxic substances such as heavy metals. Increasingly, wetlands have been found to be filters of many wastes of the economy which can be retained in the landscape to mitigate the impact of economic activity. This chapter is an evaluation of a Florida wetland system which filtered large quantities of lead from the discharge of a battery processing plant. Field methods were used to compare treatment wetlands with reference wetlands. Comparison was made between emergy analysis (spelled with an "m") and mainstream economic analysis methods in evaluation of wetland services in filtering a toxic metal. An estimate is made of the potential value of wetland filtration of lead to the state and nation.

Two systems for recovery of lead batteries and lead-contaminated waters were evaluated: (1) wetland filtration of wastewaters from acid washing of batteries; and (2) chemical treatment of wastewaters at a lead reprocessing smelter. Both systems were evaluated with *economic* and *emergy* methods. From 1970 to 1979 the Sapp Swamp, Steele City Bay, in Jackson County, FL, received acidic, lead-contaminated wastewaters from a battery reclamation operation. The 29-ha cypress-tupelo wetland is the Superfund site described in Chapters 1 and 5. The technological battery operation is a lead smelter–chemical treatment operation in Tampa, FL. The process of producing lead batteries was also evaluated using emergy, obtaining the transformity of lead batteries (Figure A11B.7) needed in calculations.

EVALUATION CONCEPTS

This chapter uses two concepts of evaluation: (1) environmental value based on the work of nature and humans in generating a product; (2) economic values based on human perceptions and market prices for a product.

Emergy Evaluation

Emergy evaluation provides common units for comparison of environmental and economic goods and services. After all the inputs are identified with systems diagramming (example: Figure 11.1), each is evaluated in *emergy* units and summed. Emergy is the energy of one kind required directly or indirectly for their production.

For instance, production of a bushel of corn may require many kinds of available energy from sunlight, wind, rain, fertilizer, equipment, and human labor, but with emergy evaluation each is expressed in units of one kind of energy previously used up. The production of wind energy requires solar energy, and the production of rain requires solar energy and wind energy (which requires solar energy). Fertilizer, equipment, and human labor are transformations of fossil fuel energy (the production of which required solar energy and geologic energy). The amounts of solar emergy necessary are back-calculated.

Emergy is thus a measure of environmental work (Odum, 1986, 1988, 1996) contributing to production. Its unit is the solar emjoule (sej — see Chapter 4). By measuring the emergy previously required per unit energy, the method recognizes differences in energy quality of environmental and economic inputs. The emergy per unit energy is called transformity (sej/J). With each successive transformation process, the transformity increases, thus measuring the position of an item in an energy hierarchy.

Emergy flow per time is called empower. The higher the empower the greater is the economic and ecological value of production (as defined in emergy units).

Emergy/mass ratios are convenient for calculating the emergy of materials which are often more easily measured in mass rather than energy terms.

Although national emergy use and gross national product are partially independent (see Discussion), the emergy/money ratio of the overall economy in a given year (emergy use/gross national

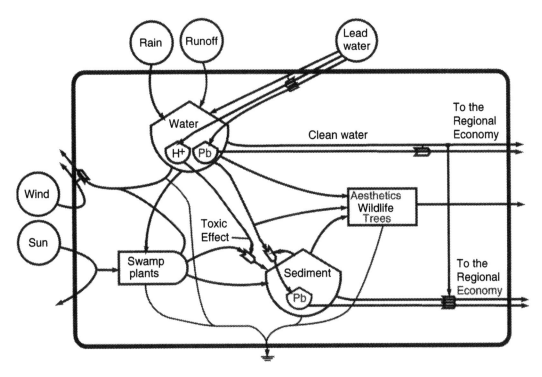

Figure 11.1   Energy diagram of swamp receiving lead-polluted water.

product) can be used to estimate the average emergy behind purchased services for which detailed energy information is lacking (Odum, 1991).

For clarity, Table 11.1 provides a summary of definitions for emergy evaluation terms.

The emergy value of a wetland depends on the energy captured and used in biological production (often measured by gross primary production). Calculation of stored emergy evaluates storages and structure, for instance, peat in cypress swamps or tidal channels in salt marshes (Odum and Hornbeck, 1996).

As systems diagrams show, the ecological goods, services, and storages considered economic amenities are all direct or indirect products of the input energy flows. Figure 11.1 is an energy diagram of the wetland receiving lead-polluted acid water. It shows the ecological system generating a storage of lead-adsorbing sediments but experiencing a toxic effect from acid waters. The sediments adsorb lead from the water column and return lead to the water column as they decay.

Table 11.1   Definitions of Emergy Evaluation Concepts

Emergy	The energy of one type required directly or indirectly in transformations to generate a product or service
Solar emergy	Solar energy required directly and indirectly to produce a product or service (units are solar emjoules — sej)
Transformity	Emergy per unit energy for a given product or service in a system
Solar transformity	Solar emergy per unit energy (units are solar emjoules/joule — sej/J)
Emergy per unit mass	Energy of one type required to generate a flow or storage of a unit mass of a material (units are sej/g)
Empower	Emergy flow per unit time (units are sej/year)
Emergy/dollar ratio	Ratio of emergy flow to dollar flow, either for a single pathway or, more commonly, for a state or a nation, where annual emergy use is divided by the gross economic product (units are sej/$)

Sources: Odum, 1988; Odum, 1991.

Flows out of the system include some lead in water (though a lower concentration than the inflow) and some lead in suspended sediment. The ecological system also generates aesthetic, wildlife, and timber value to the regional economy.

Economic Valuation

In mainstream economics, goods and services are valuable to the extent that they are useful to consumers. Some wetland contributions to people are directly useful (bird watching, boating, recreational fishing) and are called "final" goods (Scodari, 1990). *Intermediate* wetland goods are valuable to consumers because they serve as factors of production for goods which are, in turn, enjoyed directly (for example, wetland trees may be a factor in the production of wood for fuel or pulp for paper and wetland peat may be a factor of production in electricity). Where well-developed markets exist for wetland final and intermediate goods, it is argued that prices reflect their value to society. Where markets do not exist, economic value must be determined in other ways.

The *replacement cost* (or *substitution cost*) method is one such way. This method is an attempt to measure the value to society of nonmarketed wetland services such as heavy metal retention by using the cost of a substitute for that service. If society is willing to bear that cost, the value of the service must be greater than or just equal to the cost. (A different concept of replacement cost involves measuring the cost of actually replacing a natural wetland and its functions with a constructed wetland [Anderson and Rockel, 1991].)

The ability of wetlands to retain heavy metals such as lead is an intermediate wetland good. In this economic paradigm, people do not actually value the intermediate good of heavy metal retention; they value the final goods of clean water or refined lead (or the output of the industries using heavy metals). The demand for wetland retention of heavy metals (which represents the value they place on that service) is a "derived demand" — derived from the value of final goods by a firm which will use the intermediate good to satisfy the demands of consumers (McCloskey, 1985, p. 450).

To use the replacement cost method of valuing wetland service, the derived demand is assumed to be perfectly *inelastic,* which means that even with a higher-cost substitute, a firm (or society) would demand the same amount of lead retention as with "free" treatment by a natural wetland. For a discussion of why this assumption is made, see Appendix A11A.

Wetland products and services such as lead retention are called *positive externalities*. These are societal benefits that arise from wetlands which cannot be captured by the wetland property owner. Wetland production of waterfowl, for example, benefits society (especially hunters and bird watchers), but this benefit is external to the private property owner's decision-making boundary. If the owner is not personally interested, then he or she is likely to sell or convert the wetlands into other uses. Likewise, negative externalities (also called *social costs*) are costs "falling beyond the boundary of the decision-making unit that is responsible for those costs" (Bromley, 1986), such as wetland destruction from pollution. If decisions are to be made about the socially efficient provision of wetland products and services, the magnitude of these externalities must be ascertained.

In summing benefits and costs over time, mainstream economics stresses the importance of the time value of money. It is common practice in economic analysis to discount the value of future benefits and costs. Discounting has been extensively criticized and defended in the literature (see, for example, Pearce and Turner (1990, pp. 211–225), but the idea is that real current benefits (and costs) are given more weight than prospective future benefits (and costs). Several reasons exist for such discounting: inflation erodes the value of benefits over time, there may be some risk that the future benefits will not materialize, and individuals are impatient (Randall, 1987, p. 239; Pearce and Turner, 1990). Because the calculations here are in constant dollars, because the concern is with the longer run, in which case risk has less meaning, and because societal rather than individual values are considered, these three components (inflation, risk, and impatience) should not influence our analysis.

Another reason for discounting is that, on average, the economy is growing, and investments yield a positive return. If the rate of return is 4% per year, then to receive $100 a year from now, one would need to invest about $96 today. So it can be said that the *present value* of a promise of $100 a year from now is $96. The rate of real growth in gross national product over the past 20 years has been about 4% (U.S. Department of Commerce, 1990), which may be taken to represent an appropriate discount rate.

The discounted sum of annual net benefits over time is the *present value* of those benefits. For longer periods of time it is usually called the *capital asset value* of those annual net benefits. For such longer time periods, it is equivalent to the amount of money which would need to be invested at a rate of interest equal to the given discount rate such that the return would equal the net benefit. At a zero discount rate, the present value is the sum of expected net benefits. The discounting/income capitalization formula is given in Appendix A11.

METHODS

The first step in valuing the work of the wetland in retaining lead was to quantify the amount of lead actually held in the wetland sediments. The cost of providing this service was then estimated using the emergy of lost wetland productivity. The cost of replacing the wetland service with a technological treatment alternative was then calculated first in emergy terms and then in dollars.

Lead Filtered by the Wetland

The amount of lead retained in on-site wetlands and in Steele City Bay was estimated based on data in Watts (1984) and Mundrink (1989). The total lead released over the lifetime of the plant was calculated using the estimated number of batteries processed per year and the average concentrations of lead in the electrolyte.

Measurements of Wetland Status

To evaluate the loss of wetland productivity in the field, the gross primary production was estimated in 1991 for each of three ecosystem components — trees, water lilies, and aquatic producers. The swamp was divided into productivity classes based on the vegetation structure, and the various productivity values found in 1991 were used to estimate swamp production in 1981 based on observations of vegetation structure in Lynch (1981). This provided another data point to crudely estimate swamp recovery rates. To consider ecosystem effects other than productivity loss, benthic macroinvertebrate species diversity was measured, and a bioassay of toxicity was made with tree seedlings.

The emergy value of the trees killed was evaluated, but was not included in the value of wetland damage, because for the most part dead trees were not lost to the system but rather were ecologically recycled. The emergy value of the loss of wetland production over time represented the real ecological damage. The transformity of wetland gross primary production for an undamaged forested wetland in the Florida Panhandle was estimated based on average environmental inputs of solar energy, wind energy, rain, and runoff. This transformity was applied to the loss of gross primary production for the wetland system draining the Sapp Battery site, based on measurements for 1991, estimates for 1981, and a number of projected recovery rates.

Energy and Emergy Evaluation

Energy–emergy evaluation (Odum and Arding, 1991; Huang and Odum, 1991; Odum, 1996) was used to evaluate the lead battery production, the wetland filtration system, and the chemical

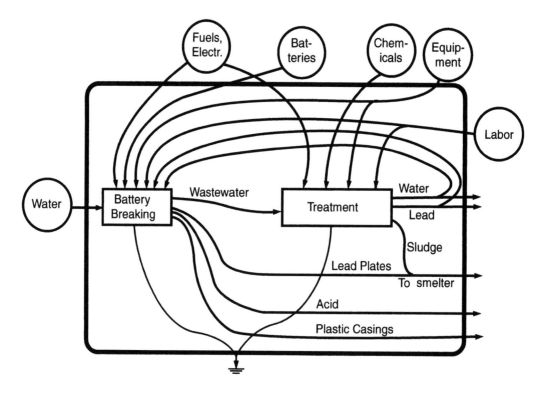

Figure 11.2   Energy diagram of a lead wastewater treatment plant showing inputs of batteries, water, energy, chemicals, equipment, and labor.

recovery system. First an energy systems diagram was constructed (Figures 11.1 and 11.2), detailing the system boundaries, important sources, components, flows, and interactions. These were arranged to show the hierarchy of components (more dilute energies converge to concentrated energies from left to right) and the quality of sources (arranged outside the boundary from left to right in order of quality). Pathway lines show flows of energy, materials, and information.

Emergy evaluation tables with five columns were prepared from the diagram:

Column 1: Item number indicating the table footnote detailing calculations
Column 2: Item from the diagram to be evaluated
Column 3: Data in typical units (joules, grams, or dollars)
Column 4: Solar emergy per unit (solar emjoules per unit: sej/J, sej/g, or sej/$)
Column 5: Solar emergy, in solar emjoules; the product of columns 3 and 4

For example, Table 11.2 evaluates annual flows of inputs and products and Table 11.3 evaluates stored quantities. From the emergy evaluation table emergy indices were calculated to help interpretations.

Economic Analysis

The value of the wetland for treating aquatic lead pollution was estimated using the mainstream economic concept of replacement cost (Scodari, 1990). This involved three steps. First, the level of treatment provided naturally was quantified. Second, the financial cost of a technological substitute was calculated that would have provided the same level of treatment. Third, evidence that the technological substitute would actually have been chosen in the absence of natural treatment was supplied to confirm the replacement cost approach.

Table 11.2 Emergy Evaluation of Yearly Flows in a Northwest Florida Swamp (29.2 ha) with and without Lead/Acid Discharge, 1991

Note	Item	Raw Units	Transformity or Emergy/Mass (sej/unit)	Solar Emergy (sej)
Without discharge				
	Energy inflows			
1	Sunlight	3.07 E12 J	1 sej/J	3.07 E12
2	Wind	1.02 E11 J	623 sej/J	6.37 E13
3	Rain, chemical	2.18 E12 J	15,444 sej/J	3.37 E16
4	Run-in	9.09 E11 J	41,068 sej/J	3.73 E16
	Ecosystem processes			
5	GPP (undamaged)	5.39 E13 J	1,317	7.10 E16
With discharge				
	Energy inflows			
1	Sunlight	3.07 E12 J	1 sej/J	3.07 E12
2	Wind	1.02 E11 J	623 sej/J	6.37 E13
3	Rain, chemical	2.18 E12 J	15,444 sej/J	3.37 E16
4	Run-in	9.09 E11 J	41,068 sej/J	3.73 E16
6	Lead inflow	3.38 E5 g	7.30 E10 sej/g	2.47 E16
7	Lead outflow	1.11 E5 g	7.30 E10 sej/g	8.07 E15
	Ecosystem Processes			
8	GPP (damaged)	1.67 E13 J	1,317	2.20 E16

Notes

1. Solar input. 29.2 ha, 1.5 E7 J m²y⁻¹ (Fernald, 1981), albedo 30%.

    $(29.2\ E4\ m^2)(1.5\ E7\ J\ m^2y^{-1})(1 - 0.30) = 3.07\ E12\ J/y.$

    Transformity = 1.0 (by definition).

2. Wind energy. Diffusion and gradient values for Tampa, FL; Odum et al., 1987, p. 25 ff.

    Winter:
    (1 E3m height)(1.23 kg/m³ air dens)(2.82 m³/m/s)(2.26 E–3 s⁻¹)
    (1.577 E7 s/half y)(29.2 E4 m²) = 8.09 E10 J/half y

    Summer:
    (1 E3 m height)(1.23 kg/m³ air dens)(1.66 m³/m/s)(1.51 E-3 s⁻¹) (1.577 E7 s/half y)(29.2 E4 m²) = 2.13 E10 J/half y

    Total = Summer + Winter = 2.13 E10 + 8.09 E10 = 1.02 E11 J/y.
    Transformity = 623 sej/J (Odum et al., 1987, p. 4).

3. Rain, chemical. 1.51 m/y (Fernald, 1981). Gibbs free energy of rain relative to seawater, 4.94 J/g.

    (29.2 E4 m²)(1.51 m/y)(1000 kg/m³)(4.94 E3 J/kg) = 2.18 E12 J/y.

    Transformity = 1.54 E4 sej/J (Odum et al., 1987, p. 4).

4. Run-in. Drainage area estimated equal to wetlands area from USGS map. Annual runoff rate for Northwest Florida 0.63 m/y (Kenner, 1966 in Fernald, 1981).

    (29.2 E4 m²)(0.63 m/y)(1000 kg/m³)(4.94 E3 J/kg) = 9.09 E11 J/y.

    Transformity = 41 E4 sej/J (Odum et al., 1987, p. 4).

5. Gross primary production (undamaged). Reference forest production from Table 5.4: 1.85 E8 J/m²/y

    (1.85 E8 J/m²/y)(29.2 E4 m²) = 5.39 E13 J/y.

    Transformity calculated from sum of emergy of 3 and 4 above divided by energy of gross primary production.

    (7.10 E16 sej/y)/(5.39 E13 J/y) = 1317 sej/J.

6. Lead inflow. Lead in process wastewater 0.27 g/battery. 1.25 E7 batteries processed by Sapp in 10 years.

    Total wetland area 29.2 ha.

    (1.25 E7 batteries/10 years)(0.27 g Pb/battery) = 3.38 E5 g/y.

    Emergy/mass of lead = 7.3 E10 sej/g.

continued

Table 11.2 (continued)　Emergy Evaluation of Yearly Flows in a Northwest Florida Swamp (29.2 ha) with and without Lead/Acid Discharge, 1991

7.	Lead outflow. Lead inflow over 10 years = (0.27 g/battery)(1.25 E7 batteries) = 3.38 E6 g. Lead retained in wetland = 2.28 E6 g (this study).
	(3.38 E6 g − 2.28 E6 g)/10 years = 1.11 E5 g/y.
8.	Gross primary production (damaged). Weighted average of wetland production from Table 5.4: 5.73 E7 J/m²/y.
	(5.73 E7 J/m²/y)(29.2 E4 m²) = 1.673 E13 J/y.
	Transformity = 1317 sej/J (calculated in note 5 above).

The technological substitute was an existing wastewater treatment plant operated by a secondary lead smelter in Tampa, FL. Operation and maintenance cost data were supplied by the firm (Neil Oakes, personal communication), and capital costs for the treatment plant were estimated using a component cost approach (James M. Montgomery Consulting Engineers Inc. 1985, p. 661) for specific wastewater treatment processes and equipment.

The financial cost of replacing the wetland's work in lead retention was calculated as the sum of the capital cost of building a treatment facility and the operating costs to treat an amount of lead equal to that which was retained by the wetland. The operating cost was obtained by multiplying the unit operating costs (dollars/kilogram) to treat lead in the treatment plant by the amount of lead (kilograms) retained in the swamp.

While the total benefit of allowing the wetland to treat lead waste was calculated from its replacement cost, there were some economic costs incurred in this wetland use. The financial cost of the loss of standing timber from the wetland was calculated from the amount of wood in tree boles in the standing stock in the reference forest (Location RF, Figure 1.3) and from current market stumpage values for wetland trees. The data from the 5 × 20-m tree plots were converted to aboveground stem mass using the following regressions from Day (1984):

$$\log_{10} \text{dry weight cypress (kg)} = -0.99 + 2.426 \log_{10} \text{dbh (cm)}$$

$$\log_{10} \text{dry weight hardwoods (kg)} = -1.0665 + 2.4064 \log_{10} \text{dbh (cm)}$$

where dbh is the diameter at breast height.

The financial cost of the loss of timber production (as distinct from the loss of standing timber) was estimated from the same market stumpage values of wetland wood multiplied by the estimated annual wood production from Johnson (1978).

With wetland treatment, lead that with chemical treatment would be precipitated and recycled to the economy was instead bound up in wetland sediments. This economic loss of lead metal was calculated according to the market value of lead (Woodbury, 1988).

The costs of wetland treatment (loss of timber, timber production, and lead metal) were subtracted from the benefits of wetland treatment (the replacement cost) to calculate the net benefit of using wetland treatment. Since the stream of benefits and costs occurred over time, the mainstream economic values of future benefits and costs were discounted at 4%. The formula used was

$$PV = \sum_{t=0}^{N} \frac{NB_t}{(1+i)^t}$$

where PV is present value, NB is annual net benefit, i is the discount rate, and t is the number of years in the future. See Appendix A11.

Table 11.3   Emergy Evaluation of Storages in a Northwest Florida Swamp (29.2 ha) with and without Lead-Acid Discharge, 1991

Note	Item	Raw Units (sej/unit)	Transformity or Emergy/Mass (sej)	Solar Emergy
Without discharge				
1	Water	1.38 E12 J	41,000 sej/J	5.67 E16
2	Lead in water	2.92 E1 g	7.30 E10 sej/g	2.13 E12
3	Wood	1.47 E14 J	32,000 sej/J	4.71 E18
4	Peat	3.26 E14 J	17,000 sej/J	5.55 E18
5	Lead in peat	2.73 E5 g	7.30 E10 sej/g	2.00 E16
With discharge				
1	Water	1.38 E12 J	41,000 sej/J	5.67 E16
6	Lead in water	8.47 E3 g	7.30 E10 sej/g	6.18 E14
7	Wood	3.02 E13 J	32,000 sej/J	9.65 E17
4	Peat	3.26 E14 J	17,000 sej/J	5.55 E18
8	Lead in peat	2.28 E6 g	7.30 E10 sej/g	1.66 E17

Notes:

1. Water. Depth above peat = 0.5 m. Depth of peat = 0.5 m. Percent moisture = 89.6%. Density of wet peat = 1.0 E6 g/m³. Gibbs free energy of water = 4.94 J/kg.

    Water in peat = (29.2 E4 m²)(0.5 m)(1.0 E6 g/m³)(0.896).

    (4.94 J/g) = 6.54 E11 J.

    Water above peat = (29.2 E4 m²)(0.5 m)(1.0 E6 g/m³).

    (4.94 J/g) = 7.30 E11 J.

    Total water above and in peat = 1.38 E12 J.

    Transformity = 4.1 E4 sej/J (Odum, 1992b).

2. Lead in water (background). Pb conc = 2.0 E-10 g Pb/g water (Förstner and Wittmann, 1983, p. 87, avg for freshwater).

    (29.2 E4 m²)(0.5 m depth)(1 E6 g/m³)(2.0 E-10 g Pb/g water) = 29.2 g Pb.

    Emergy/mass = 7.3 E10 sej/g (Table A11B.6).

3. Wood. Mass from reference forest tree plots = 34.4 kg/m². Wood energy 3500 kcal/kg (Chapman & Hall, 1986, p. 467).

    (29.2 E4 m²)(34.4 kg/m²)(3500 kcal/kg)(4186 J/kcal) = 1.47 E14 J.

    Transformity = 3.2 E4 sej/J (Odum, 1992b, p. 27).

4. Peat. Depth 0.5 m. Moisture 89.6%. Density of wet peat 1.0 E6 g/m³.

    Peat energy 2.15 E4 J/g (Odum, 1992b, p. 27).

    (29.2 E4 m²)(0.5 m)(1 − 0.896)(1.0 E6 g/m³)(2.15 E4 J/g) = 3.26 E14 J.

    Transformity = 1.7 E4 sej/J (Odum, 1992b, p. 27).

5. Lead in peat (background). Pb conc = 1.8 E-5 g Pb/g sediment (Okefenokee Swamp, GA; Nixon and Lee, 1986, p. 116).

    (29.2 E4 m²)(0.5 m)(1 − 0.896)(1.0 E6 g/m³)(1.8 E-5 g Pb/g peat) = 2.73 E5 g Pb.

    Emergy/mass (see note 2).

6. Lead in water (contaminated). Pb conc = 5.8 E-8 g Pb/g water (Ton, 1990).

    (29.2 E4 m²)(0.5 m depth)(1 E6 g/m³)(5.8 E-8 g Pb/g water) = 8.47 E3 g Pb/ha.

    Emergy/mass (see note 2).

7. Wood. Mass (weighted average of tree plots) = 7.05 kg/m². Wood energy 3500 kcal/kg (Chapman & Hall, 1986, p. 467).

    (29.2 E4 m²)(7.05 kg/m²)(3500 kcal/kg)(4186 J/kcal) = 3.02 E13 J.

    Transformity (see note 3).

8. Lead in peat (contaminated). Total Pb in peat estimated at 2276 kg for 29.2 ha (Figure 5.1).

RESULTS

Lead Retained by the Swamp

Lead retained in on-site wetlands was estimated to be about 1000 kg using Watts' data (1984). Lead retained in Steele City Bay was estimated at 1354 kg. The average of lead concentrations reported by Ton (1990) for sediments in Steele City Bay was 52.1 kg/ha, or 1198 kg in all 23 hectares. The average of the two values for Steele City Bay was 1276 kg. Thus, the total lead retained in on- and off-site wetlands was estimated at 2276 kg. For details on calculations, see Pritchard (1992, Appendix E).

Assuming a linear rate of increase in battery processing by Sapp Battery from zero at the beginning of 1970 (when the plant opened) to a peak of 50,000 batteries per week in 1979 (Watts, 1984), the total number of batteries processed was estimated at 12,525,000. Cumulative lead release from those batteries over 10 years was estimated to be between 1528 and 6162 kg of particulate and dissolved lead, based on data on electrolyte content from Watts (1984) (calculations in Pritchard, 1992, Appendix E). At a secondary lead smelter in Tampa, FL, process wastewater contained about 0.27 g lead for every battery. This, multiplied by the estimated 12,525,000 batteries processed at Sapp, would put cumulative lead releases at 3382 kg.

The removal rate is the percentage of lead released that was retained in the wetland system. Using the range of concentration given by Watts (1984), the removal rate for lead by the wetland system was between 37 and 100%, with a middle value of about 67% based on data on wastewater lead concentrations at a secondary lead smelter in Tampa, FL.

Since it is likely that the rate of battery processing at Sapp Battery increased exponentially rather than linearly to its peak rate as was assumed, 12 million batteries processed is probably an overestimate, making the actual removal rate higher than the calculated removal rate. However, emergy and economic calculations that follow are based on the amount of lead retained rather than on the removal rate. We evaluated the work the wetland did, not the work it did not do (i.e., lead not absorbed from the waters flowing out).

Emergy Evaluation of Impacted Wetlands

The emergy per gram of lead metal was calculated by summing environmental work and human work in extracting, refining, and processing in Appendix A11 and was 7.3 E10 sej/g.

The emergy evaluation of the wetlands which received wastewater from the Sapp Battery plant based on the diagram in Figure 11.1 is given in Tables 11.2 (flows) and 11.3 (storages). The two largest sources, rain and run-in, were taken as the main annual emergy input to the swamp system. The reference forest wetland (Location RF) was used to represent the productivity of the impacted wetland before damage began (line 5 in Table 11.2). The transformity of gross primary production based on productivity values from the reference forested wetland was about 1300 sej/J.

Productivity data (Table 5.6) were used to estimate the actual level of energy processing in the swamp in 1991 (line 8 in Table 11.2). Remote sensing information from a previous study (Lynch, 1981) was used to estimate the energy flows and transformations in the local wetlands system for two points in time. Table 5.4 shows the reduction in empower per square meter due to the acidic discharge for 1991. In Table 11.4 these estimates are multiplied by the appropriate areas to convert them to total empower for the wetland complex (calculations in Pritchard, 1992, Appendix F). Figure 11.3 is a time graph of wetland empower based on the above calculations.

In the absence of detailed time-series data, several simplifying assumptions were made. It was assumed that the decrease in empower of the system was linear from the beginning of operations at Sapp Battery to its closure, a period of 10 years. Lynch's measurements were made in 1981, soon after the cessation of operations; the data he cites are taken as the minimum productivity of the system. Linear recovery is also assumed up to 1991.

Table 11.4   Empower of Wetlands after Cessation of Lead/Acid Discharge

	1981	1991
Empower of wetland system[a]	1.0 E16 sej/year	2.2 E16 sej/year
Potential empower[b]	7.1 E16 sej/year	7.1 E16 sej/year
Empower loss[c]	6.1 E16 sej/year	4.9 E16 sej/year
Ratio of observed to potential empower	0.14	0.31

[a] 1981 value is based on remote sensing data from Lynch (1981); 1991 value is
   based on the results of this study. Calculations in Pritchard (1992, Appendix F).
[b] Based on the productivity of the reference forest (2.4 E11 sej/m²/year on 29.2 ha
   of wetlands).
[c] Difference between potential and observed empower.

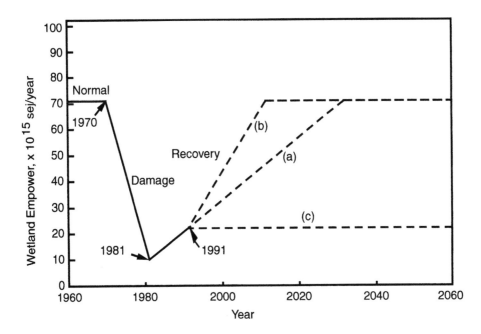

Figure 11.3   Empower graph of destruction and recovery of wetlands which received lead/acid discharge
between 1970 and 1980, based on Table 5.4. Three recovery rates are projected: (a) present
rate, (b) double the present rate, and (c) no further recovery.

Figure 11.3 shows three projections of future recovery, based on (1) the rate of recovery from
1981 to 1991, (2) double the rate of recovery from 1981 to 1991 (assuming that this may be the
effect if water levels were managed to allow regeneration), and (3) no further recovery past 1991.
The lost ecosystem production is the difference between potential and actual production.

The empower loss over time was integrated to obtain total emergy loss. Table 11.5 gives the
value of the integration using each of the projected recovery rates. Notice that if it is the case that
permanent ecological damage has occurred and that recovery time is infinite, the damage is also
regarded as infinite since it continues in perpetuity. This is in contrast with the mainstream economic
notion of discounting future benefits and losses discussed below.

The emergy in the storage of wetland trees that either died in place and fell or were cut down
and land-filled was substantial. Table 11.3 showed an emergy of wood (in stems only) of 1.61 E17
sej/ha for a typical Northwest Florida swamp. There were 25 ha of total tree mortality (Lynch,
1981), so the total emergy of dead trees was 4.03 E18 sej. Neither fallen trees nor land-filled trees
have completely oxidized, so much of the stored emergy has merely been transferred from standing
stock to the detrital pool or to a landfill. This stored emergy is therefore not counted as a loss in
the evaluation of wetland lead retention.

Table 11.5   Emergy Loss during Wetland Destruction and Recovery

Recovery Projection[a]	Recovery Time (years)	Loss (sej[b])
(a)	51	1.9 E18
(b)	30	1.4 E18
(c)	Infinite	Infinite

[a] For explanation of projections see Figure 5.5.
[b] Integration of empower reduction in Figure 5.5.

Emergy Evaluation of Lead Smelter-Chemical Recovery System

The wastewater treatment plant of a secondary lead smelter in Tampa, FL, is diagrammed in Figure 11.3, and the emergy evaluation is given in Table 11.6. Several new transformities were calculated for chemicals used in the treatment system by making an emergy evaluation of their systems of production (Appendix A11).

Table 11.6   Emergy Evaluation of a Water Treatment Plant (All Flows Are per Year)

Note	Item	Raw Units	Transformity or Emergy/Mass (sej/unit)	Solar Emergy (E16 sej)	Solar Emergy (sej)
Junk batteries		2.2 E6 batteries			
1	Plastic battery casings	6.3 E13 J	6.6 E4 sej/J	413.5	4.1 E18
2	Battery acid	6.8 E9 g	9.1 E7 sej/g	62.0	6.2 E17
3	Lead in plates	2.7 E10 g	7.3 E10 sej/g	198,639.5	2.0 E21
Trucking cost for acid removal					
4	Fuel	5.1 E11 J	6.6 E4 sej/J	3.3	3.3 E16
5	Services	5.8 E4 $	1.6 E12 sej/$	9.0	9.0 E16
Process wastewater					
6	Water	2.5 E11 J	4.8 E4 sej/g	1.2	1.2 E16
7	Lead	6.0 E5 g	7.3 E10 sej/g	4.4	4.4 E16
Energy inputs to treatment					
8	Caustic	9.7 E7 g	6.7 E9 sej/g	65.2	6.5 E17
9	Lime	3.0 E7 g	1.6 E9 sej/g	4.9	4.9 E16
10	Ferrous sulfate	3.3 E7 g	1.0 E9 sej/g	3.3	3.3 E16
11	Diatomaceous earth	1.6 E7 g	2.0 E9 sej/g	3.3	3.3 E16
12	Electricity	1.4 E12 J	2.0 E5 sej/J	29.0	2.9 E17
13	Labor	6.3 E0 p/year	3.4 E16 sej/p	21.3	2.1 E17
Plant capital value					
14	Equipment	3.2 E4 $	1.6 E12 sej/$	5.0	5.0 E16
Effluent					
15	Water	6.2 E10 J	4.8 E4 sej/J	0.3	3.0 E15
16	Lead	5.0 E2 g	7.3 E10 sej/g	0.004	3.7 E13
Recovered metals					
17	Lead	6.0 E5 g	7.3 E10 sej/g	4.4	4.4 E16

Notes:

1. Plastic battery casings. 3.0 E6 lb (plant estimate); 4.6 E7 J/kg (Hall et al., 1986, p. 4).
   (3.0 E6 lb)(0.454 kg/lb)(4.6 E7 J/kg) = 6.3 E13 J.
2. Battery acid. 7490 short tons (plant estimate) of 15 to 20% sulfuric acid (Watts, 1984).
   (7.5 E3 tons)(9.07 E5 g/ton) = 6.79 E9 g.
   Emergy per gram = 9.1 E7 sej/g (Table A11B.5).
3. Lead in plates. Lead per battery = 20.0 lb (Watts, 1984).
   (2.22 E6 batteries)(20 lb/battery)(454 g/lb) = 2.02 E10 g.
   Emergy per gram = 7.3 E10 sej/g (Table A11B.6).

Table 11.6 (continued)   Emergy Evaluation of a Water Treatment Plant (All Flows Are per Year)

4. Fuel. 312 trips per year at 100 mi per trip (plant estimate).

    Truck mileage about 8 mi/gal. Gasoline = 3.42 E7 J/l (Cervinka, 1980, p. 15).

    (312 trips)(100 mi/trip)(1 gal/8 mi)(3.79 l/gal)(3.42 E7 J/l) = 5.05 E11 J.

5. Human services in trucking. Payment for hauling = $58,000.

    1990 Emergy/money ratio = 1.55 E12 sej/$ (Odum, 1992b).

6. Water used to cool saws, clean battery casings, equipment, and workplace.
    1.3 E7 gal (plant estimate) = 5.0 E7 l. Gibbs free energy of freshwater = 4.94 J/g (Odum, 1992b).
    (5.0E7 l)(1000 g/l)(4.95 J/g) = 2.49 E11 J.

7. Lead in process wastewater. Average concentration = 12 mg/l (plant estimate).

    (12 mg/l)(0.001 g/mg)(5.0 E7 l) = 6.0 E5 g.

    Emergy per gram = 7.3 E10 sej/g (Table A11B.6).

8. Caustic soda. 214 tons of liquid (50%) caustic (plant estimate).

    214 tons * 50% = 107 tons.

    (107 tons)(2000 lb/ton)(454 g/lb) = 9.7 E7 g.

    Emergy per gram = 6.7 E9 sej/g (Table A11B.3).

9. Hydrated lime. 66,000 lb (plant estimate).

    (66,000 lb)(454 g/lb) = 3.0 E7 g.

    Emergy per gram = 1.6 E9 sej/g (Table A11B.2).

10. Ferrous sulfate. 1440 bags (plant estimate) at 50 lb/bag.

    (1440 bags)(50 lb/bag)(454 g/lb) = 3.27 E7 g.

    Emergy per gram = 1.0 E9 sej/g (value for sedimentary rocks, Odum, 1992b).

11. Diatomaceous earth. 36,000 lb (plant estimate).

    (36,000 lb)(454 g/lb) = 1.63 E7 g.

    Emergy per gram = 2.0 E9 sej/g (Table A11B.4).

12. Electricity. Used 4.03 E5 kwh (plant estimate).

    (4.03 E5 kwh)(3.6 E6 J/kwh) = 1.45 E12 J.

13. Labor. The treatment plant operation involves the following amount of labor, measured in person-years (1 person working 8 to 9 h/day for 1 year): 3 operators, 1 mechanic, 1 foreman, 0.5 lab tech, and 0.75 plant engineer = 6.25 person-years.

    Emergy/person = GNP/population = 3.4 E16 sej/person (Odum, 1992b).

14. Human services embodied in plant equipment. Capital cost = 3.16 E5 $(1988) (estimated in this study). Assumed plant lifetime = 10 years.

    Annualized cost = ($3.16 E5)/(10 years) = $16,000/year.

15. Effluent water. 25% of inflow (75% is recycled; plant estimate).

    (0.25)(5.0 E7 l)(1000 g/l)(4.94 J/g) = 6.2 E10 J.

16. Effluent lead. Average concentration = 0.04 mg/l (plant estimate).

    (0.04 mg/l)(0.001 g/mg)(1.3 E7 l) = 504 g.

    Emergy per gram = 7.3 E10 sej/g (Table A11B.6).

17. Recovered lead. Influent lead = 6.05 E5 g (note 8 above).

    Influent – effluent = 6.05 E5 g – 504 g = 6.04 E5 g.

    Emergy per gram = 7.3 E10 sej/g (Table A11B.6).

The Tampa facility receives whole batteries which are sawn and drained. The lead plates are sent to the smelter on site. The plastic casings are washed and sold to a recycler. The electrolyte is collected, allowed to settle, and shipped to phosphate fertilizer plants near Tampa to obviate the need for expensive neutralization.

Water is used to cool the saws and to wash the plastic casings and the work area. This and other process wastewater from the plant is collected and treated prior to release into the Tampa sewer system. Storm water is also collected and treated in a parallel but separate process which is not considered here.

The treatment system involves hydroxide precipitation of dissolved metals using caustic soda and hydrated lime. Ferrous sulfate is added as a coagulant. The wastewater is passed through a filter of diatomaceous earth. The metal-rich precipitate from the chemical process and the spent diatomaceous earth are used in the smelter as sinter. Water from the filter is sent to a settling pond from which it drains into the municipal sewer system of Tampa, which leads to a standard sewage treatment plant.

Comparison of Treatment Systems

Table 11.7 is a summary table for treatment of lead by the two alternatives evaluated — wetland and chemical treatment. The "emergy necessary" is a "cost" to society since it cannot be used for other purposes. The emergy necessary for wetland treatment was evaluated as the total lost emergy production by the wetland complex during the Sapp Battery plant operation and during ecosystem recovery. The emergy necessary for chemical treatment was evaluated as the emergy of services in treatment plant construction an
d materials (the money cost of construction multiplied by the U.S. emergy/$ ratio) and the treatment emergy per kilogram of lead removed from the waste stream, which includes chemicals, electricity, labor, and fuel, and services in acid removal.

Table 11.7 Summary Table of Emergy Indices for Lead Treatment Alternatives

Description	Units	Wetland Treatment (30- to 50-year recovery)	Chemical Treatment
Emergy necessary for operation/lead retained	sej/kg Pb	6.15 E14 to 8.35 E14[a]	2.3 E15[b]
Lead retained in swamp	kg Pb	× 2276	× 2276
Subtotal	sej	1.40 E18 to 1.90 E18	5.23 E18
Emergy in equipment and building	sej	+ 0.00	+ 0.50 E18[c]
Total emergy necessary	sej	1.40 E18 to 1.90 E18	5.73 E18
Treatment efficiency[d]	% retention	75	99
Emergy/emergy lead[e]	sej/sej Pb	8.42	31.75

[a] Based on emergy of wetland production loss of 1.4 E18 to 1.9 E18 sej from Table 11.5 divided by the lead retained in the wetland, 2276 kg.

[b] Based on operating emergy for lead wastewater treatment plant (the sum of rows 4, 5, and 8 to 13 in Table 11.6) divided by the lead retained by the treatment plant, 604 kg.

[c] Based on financial capital cost from Table 11.9 divided by 1990 emergy/money ratio of 1.55 sej/$.

[d] Treatment efficiency = lead retained/lead input.

[e] Emergy of lead = (mass of lead retained) * (emergy/mass of lead), 7.3 E10 sej/g Pb.

As shown in Table 11.7, the emergy cost of chemical treatment was four to five times higher than the emergy cost of wetland treatment, whether calculated per battery, per gram of lead retained, or per emergy of lead retained. The ratio of emergy in treatment to the emergy of lead retained is a kind of investment ratio, and had a value of 8 to 11 for wetland treatment and 32 for chemical treatment.

Table 11.8 shows the net benefit in emergy terms of using wetland treatment rather than chemical treatment. The value of lead not recycled was included since it was a loss to the economy, and, unlike dead wetland trees, it could not be used by the ecological system.

The net benefit from using wetland treatment was between 3.8 E18 and 4.29 E18 sej. The average value of the swamp (29.2 ha) was 1.29 E17 to 1.47 E17 sej/ha.

Table 11.8 Emergy Value of Wetland Treatment Using the Replacement Cost Method

Zero Discount Rate	Emergy Value (sej)
Benefit	
Replacement cost	5.73 E18
Emergy cost	
Timber destroyed	N/A
Loss of production[a]	1.40 E18 to 1.90 E18
Lead not recycled[b]	4.4 E16
Total cost (emergy loss)	1.44 E18 to 1.94 E18
Net benefit[c]	3.79 E18 to 4.29 E18
B/C ratio[d]	2.9 to 4.0

[a] Based on recovery projections of 25 and 50 years.
[b] 2276 kg lead at 7.3 E10 sej/g (Pritchard, 1992, Appendix D).
[c] Benefit minus total losses.
[d] Benefit divided by total losses.

Economic Analysis Using Money

Capital costs for treatment plant construction were estimated according to the methods and data in Gumerman et al. (1979a and 1979b) and James M. Montgomery Consulting Engineers Inc. (1985). Values for unit processes are shown in Table 11.9. The sum of these processes were updated to present costs using the *Engineering News-Record* (ENR) Building Cost Index (BCI) for the years concerned, according to the following formula:

$$\text{Updated Cost} = 1978 \text{ Construction Cost} \left(\frac{\text{Current BCI}}{1978 \text{ BCI}}\right)$$

Operating costs for the treatment plant are given in Table 11.10.

Table 11.9 Capital Costs of 36,000-gpd Water Treatment Plant, 1990 Dollars

Item	Unit Design Value	Construction Cost, 1978	Construction Cost, 1990[a]
Lime feed system[b]	7.5 lb/h	$15,000	$23,258
NaOH feed system[b]	1,172 lb/day	25,000	38,764
Ferrous sulfate feed system[b]	8.2 lb/h	15,000	23,258
Storage tanks[c]	12 m³ (3)	15,900	24,654
Circular clarifier[b]	100 ft²	30,000	46,516
Rapid mix basin[b]	100 ft³	10,000	15,505
Pressure diatomite filter[c]	36,000 gpd	45,000	69,775
Dewatering lagoon[b]	40,000 ft³	9,000	13,955
Subtotal		$164,900	$255,685
Contractor overhead[c]	12%	19,788	30,682
Engineering	11%[d]	18,139	28,125
Subtotal		$202,827	$314,493
Legal, fiscal, admininistrative[c]		6,200	9,613
Total capital cost		$209,027	$324,107

[a] Based on *Engineering News-Record* Building Cost Index (1913 = 100; 1978 = 1731; 1990 = 2684).
[b] Gumerman et al., 1979.
[c] Hansen et al., 1979.
[d] J.M. Montgomery Consulting Engineeers, Inc., 1985.

Table 11.10 Annual Operating Costs of a 36,000 gpd Water Treatment Plant, 1990 Dollars

Item	Quantity ($/unit)	Price ($)	Value
Energy and materials input to treatment			
Caustic soda	214 tons	346	74,000
Hydrated lime	33 tons	85	2,805
Ferrous sulfate	36 tons	300[a]	10,800
Diatomaceous earth	18 tons	450	8,100
Electricity	403 mwh	66	26,498
Subtotal (per year)			$122,203
Labor and management	6.25 p · year	40,000[b]	$250,000
Opportunity cost of capital, $162,000[c]	@4.0%		6,480
Total plant operating costs (per year)			$378,683
Trucking cost for acid removal			58,300
Total (per year)			$436,983
Total (per gram lead retained)[d]		$0.72	
Total (per battery)[e]		0.20	

Note: All data are from Tampa lead treatment plant except as noted.

[a] Traylor Chemical and Supply, Orlando, FL, quote over phone, 3-12-92.
[b] Average 1990 skilled labor wage from *Engineering News-Record*.
[c] Average value of physical capital taken to be one half of capital cost from Table 11.9.
[d] Based on 6.04 E5 g Pb retained.
[e] Based on 2.22 E6 batteries processed.

The net present value of the total benefit (replacement costs) is given in Table 11.11. Details of the calculation, including the time stream of benefits and costs, are given in Appendix A11

Table 11.11 Net Present Value of Benefits and Costs Associated with Wetland Lead Retention at Discount Rates of 4 and 0% (Details in Appendix A11A)

4% Discount Rate	Net Present Value
Benefit	
Replacement cost	$1,617,375
Emergy cost	
Timber destroyed[a]	20,478–58,764
Loss of production[b]	4,740–13,203
Lead not recycled[c]	1,563
Total cost (emergy loss)	26,781–73,530
Net benefit[d]	$1,543,845–1,590,594
B/C ratio[e]	22–60
0% Discount Rate	**Net Present Value**
Benefit	
Replacement cost	$1,962,827
Emergy cost	
Timber destroyed[a]	23,000–66,000
Loss of production[b]	10,500–29,251
Lead not recycled[c]	1,980
Total cost (emergy loss)	35,480–97,230
Net benefit[d]	$1,865,597–1,927,347
B/C ratio[e]	20–55

[a] 25 ha destruction at 344,000 kg/ha and range of market values for timber of $0.0027 to $0.0077 per kg.
[b] Based on market values for timber and a 50-year recovery period.
[c] 2276 kg lead at $0.87/kg (Woodbury, 1988).
[d] Benefit minus total cost.
[e] Benefit divided by total cost.

(Table A11.1). Results are given for a 4% discount rate and a 0% discount rate (for comparison with emergy analysis results which are not typically discounted). The capital costs were assumed to occur in the first year of operation. Unit operating costs (dollars/kilogram lead retained) were multiplied by the amount of lead retained in the wetland during each time period, which increased over the 10 years of operation.

The economic loss due to timber destruction was calculated from the market stumpage value of trees in the reference forest (Location RF on map [Figure 5.2]). The mass of tree stems calculated from the Day (1984) regression was 344,000 kg/ha (standard error = .148,000 kg/ha). The market value of black gum on the stump was $0.0027/kg for use as pulpwood (Earl Clark, Georgia-Pacific Corporation Timber Department, personal communication) and $0.0077/kg for veneer (David Cannon, Cross City Veneer, personal communication). The market value per hectare of wetland wood was thus between $929 and $2650 per hectare. Lynch (1981) estimated the area of wetland destruction based on absence of trees at 25 ha. The economic loss of the standing crop of trees was therefore between $23,000 and $66,000. For the purposes of discounting, this loss was assumed to be spread equally over the first 5 years of operation.

In addition there was a loss in production of timber over the period of destruction and recovery. Johnson (1978) estimated that Taxodium/Nyssa stands produce over 100 ft^3 of wood per acre annually (over 5075 kg/ha^{-1} year^{-1}). Using the market values given above for wetland wood, the annual production would be valued at $14 to $39/ha^{-1}year^{-1}, or $350 to $975 (25 ha)$^{-1}$year^{-1}. For purposes of discounting it was assumed that the wetland production fell at a constant rate from 1970 to 1980, and that recovery would be over a period of 50 years.

The lead that was retained by the wetland represents a loss to the economy, because in the replacement cost scheme it would be recycled when the chemical precipitate is returned to the smelter. The value of the lead could either have been added to the cost side, as was done here, or subtracted from the benefit side (i.e., it would have partly offset the treatment cost). If the 2276 kg of lead had been returned to a smelter, as it is at the secondary lead smelter in Tampa, FL, the value of the lead would have been about $2000 (2276 kg * $0.87/kg [Woodbury, 1988]).

The discounted net value of wetland treatment was about $1,550,000 (Table 11.11), or $53,000/ha^{-1} ($21,000/acre^{-1}).

DISCUSSION

Emdollar Evaluation of Wetland Lead Retention

If the U.S. economy with a 1990 gross national product of $5.4 trillion is assumed to be emergy estimated at 8.6 E24 sej, then the portion of the economy accounted for by the swamp for lead treatment (about 4.0 E18 sej from Table 11.8) is comparable to $2,500,000, or about $85,000 per hectare ($35,000 per acre).

Economic Valuation of Wetland Lead Retention

Since the total costs of wetland damage were known, the areas of producer surplus and social cost could be approximated (Appendix A11).

The value of the sulfuric acid reclaimed under chemical treatment could have been considered a loss under wetland treatment. In the chemical treatment alternative, the sulfuric acid was shipped to local fertilizer plants for simple dilution of higher quality sulfuric acid, and had a minimal value as acid. This practice was largely a useful disposal means obviating the need to treat the acid (Neil Oakes, personal communication). The value of the acid was therefore not considered further.

Using only the market value of wetland timber to represent the social loss from wetland damage is incomplete. There are no doubt other losses to society incurred by damaging wetlands, such as loss of aesthetic value, loss of fishery and wildlife support, and so on. These would be important to consider in a more comprehensive analysis. From a mainstream economics perspective, these losses may have had a low marginal value considering the size of the wetland relative to the total wetland area of Florida or the nation. Were wetlands to come into use for heavy metal retention on a larger scale, the value of each additional unit of these other losses would become more and more important, making it difficult to scale up from the value for this wetland to a potential value for all similar wetlands.

The calculated net benefit to society from using the wetland was not very sensitive to the discount rate used (see Table 11.11), because the large total benefits (replacement cost) accrued within 10 years, and hence were little discounted, whereas the much smaller social costs (value of timber lost) stretched forward 25 to 50 years, and were even smaller once discounted.

For fixed wetland size, very high loading levels could be more expensive than chemical treatment, taking into account the level of ecological damage and the lower removal rate that would result. The cost-effectiveness of chemical treatment at high loading levels was also found by Baker et al. (1991) in comparison with wetlands treating acid mine drainage.

Comparison of Emergy and Economic Evaluations

Table 11.12 puts the two measures of value (emergy vs. mainstream economic value) on a similar scale by showing the proportion of the national economy represented by each. The ecological economic value calculated in emergy is shown as a fraction of estimated 1990 national emergy use, and the mainstream economic value (undiscounted) is shown as a fraction of gross national product. The undiscounted sum of costs and benefits is used since it is unclear what discount rate could be used for emergy.

Table 11.12 Comparison of Ecological Economic and Mainstream Economic Value of Wetlands for Lead Treatment

0% Discount Rate	Ecological economic value[a] Fraction of National Emergy Use Emergy/National Emergy Use (times E-10)	Mainstream economic value[b] Fraction of Gross National Product Value/GNP (times E-10)
Benefit		
Replacement cost	6702	3592
Costs		
Timber destroyed	N/A	42–121
Loss of production	1637–2222	19–54
Lead not recycled	51	4
Total costs	1680–2274[c]	65–178[d]
Net benefit[e]	4428–5013	3414–3527
B/C ratio	3–4	20–55

[a] Emergy values in Table 11.4 divided by 1990 U.S. estimate of emergy use of 8.55 E24 sej.
[b] Dollar values in Table 11.7 divided by 1990 U.S. GNP of $5.465 E12 (1990 dollars).
[c] Two cost estimates based on 30- vs. 50-year recovery projections.
[d] Two cost estimates based on range of market values for wetland timber.
[e] Benefit (replacement cost) minus total costs.

The cost of replacing the wetland service with chemical treatment as measured with emergy is twice as large as when measured in dollars. However, the emergy cost of using wetlands is also larger than the dollar cost, so that although the net benefit measured in emergy is larger than in dollars, the benefit/cost ratio is much smaller.

As has been noted, a more comprehensive mainstream economic analysis would attempt to measure more of the costs of using wetlands than timber loss. The difference in the fraction of the economy accounted for as wetland damage in each of the two methods is, however, almost two orders of magnitude, so adding in a few more items in the mainstream analysis would change the comparison only a little.

If emergy is the driving force in the economy, then mainstream economics underestimates the cost of both treatment alternatives, because both are resource intensive (one in nature and the other in fossil fuel), and it ignores costs that are not recognized or borne directly by humans. There may be a greater direct and indirect impact on social welfare with both treatment systems than is shown by the money circulating in the process.

While there are substantial differences between the emergy method of analysis and mainstream economics, both agree that value is a systems property, determined at system steady state (which the mainstream economists call "general equilibrium"). Both agree that something is being maximized by the system (maximum power of the system in energy theory, maximum profit, or maximum utility in mainstream theory). They fundamentally disagree on appropriate system boundaries and the appropriate scale of analysis.

Wetland Potential for Lead Filtration in the Nation

The 29.2 ha of wetlands draining the Sapp Battery site retained 2276 kg of lead over 10 years. The lead removal rate was 7.8 kg/ha/year. The total release of lead to air, surface water, and public sewers reported in the U.S. EPA Toxics Release Inventory (1989) for 1987 was 1.5 million kg. If the great majority of atmospheric lead is rained out over land, then at 7.8 kg/ha/year, it would take 192,000 ha of suitable wetlands to treat that level of lead emission, or approximately 0.5% of the nations wetlands. It is surely the case that wetlands are even now mitigating the impact of heavy metal pollution by providing retention, albeit at much lower levels than in the Sapp Battery wetlands.

In 1985 51 million batteries were recycled in the U.S. (Putnam Hayes and Bartlett Inc., 1986, 1987). Breaking and smelting generated 14,000 kg of lead in process wastewater. If all this waste had been chemically treated it would have cost $10 million in money and used resources worth 3.2 E22 sej (from data in Tables 11.3 and 11.6). If the lead had been treated with natural wetlands at the same loading as at the Sapp site, 1800 ha of wetlands would have been used, the financial cost would have been zero (although there would have been opportunity costs), and the emergy loss from ecological damage would have been about 1.0 E22 sej. With a greater wetland area and lower loading rate, the ecological damage would have been much less.

Lead recycling rates are extremely susceptible to the vagaries of virgin lead prices. During the fall in lead prices in the early 1980s, 6% of the secondary smelters in the U.S. went out of business (U.S. Congress, 1991). In 1980 only 10 million batteries (of 60 million spent) were not recycled. In 1985 22 million batteries (of 73 million spent) were not recycled (Putnam Hayes and Bartlett Inc., 1987). These were land-filled, legally and illegally, and represent 200 million kg of lead released into the environment or buried (at 9.1 kg lead/battery). Twenty-five sites on the National Priority List of Superfund are now related to battery collection or recycling (U.S. Congress, 1991). If those batteries had been recycled, even with no treatment of wastes, the amount of lead released to the environment would have been only 6000 kg. If using wetlands for some wastewater treatment or final polishing of wastewater can lower the costs of treatment, thereby making lead recycling more profitable, more lead will be recycled and less will be land-filled.

Implications for Environmental Policy

The fact that society restricts pollution with effluent standards implies that society is willing to pay a substantial amount for minimum level but is not willing to pay much extra for high levels of abatement.

For many years wetlands were undervalued by individuals and society. Now wetlands protection laws have "institutionalized" a recognition that the real value of wetlands far exceeded previous values. Yet the dichotomy of preservation or conversion as the only two uses for wetlands may at times stifle innovative attempts to tie wetlands into the human economy through their biogeochemical cycles. Much of the potential value of wetlands to the economy lies in their careful use.

It is often thought that the damages to society in using natural systems as waste absorbers is underestimated. As was shown here benefits may be larger than usually believed. Also, there are indirect environmental consequences to using technological substitutes that are directly and indirectly fossil fuel intensive. In other words, wetlands may contribute even more to the economy than can be estimated by dollar values, because of the environmental impacts and the reduced use of fossil fuel-intensive technology.

Despite the fact that there appears to have been a net benefit, the case of Sapp Battery was surely not optimal. Because battery casings and sulfuric acid were not recycled, as they are now, wetland damage was more than it should have been. Also, because the savings in treatment cost went largely to Sapp while the cost of wetland damage was suffered by society, more damage occurred than would have been optimal. As humans organize to understand the potential and existing role of wetlands in the economy, the scale of appropriate wetland use can be determined.

The ecological destruction of much of Steele City Bay and the pollution of its sediments with lead were unfortunate and a misuse of nature. We suspect, however, that the ecological and economic damage to nature and society would have been worse had this wetland not been there to intercept the effluent from the Sapp Battery plant. Retaining wetlands on the landscape can buffer the effects of willful environmental negligence. Although an abundance of wetlands may be somewhat redundant from only a "heavy metals retention" viewpoint, they provide important insurance against environmental accidents and catastrophes, as well as a multitude of other services.

Ecological engineering is a way to fit heavy metal industries into a prosperous society by tying into nature's cycles of elements, as shown here and in the Polish companion study (see Chapter 9; Wójcik and Wójcik, 1989; Wójcik et al., 1990).

Further research on transformities of various forms of chemicals and elements is needed. If the EPA Toxics Release Inventory weighted toxic releases by their transformity, an estimate of emergy loss through pollution could be calculated. If transformity is in some way proportional to toxicity, a ranking of hazardous potential would be possible, since the amounts released are known.

SUMMARY AND CONCLUSIONS

A wetland case study was used to derive an economic value for wetland lead retention using both ecological economic and mainstream economic methods. A 29.2-ha cypress/black gum wetland received lead/acid discharge from a reclamation facility that processed an estimated 12.5 million automobile batteries in 10 years. The wetland retained about 2276 kg of lead, an average removal rate of about 7.8 kg/ha^{-1}year^{-1}.

The ecological damage caused by the lead/acid discharge was substantial. Gross primary production in the damaged wetland was 70% less than gross primary production in a nearby undamaged wetland. Leaf area index of trees was less than 10% in the damaged wetland. Benthic macroinvertebrate diversity was normal. Regeneration of canopy species was probably inhibited not by toxicity, but by constant high water levels, since tree seedlings planted in contaminated areas survived.

Both emergy and mainstream economics show a net benefit from the use of the wetland for lead retention (with some ecological damage) compared with use of a chemical treatment system (with no wetland damage). Lead retention by the wetland saved the economy resources valued at about 4.0 E18 sej and about $1.5 million.

Replacement cost was defined as the resources necessary (in emergy or dollar terms) to replace the lead retention function of the study wetland with a chemical treatment system. Both emergy and mainstream economics showed a net benefit from the use of the wetland for lead retention (with some ecological damage) compared with use of a chemical treatment system (with no wetland damage). Using the wetland resulted in net savings to the economy of resources valued in emergy terms at 3.8 E18 to 4.3 E18 sej, and resources valued in dollar terms at $1.5 to $1.6 million (over 10 years of wetland use, 50 years of wetland damage recovery, and with a discount rate of 4%).

The benefit/cost ratio was lower using emergy evaluation (B/C ratio 3:4) than with mainstream economics (B/C ratio 20:55), because the value of wetland damage was larger using emergy evaluation. Since the gains from wetland use went largely to the polluting firm and the losses from wetland damage were suffered mainly by others, wetland damage may have been excessive.

Emergy and mainstream economic methods were compared by expressing each as their fraction of the national economy comparing emdollars and dollars. The emergy method valued the emdollars of work of the wetland nearly twice as high as the dollars from the mainstream economic method (6.7 E-7 vs. 3.5 E-7). However, the emdollars of wetland damage was also greater (by a factor of 35 to 9) than the market value of the damage. As a result, the net benefit of using the wetland for lead retention was close using the two methods; the ecological economic value as a fraction of national emergy use was 4.5 E-7 to 5.0 E-7 and the mainstream economic value as a fraction of gross national product was 3.4 E-7 to 3.5 E-7.

When expressed as a fraction of the economy, values derived from emergy evaluation are nearly always higher than mainstream economic values because unpaid inputs from the environment are included. At present both methods reveal valuable information about the structure of the economy and its dependence on natural systems. They should both be used in gathering information for use at the public policy level. Mainstream economic methods reveal valuable information about human behavior and preferences, whereas ecological economic methods reveal more about the real dependence of the human economy on natural systems and resources.

Capacity for retention of toxic substances may be one of the highest values of wetlands, but the standard economics method used (based only on money values) may underestimate both the replacement value of the wetland and the value of ecological damage.

ACKNOWLEDGMENTS

This work, based on research conducted for a master's degree at the University of Florida (Pritchard, 1992), was made possible by a grant for the study of Heavy Metals in Wetlands from the D.T. Sendzimir Foundation to the Center for Wetlands, University of Florida, with H.T. Odum as Principal Investigator. Thanks are in order to my master's committee, H.T. Odum, Gary Lynne, and J.J. Delfino, and to my collaborator, Shanshin Ton, whose thesis "Natural Wetland Retention of Lead from a Hazardous Waste Site" provided much of the data on lead analyses and processes. Other assistance was provided by Robert Woithe, Jan Sendzimir, Karen Pritchard, Joanna Pritchard, Peter Keller, Clay Montague, and Kathleen Dollar. Peter Wallace graciously donated the tree seedlings used in the toxicity test. William Dunn of CH2M-Hill provided access to data and reports on the site. Neil Oakes of Gulf Coast Recycling provided the data for replacement cost analysis.

PART IV

Value and Policy

Based on field and laboratory work on leaded swamps and extensive new literature on heavy metals, Part IV evaluates wetlands for heavy metal filtration, the state of relevant environmental laws, and suggested policies. Chapter 11 by Lowell Pritchard, Jr. compares economic and EMERGY evaluation of the Steele City Swamp in Florida. Chapter 12 by Wlodzimierz Wójcik evaluates wetland lead filtration in Poland. Chapter 13 by Jay D. Patel reviews the history of environmental law in the U.S. relevant to lead and wetlands. Finally, Chapter 14 summarizes, with suggestions for policy on the industrial ecology of lead.

Emergy Evaluation of Treatment Alternatives in Poland

Wlodzimierz Wójcik, Slawomir Leszczynski, and Howard T. Odum

CONTENTS

Two options for treatment of mine wastewater were compared with emergy evaluations. The first option is a conventional physicochemical method with coagulation and filtration proposed by a Swedish company. The second option utilizes the natural filtration capacity of the Biala River wetland.

METHODS

Real wealth requirements and contributions of treatment were evaluated by estimating flows and storages of emergy in inputs and outputs from the treatment systems as explained briefly in Chapter 2 and applied in Chapter 11. Energy systems diagrams were prepared to identify the most important flows (Figures 12.1 and 12.2). Then an emergy analysis table was prepared with each of the important items as a line item. Data expressed in energy, mass, and money units were multiplied by emergy per unit to obtain emergy flow values. Emergy/money ratio was obtained from an emergy analysis of Poland (Appendix A12).

Treatments are best that use less emergy resources from the economy while diverting more emergy of toxic waste from harmful impact and converting more waste emergy into useful or potentially useful products or storages.

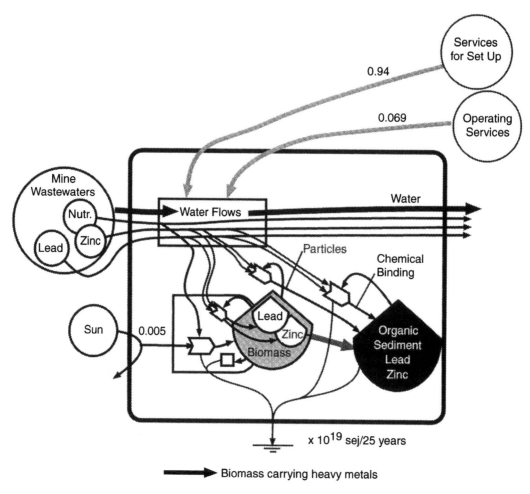

Figure 12.1   Emergy diagram of natural treatment by the wetlands on the Biala River.

EMERGY EVALUATION OF WETLAND TREATMENT

Using information obtained from the Biala River wetland studies (Chapter 9), an ecological engineering design for the most efficient wetland treatment of heavy metal wastes was prepared. To improve the processing, a reconstruction of the wetland was proposed to change the hydraulics of the water flow. For this purpose several dikes and barriers could be built across the wetland as explained in Chapter 9 (Figures 9.18 and 9.19). Moreover, additional planting of the marshy vegetation might accelerate the self-organization of the vegetation to the new condition.

The analysis was started by preparing a diagram with all external sources of energy, components, and connections describing the flows of mass and energy (Figure 12.1). This phase of research helped us understand how the system is functioning and what the connections are between the system components. The interior of the systems diagram was simplified to include a water flow unit, a biomass production unit, and a tank or deposit of organic sediments.

The summary diagram includes the inflows from external sources into the treatment system for emergy evaluation (Table 12.1). Environmental contributions were those of the land and sunlight. The rain was small relative to the wastewater inflow and not evaluated.

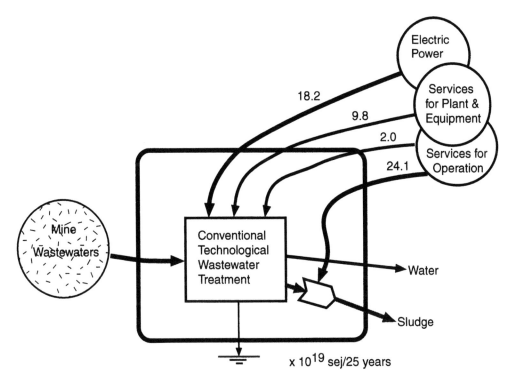

Figure 12.2   Values of main flows in the model of wetland treatment in Figure 12.1.

Energy of solar radiation reaching the surface was calculated with a function representing changes with season:

$$f(t\text{-time}) = a + b*(\sin(t/c))2*1000*3600$$

This function was worked out based on data collected by Olecki (1991) using the EUREKA computer program. Coefficients were as follows: a = 7.3274, b = 107.8975, c = –4.2201.

Wastewater Contributions

Wastewater inflow in cubic meters per second was described by a function expressed by the equation

$$f(t) = a + b^{ct}$$

where t = time, a = 1.7511, b = 0.9919, c = 0.5098. For the 25-year evaluation the total wastewater processed was estimated to be 1.7 E9 m³/25 years.

The inflowing waters contained emergy of the water, the nutrients, lead, and zinc transported together. The waters were partially used by transpiring plants, and this emergy contributed to the treatment work. The rest of the water flowed out, a contribution to downstream users.

Dilute concentrations of nutrients, lead, and zinc were estimated for the inflow waters to evaluate their emergy content (Table 12.1). This emergy inflow was mainly retained in the system as biomass and sedimentary deposits.

Table 12.1 Emergy Evaluation of Wetland Treatment Flows per 25 Years; Area, 74 Hectares

Note	Item	Data (raw) (units)	Transformity (sej/unit)	Solar Emergy (sej)
Environmental contribution				
1	Sun, joules	5.402 E16 J	1 sej/J	5.402 E16
2	Land	1.295 E14	3.45 E4 sej/J	4.468 E18
3	Total			4.522 E18
Mine wastewater inflows				
4	Water used, joules	8.5 E15	4.8 E4	4.08 E20
5	Nutrient nitrogen, grams	5.1 E9	1.05 E9	5.355 E18
6	Lead (dilute), grams	8.5 E8	3 E8	2.55 E17
7	Zinc (dilute), grams	2.55 E9	6 E8	1.53 E18
8	Total			4.151 E20
Purchased from the economy				
9	Setup costs	1.57895 E5 $	6.0 E12	9.477 E17
10	Operational costs	0.11579 E6 $	6.0 E12	6.94737 E17
11	Total	1.736 E6 $		1.645 E18
Products				
12	Usable water, joules	8.5 E15	4.8 E4	4.08 E20
13	Organic sediment, grams dry	7.82 E10	0.36 E9	2.817 E19
14	Total			4.362 E20

Notes:

1. 2.161 E15 J/year (Olecki, 1991) * 25 years = 5.402 E16 J/25 year.

2. Area share of global continental land cycle (Odum, 1996, p. 303).

    [a](7.0 E6 sej/m²/year)(25 years)(74 E4 m²) = 1.295 E14.

3. Sum of items #1 and #2 = 4.522 E18.

4. Water transpired: (1.7 E9 m³/25 years)(1 E6 g/m³) (5 J/g free energy) = 8.5 E15. Transformity of stream water (Odum, 1996, p. 309).

5. Nutrients used: (3 g nitrogen/m³ water)(1.7 E9 m³/25 years) = 5.1 E9.

    Transformity of dilute nitrogen (Odum, 1996, p. 309).

6. Lead: (0.5 g/m³)(1.7 E9 m³/25 years) = 8.5 E8 g/25 years.

    Transformity of dilute metal — see Chapter 4.

7. Zinc: (1.5 g/m³)(1.7 E9 m³/25 years) = 2.55 E9 g/25 years.

    Transformity of dilute metal — see Chapter 4.

8. Sum of items #4, #5, #6, and #7 = 4.151 E20.

9. 1.5 billion Polish zlotys: 9500 Zl/$ = 1.57895 E5 $/25 years.

10. 44 million Polish zlotys/year: 9500 Zl/$ * 25 years = 1.1579 E5 $.

11. Sum of items #9 and #10 = 1.645 E18.

12. Usable water outflow:

    (1.7 E9 m³/25 years)(1 E6 g/m³) (5 J/g free energy) = 8.5 E15 J.

    Transformity of stream water (Odum, 1996, p. 309).

13. Organic deposits including bound lead and zinc:

    (846 dry g/m²/year)(25 years)(74 E4 m²)(5 J/g) = 7.82 E10 g.

    [b] Lead deposited: (1.034 E3 g lead/m²/year)(25 years)(74 E4 m²) = 7.65 E8 g.

    [c] Zinc deposited: (3.274 E3 g zinc/m²/year)(25 years)(74 E4 m²) = 2.42 E9 g. Transformity of peat (Odum, 1996, p. 311).

14. Sum of items #12 and #13 = 4.362 E20.

[a] The value of land for Biala River wetland calculated as for rapid orogenezic cycle.

[b] (1.7 E9 m³/25 years)(0.5 g/m³)(0.9)/(74 E4 m²) = 1.034 E3 g/m²/25 years   → 2.295 E17 sej

[c] (1.7 E9 m³/25 years)(1.5 g/m³)(0.95)/(74 E4 m²) = 3.274 E3 g/m²/25 years   → 1.452 E18 sej

Σ 1.681 E18 sej

Money Flows for Costs and Investments

Investment costs were assumed to be 1.5 billion Polish zlotys (for 1992 year) including: cost of the land, designing costs, costs of materials and machine work, and labor costs. To define the emergy corresponding to the costs (services and labor), Polish zlotys were first converted into dollars at the exchange rate of 9500 Zl/$. Emergy/money ratio = 6.0 E12 sej/$ was applied as calculated in the analysis of Poland in Appendix A12. Operations costs were assumed to be equal to 44 million zlotys per year, covering the following costs:

Payment for manual work, 7,200,000 Zl/year
Payment for scientific work, 25,000,000 Zl/year
Machine-hours, 12,000,000 Zl/year

Flows through Main System Compartments

As shown in Figure 12.1, the inflowing waters, nutrients, lead, and zinc are used and processed by more than one pathway, and the emergy flow of each can be calculated as a proportionate "splitting" of the input emergy. However, for the purposes of this overview analysis, these details are not necessary except to determine how much of the input emergy remains stored on site and how much passes downstream (Table 12.1). Because most of the plants are 1-year plants, it was assumed that all of the biomass flows to the deposit tank each year.

Value of Products

These two systems generate a valuable flow of usable water. Table 12.1 shows this product to be a large emergy contribution, which can be compared with the emergy of the costs from the economy. The value of the contributed water (to the cost) is 4.08 E20 sej, so that when divided by the emergy/money ratio, we find the contribution in 25 years is 6.777 E7 emdollars.

The other main product is the deposited sediment containing the heavy metals. The emergy accumulated in this deposit is a measure of the environmental protection achieved and potential value when some use may be found for these sediments in the future. When 1.681 E18 sej is divided by the emergy/money ratio, a value of 2.793 E5 emdollars is found.

EMERGY EVALUATION OF CONVENTIONAL TREATMENT

A conventional technological treatment uses sand filtration with sodium sulfide and polymers as a flocculant. The chemicals are dissolved in special tanks and introduced into pipes that feed to the filters immediately before the pumps supply wastewater from an equalization reservoir. The sand filter units are flushed periodically and sludge transported by pumping into sedimentation tanks. Inputs are evaluated in Table 12.2. Figure 12.2 summarizes the emergy flows.

Total required input of emergy for this method is 3.925 E18 sej/year or 9.813 E19 sej/25 years, while input of energy for operation is 8.64 E18 sej/year or 2.016 E20 sej/25 years (Table 12.2).

COMPARISON OF EMERGY FLOWS OF WETLAND
AND TECHNOLOGICAL TREATMENT

Where the flows of water are large and similar in both systems, a partial but important analysis can be made by examining only what has to be purchased from the economy. The system that requires less for the same task is the best one energetically and economically (Table 12.3).

Table 12.2   Emergy Evaluation of Conventional Treatment Method (Flows per 25 Years; Area, 5.0 ha)

Note	Item	Data (Units)	Emergy/Unit	Solar Emergy
	Environmental contribution			
1	Land	5.0 ha	6.29 E10	3.145 E15
	Mine wastewater inflows			
2	Total water, lead, zinc			4.151 E20
	Purchased from the economy for setup			
3	Hydraulic installation	8.125 E6 $	6.0 E12	4.875 E14
4	Buildings and roads	1.625 E7 $	6.0 E12	9.75 E19
5	Cost of land	1.0526 E5 $	6.0 E12	6.31578 E17
6	Total setup			9.813 E19
	Purchased from the economy for operations			
7	Electrical energy for operation	1.1455 E15 J	15.9 E4	1.82134 E20
8	Labor for operation	6.06315 E5 $	6.0 E12	3.6398 E18
9	Chemicals	2.64677 E6 $	6.0 E12	1.58806 E19
	Total operations			2.016 E20
10	Sludge disposal	4.01786 E7 $	6.0 E12	2.41071 E20
11	Total operation			4.427 E20
	Products			
12	Usable water	8.5 E15	4.8 E4	4.08 E20
13	Sludge	2.287 E12	0.5 E9	1.143 E21
14	Retrieved metals	2.04 E9	0.5 E9	1.02 E18
15	Total product			1.552 E21

Notes:

1.   (5 E4 m²)(6.29 E10 sej/m²/year (Odum, 1996, p. 110)(25 years).

2.   Item #8 in Table 12.1 = 4.151 E20.

3.   Costs of hydraulic installation:

          $8.125 million.

4.   Building and roads:

          $16.25 million.

5.   50,000 m² * 20,000 Zl/m²: 9500 Zl/$ = 1.05263 E5 $.

6.   Total setup = sum of items #3, #4, and #5 = 9.813 E19.

7.   Electrical emergy for operation:

      4.582 E13 J/year * 25 years = 1.1455 E15 J/25 years.

8.   Labor for operation:

      4 persons * 4.8 million Zl/month * 12 months: 9500 Zl/$ = 24,252 $/years

      24,252 $/year * 25 years = 6.06315 E5 $/25 years

9.   Chemicals:

      Sodium sulfide

              12,614.4 kg/year * 5.357 $/kg = 67,577.1 $/year.

              67,577.1 $/year * 25 years = 1.68943 E6 $/25 years.

      Polymer

              6307 kg/year * 6.071 $/kg = 38,293.7 $/year.

              38,293.7 $/year * 25 years = 9.5734 E5 $/25 years.

      Total chemicals

              105,870.8 $/year * 25 years = 2.64677 E6 $/25 years.

10.   Sludge disposal:

              1,607,143 $/year * 25 years = 4.01786 E7 $/25 years.

11.   Total operation = sum of items #7, #8, #9, #10 = 4.427 E20.

Table 12.2 (continued)   Emergy Evaluation of Conventional Treatment Method (Flows per 25 Years; Area, 5.0 ha)

12.	Water output = (1.7 E9 m³/25 years)(1 E6 g/m³)(5 J/g) = 8.5 E15 J/25 years.
13.	Sludge = (7.625 E11 g wet/25 years)(60% dry of wet)(5 J/g) = 2.287 E12.
14.	Retrieved metal = 2.04 E9 g/25 years.
15.	Total product = sum of items #12, #13, and #14 = 1.552 E21.

Table 12.3   Comparison of Requirements from the Economy for Wastewater Treatment Methods (25 Years)

Category and Units	Conventional Method	Wetland Method
Emergy evaluation[a]		
Operation	73.8 million em$	0.116 million em$
Total[b]	90.1 million em$	0.274 million em$
Economic Costs		
Operation	43.4 million $	0.116 million $
Total[b]	67.9 million $	0.274 million $

[a] Emergy values expressed as emdollars:
solar emdollars = (solar emjoules)/(6 E12 sej/$).
[b] Total = setup + operation.

Table 12.4 summarizes the emergy flows for the two treatment methods. The emergy required for installation of the wetland method is 4.151 E20 sej/25 years, while emergy to establish the conventional method is 9.813 E19 sej/25 years. Therefore conventional treatment methods would require 68.5 times more emergy from the economy. This emergy difference was even greater for operations. Emergy of conventional methods was 600 times higher than that required from the economy for the wetland treatment method.

The natural method is environmentally compatible. In the calculations several wetland contributions were neglected that would increase emergy values such as the benefits from the small impoundments created and increases in wildlife.

Table 12.4   Summary of Emergy Flows for the Two Treatment Methods

Category	Units	Conventional	Wetland
Establishing of a system	sej/25 years	9.813 E19	4.151 E20
For operation only	sej/25 years	2.016 E20	6.947 E17

CHAPTER 13

The Evolution of Environmental Law and the Industrial Lead Cycle

Jay D. Patel

CONTENTS

INTRODUCTION

In several centuries of industrial development and the technological revolution, laws and legal practice have regulated the relationship of industry and environment. The laws controlling mining, manufacture, waste, and recycle of lead may be typical of society's control of material cycles. The evolution of public attitudes and resulting laws may be analogous to developments observed when an ecological system develops using a new area. Exploitive, competitive uses of initially concentrated resources, and accumulating wastes, are followed by increased efficiency, reuse, and recycling to the environment. Largely responding to market values, the laws on materials such as lead have not yet recognized the contributions of materials carrying the prior work of nature.

In this chapter, the history of certain laws that pertain to the extraction and processing of lead is reviewed. This review will look closely at how the individual statutes measure up to the systems ecology approach to efficient human ecosystem interrelation. Upon review of each statute, the method the federal (and in some cases state) government uses to control the possible negative effects of lead will be identified. This report will then consider changes that would be necessary to make a smooth, sustainable system of lead cycle and use including nonmarket values.

STATUTORY ANALYSIS

Mining Law of 1972

The statutes that govern lead mining today date back to the General Mining Law of 1972. In passing this law, Congress intended to protect the rights of individuals to stake claims on mineral land and extract those resources. The Mining Law provides that lands that belong to the U.S. "shall be free and open to exploration and purchase" (Note 1, Appendix A13). In 1872 Congress had intended that people should extract these resources. In that time of growth and industrialization the supply of these resources would encourage more growth. To encourage these developments the statute identifies areas that can and cannot be mined, and defines requirements for the prospector to claim the land and assume the mineral rights (Note 2, Appendix A13).

The importance of the Mining Law may not lie in what it did outright, but rather in what environmental awareness resulted from the subsequent exploitation of the land. What resulted was the formation of national parks such as Yellowstone National Park, which was established in 1872, and the formation of the Sierra Club, which formed in 1892. In addition, certain federal lands were set aside and were exempt from mining activities and mining rights previously given in the Mining Law of 1872. Subsequent enactments have imposed constraints on mining operations on other federal lands with environmental protection as an explicit goal.

In 1976 the Federal Land Policy and Management Act (FLMPA) gave the Bureau of Land Management the responsibility of managing public lands for multiple purposes (Note 3, Appendix A13). The act includes, among these purposes: "recreation, range, timber, minerals, watershed, wildlife and fish, and natural, scenic, scientific and historic values" (Note 4, Appendix A13). In managing these lands, the bureau has adopted a reasonably prudent standard to determine if activities that disturb the surface are necessary and appropriate (Note 5, Appendix A13). Mining operations are then subdivided into three categories based on the extent of land used. Where mining involves more than 5 acres the Bureau of Land Management requires that a plan of operations be submitted (Note 6, Appendix A13), and that the plan be subject to an environmental assessment (Note 7, Appendix A13). In addition, an Environmental Impact Statement (EIS) may be required (Note 8, Appendix A13). However, the regulations do not clearly grant the BLM authority to refuse a plan of operations.

The Forest Service requires a similar plan when permission is sought for mining in national forests, if there is intent to use mechanized equipment (Note 9, Appendix A13). The Forest Service may require an environmental assessment and may determine if an EIS is needed. Although the regulations do not include the possibility of denying a permit, the Forest Service can defer granting a permit by requiring changes in the plan, or requiring that an EIS be completed (Note 10, Appendix A13).

National Environmental Policy Act of 1969 (NEPA)

In 1969, the National Environmental Policy Act (NEPA) was enacted to allow the Federal Government to "create and maintain conditions under which man and nature can exist in productive harmony" (Note 11, Appendix A13). On its face the act would appear to be a very progressive statement about the environmental goals of the Federal Government. Substantively, however, the statute makes the protection of the environment merely an element for consideration in all major federal actions.

Section 4321 is the declaration of purpose that establishes why the Federal Government has enacted this statute. Using words such as "productive and enjoyable harmony," and stating a goal to "enrich the understanding of the ecological systems," the first paragraph of NEPA opens the door on a new attitude about environmental issues.

Section 4331 of NEPA appears on its face to be a dream come true for some environmentalists. The policies and goals that are in print in the statute demonstrate a profound understanding of the need for governmental and individual contribution to the preservation of the environment and its resources. The statute foresees the need to protect the environment for the next generation and to minimize health risks and "undesirable and unintended consequences." NEPA mandates that the Federal Government coordinate plans that "approach the maximum attainable recycling of depletable resources." All of the goals described in the section would require governmental and individual participation, and Section 4331 expresses the intention to achieve these goals.

As profound as the purpose of NEPA might seem, the policy goals themselves (Note 12, Appendix A13) are problematic and flawed when viewed from a systems ecology standpoint. The most important fault in these policies is the adoption of a view that the environment is for man's productive use. Most glaringly the statute assures "healthful and *productive* ... surroundings." I doubt that the legislators had the intent to define productivity in terms of environmental productivity. It is more likely they are concerned about how productive humans can be in that environment rather than how efficiently humans can interact with their environment. The statute also seeks to "attain the widest range of beneficial uses of the environment." But nowhere in the goals of this statute is there any mention of the trade-off between what is taken from the environment and what is given to the environment in return. A balance between use and feedback must be another primary goal for environmental policy.

What then is the method the Congress adopted to achieve its goals? Primarily, NEPA serves to make the environment an issue in any federal action that is of major consequence. Secondarily, the statute serves to empower the president with more information about the environment through the Council on Environmental Quality (CEQ).

The thrust of NEPA is to require the federal agencies to prepare an EIS, and to conform their actions to the policies of NEPA (Note 13, Appendix A13). The EIS is described by the statute as a "detailed statement by the responsible official on the environmental impact of the proposed action, any adverse environmental effects should the proposal be implemented, alternatives to the proposed action, the relationship between local short-term and long-term productivity, and any irreversible and irretrievable commitments of resources." The detailed content requirements of the EIS have produced many lengthy reports with questionable effects on final government actions.

However, NEPA is important in that it applies an environmental element to all agency decisions that are considered major federal actions (Note 14, Appendix A13). The absence of a statutory definition of "major federal action" led to many legal challenges where environmentalists tested the power of NEPA. The Council on Environmental Quality then summarized the case law and defined "major federal action" in their regulations (Note 15, Appendix A13). Major federal actions have been defined as actions the federal government is funding or carrying out, private projects that may require federal approval, and actions with effects that are major and are subject to federal control or responsibility. This definition is significant in reaching action beyond that which is purely federal, but requires a threshold of significant federal involvement, funding, or control before NEPA applies.

Besides producing lengthy reports about environmental impacts related to major federal agency actions, NEPA fails to achieve much measure of the goals laid out in the opening sections of the statute. Much of the case law related to NEPA demonstrates this failure. The cases do not necessarily involve questions of goal achievement. Instead, the legal points that were argued demonstrate that the importance of the goals has been overlooked.

In *Stryker's Bay* v. *Karlen* (Note 16, Appendix A13), the Supreme Court ruled that NEPA does not dictate the weight of an EIS in an agency decision. Rather, the court ruled that if the EIS were performed and considered, it would be the decision of the agency as to the importance of the EIS. This Supreme Court decision seems to undermine the importance of the goals of NEPA. Justice Marshall, in his dissent, seems to agree. It is the court's lack of support of the goals that breaks down the effectiveness of those goals.

Another way the court system has undermined the effectiveness of NEPA falls under the questions of timing and scope of the EIS. *Kleppe* v. *Sierra Club* (Note 17, Appendix A13) raised these points and in essence the court deferred to the agencies to decide the appropriate timing and scope. The court failed to balance the importance of the action itself. In its decision the court said, "even if the environmental interrelationships could be shown conclusively, practical considerations of feasibility might well necessitate restricting the scope." The court should have recognized that the scope of the EIS can drastically change the effectiveness it has on the related actions. This could have been better resolved had the court approached the issue with regard to NEPA's goals which "encourage productive and enjoyable *harmony* between man and his environment" (emphasis added) (Note 18, Appendix A13). Such an approach might include the use of those goals in determining the best time possible and best scope with which to approach the measurement of impact. To ensure productive harmony, the agency should have been required to address the long-term, regional coal productivity and its environmental impact in its EIS.

Comprehensive Environmental Response, Compensation and Liability Act of 1980 (ERCLA)

The Comprehensive Environmental Response, Compensation and Liability Act, also known as CERCLA, was signed into law in 1980 as a political response to the incident in Love Canal, New York. Unlike NEPA, CERCLA does not outline goals, nor does it place emphasis on achievement. The structure of this statute is to identify responsible parties and make those parties liable for their improper disposal techniques. This policy is intended to deter unsafe practices, and within the legal system should function effectively. But within the larger system it seems to lack a broader understanding of how to approach the problem of disposal in a system where humans and the environment productively interact.

This statute has two parts. The first role of CERCLA is to identify individuals, corporations, and any other party responsible for improper hazardous materials disposal or handling that has resulted in or will cause "releases" into the environment. Second, this statute is designed to make these responsible parties financially liable for correcting the problems that resulted from improper disposal or handling. But, somewhere within the legal mire that resulted from this statute, the idea of responsibility was lost.

This is not to say that the statute has failed to impose financial responsibility. There are many cases and articles that confirm the imposition of financial responsibility and define its controls. Rather, the big picture that the environment could be cleaned up and protected through the efficient use of available resources was lost, and instead money was spent on ineffectual cleanup with limited benefit. The high costs associated with cleanup became the issue and the environment was lost in the shuffle. Legal cases that worked out details of liability imposition did not address the interactions between man and the environment.

CERCLA costs can be defined as associated with the cleanup of a site or with the transaction costs. The transaction costs incurred in the cleanup of a Superfund site include the costs incurred to recover funds from potentially responsible parties (PRP). This process can involve extensive and expensive investigation and litigation.

High transaction costs have been a major source of criticism of CERCLA and have resulted in bipartisan efforts to amend the statute. Delays in cleanup, increasing costs of cleanup, and costs engendered by unsatisfactory cleanups are distinct problems contributing to CERCLA's high overall cost to society.

Costs associated with cleanup are those costs that encompass the engineering and execution of a cleanup plan. As the NPL list increases and sites grow older, the cost for cleaning up any given individual site increases. On one hand, quick cleanups require more Superfund money up front, but generally they make recovery of the cost easier. This is because problems are easily identified in the field rather than in the courthouse. Where litigation or investigation precedes cleanup, identification of PRPs increases transaction costs but may reduce overall cleanup costs. The cleanup costs are lower because the litigation and investigation can lead to an understanding of what in the site needs to be cleaned up.

If resources were not scarce, all sites listed would be cleaned quickly and thoroughly. But in a world of scarcity the goals for CERCLA may need reassessment. The solution for the Superfund cost problem might eventually prove to be the solution for the problem of cleaning up old toxic waste sites.

The policy of cleaning up sites to pristine condition should be replaced with the policy of getting sites to a state of neutrality quickly and with least cost approach. Neutrality simply means a state of minimized long-term external effects. Sites will not be returned to pristine conditions, but will be transformed into places that have minimal effect on the surrounding health of the environment and people. Minimum costs and minimum effects would be the primary goals, with minimum time as the driving force to complete the cleanups.

This policy differs from CERCLA in that minimizing costs should have an economic impact on industry. Placing the responsibility of paying minimal costs should stimulate PRP's voluntary cleanup. In addition, there would be economic incentive to produce a competitive market for cleanup. This market would in turn reduce cleanup costs themselves. Cleanup standards that realize pristine end results are inefficient. The cost of achieving the new goals becomes realistic. The scientific justification for lowering cleanup goals recognizes that some sites are so polluted that they are irreversibly damaged.

The Resource Conservation and Recovery Act of 1976 (RCRA)

The Resource Conservation and Recovery Act (RCRA) of 1976, originally enacted as the Solid Waste Disposal Act of 1965, is the federal legislation that regulates solid and hazardous wastes. RCRA was revised in 1984 by the Hazardous and Solid Waste Amendments. The statute serves as the federal regulatory tool for reducing unsafe waste disposal practices.

The opening section of the statute reports congressional findings with respect to solid waste. These findings show an increase in the amount of waste material that is being discarded. In addition, the findings recognize financial, managerial, intergovernmental, and technical problems that state and local governments have with handling and reducing waste.

To solve these problems the statute establishes a strong partnership between the federal and state governments. Section 1003 of the statute, the objectives and national policy section, declares that "whenever feasible, the generation of hazardous waste is to be reduced or eliminated as expeditiously as possible, and waste that is generated should be treated, stored or disposed of so as to minimize the present and future threat to human health and the environment."

With regard to hazardous waste, the statute serves to identify and define what materials are considered hazardous. These materials are either listed wastes or characteristic wastes. The EPA, in 1980, produced an original list of 100 hazardous wastes in four priority groups. Lead is listed among the Priority Group 1 list for hazardous materials. The original list was easy to generate because these hazardous materials were generally agreed-upon wastes and process by-products. Unfortunately, in the 6 years that followed, only an additional six wastes were added to those listed. The inadequate expansion of the list was blamed on lack of both funding and a systematic approach. In 1984 the Hazardous and Solid Waste Amendments (HSWA) expanded the criteria for listing to include wastes whose constituents were in levels known to endanger human health.

Wastes may also be regulated under RCRA if they are considered characteristic wastes. These are generally identified by toxicity, corrosivity, ignitability, or reactivity. These general characteristics seem broad enough to cover those wastes that may not be listed but are potentially harmful to health and the environment. But many problems exist with this strategy as well. Initially it is the responsibility of the waste handler to determine if the waste exhibits these characteristics and is then subject to RCRA regulation. Problems of unavailable and complicated test methods make the determination difficult.

Mixtures of wastes is another topic which has created practical difficulties under RCRA. Some mixtures of hazardous wastes produce materials which are more desirable, while other mixtures worsen the health and environmental consequences of disposal. The regulations allow mixtures containing characteristic hazardous wastes to be classified as nonhazardous if the mixture does not exhibit the hazardous characteristics. On the other hand, mixtures that contain listed hazardous wastes are automatically considered hazardous, and can only be delisted through the RCRA delisting procedures.

Once identified the statute takes a cradle-to-grave approach in dealing with the waste. Separate sections of the statute deal individually with generators, transporters, and treatment, storage, and disposal (TSD) facilities. Part of the cradle-to-grave approach incorporates a document trail for all hazardous wastes. In particular, a Uniform Hazardous Waste Manifest must accompany all hazardous waste from generator to TSD.

In general, facilities that are identified as hazardous waste generators must do the following: obtain an EPA identification number, prepare a manifest for transporting wastes off-site to a permitted TSD facility, package and label the waste as hazardous, accumulate the wastes for no longer than 90 days without a storage permit, maintain records of signed manifests, and issue biennial reports of waste generated.

With regard to standards applicable to generators of hazardous waste, Section 3002, every 2 years a permitted generator must report on quantities of waste produced, efforts undertaken to reduce the volume and toxicity of waste, and the changes in volume and toxicity of the waste compared to other years. Other sections of the statute have related themes. Section 5003 requires the secretary of commerce to identify markets, and encourage development of new uses for recovered materials. This mandate, though brief, could have positive effects on recycle of materials. Also, Section 5004 mandates that the secretary of commerce evaluate the commercial feasibility of resource recovery facilities and assist in selection of those facilities.

Lead is listed as hazardous waste number D008 under RCRA. In 1989 the USEPA Toxics Release Inventory listed the industry-reported releases and transfers of lead and lead components. Among this list the largest releases occurred in on-site land disposal and off-site transfer. Air, surface water, public sewage systems, and underground injection of lead each released less lead by an order of magnitude or more.

Specific standards for treatment of lead wastes vary per media being treated. When wastewaters contain lead and have characteristic toxicity, the regulations dictate an extraction procedure should achieve 5.0 mg/l. With nonwastewaters, thermal recovery in secondary smelters is required; and with radioactive lead solids, macroencapsulation of the waste must be performed (Note 19, Appendix A13). When lead is incinerated the standards vary from 4.3 to 9200 g/h for a feed rate screening limit. These standards are drawn from tables that account for stack height and terrain complexity in setting precise limits (Note 20, Appendix A13). With regard to land disposal, all lead has been banned as of May 8, 1992, when nonwastewaters before secondary treatment were added to the ban. All other forms of lead had been banned from land disposal since August 8, 1990 (Note 21, Appendix A13).

What RCRA accomplishes for the environment is unclear. RCRA's weak program for reuse and recycle is a key problem with the statute. There is no provision that forces generators of waste, hazardous or otherwise, to reduce production of those wastes. Because Congress chose to regulate disposal — not production — of hazardous waste, it created little incentive to seek better solutions for reuse and recycle of materials.

Unlike industrial hazardous wastes that are regulated under RCRA, wastes generated in small quantities either commercially or by households are not banned from municipal landfills. Hazardous materials such as batteries that are used by general consumers are not regulated directly in RCRA. Instead, RCRA regulates the way landfills are designed and sited. This type of legislation is reminiscent of end-of-pipe-type cleanup.

RCRA's mode of regulating the water at its final resting place is similar to other statutes of the era. The Clean Water Act and Clean Air Act have similar end-of-pipe cleanup regulations. There appears to be a conflict between the goal of reducing waste and the way the statute regulates the handling of waste.

The positive side of RCRA is the cooperation between the federal and the state governments. The protection of the environment is a national problem, and the support of the federal government to the next smaller scale is necessary to see that objectives are set and achieved. With respect to RCRA, the federal government is better suited to deal with major technical and financial problems, and these sections of the statute give the federal government the authority to help find solutions and enforce regulations throughout the states.

Air Pollution Prevention and Control Act of 1963 (CAA)

The Air Pollution Prevention and Control Act of 1963 (CAA) was established by Congress to respond to a growing problem of poor air quality, especially in urbanized areas. It was amended numerous times including the Clean Air Amendments of 1970, 1977, and 1990.

It is interesting that this statute is generally considered an "environmental" law. Nowhere in the statute's Findings and Declaration of Purpose Section (Note 22, Appendix A13) does it discuss natural environmental concerns. The purpose of this statute is to "protect and enhance the quality of the nation's air resources so as to promote the public health and welfare and the productive capacity of its population (Note 23, Appendix A13). This is without question a goal seeking the protection of humans, with no mention of protection of ecosystems or the natural environment in general. However anthropocentric this goal might seem, cleaning the nation's air will also have profound effects on natural systems.

The statute does much to clean up the quality of air, targeting sources that are large producers of air pollution. Many sections of the statute deal directly with certain sources or types of pollution the Congress has determined to be important. Some of these problems include ozone, motor vehicle emissions, acid deposition, particulates, carbon monoxide, and stratospheric ozone.

In the processing and manufacture of lead and lead products, smelting is an important process. The heating of lead for refining or processing, called smelting, releases many different gases. Most of these gases are released into the atmosphere. For this reason the Clean Air Act contains most of the regulatory control over these processes.

The National Ambient Air Quality Standards (NAAQS) set up by the Clean Air Act defines a maximum level allowed for certain pollutants. The NAAQS for lead is the arithmetic mean of $1.5 \ \mu g/m^3$ averaged over a calendar quarter. Under the statute, EPA develops NAAQS, and then each state is required to implement a plan (SIP) to achieve the goals. This means that states can adopt different regulations on different types of pollutant sources.

Nonattainment of the goals has been one of the major problems of the Clean Air Act. Deadlines have been repeatedly missed, and the weak tools the statute initially gave to the EPA to enforce the NAAQS proved of little value, leading to the numerous amendments to the original statute.

In addition to the NAAQSs, the EPA has generated a list of new source performance standards (NSPS). The NSPSs are technology-based standards that force technological controls over the type of new system a company can build. For lead these regulations define maximum effluent limitation guidelines that are based upon the Best Practicable Control Technology (BPT). Under these regulations there are a number of different processes where lead discharge is monitored. These include primary lead smelters and zinc smelters. Under the regulations for zinc and lead, emissions are regulated based on Best Available Control Technology (BAT).

Clean Water Act (CWA)

The Federal Water Pollution Control Act (FWPCA) was enacted in October 1972 (Note 24, Appendix A13). It was renamed the Clean Water Act (CWA) in 1977 when it was amended. The goal of the act is to "restore and maintain the chemical, physical and biological integrity of the nation's waters" (Note 25, Appendix A13).

The methods the statute employs include permitting requirements for point sources and technology forcing water quality criteria. The extraction of lead ores from the mined rock, called beneficiation, produces much mine wastewater as a major coproduct. This water, whether it enters the ground or surface waters, is heavily regulated. The CWA requires a National Pollutant Discharge Elimination System (NPDES) permit for all runoff water into waters of the U.S. (Note 26, Appendix A13). The regulations define daily and monthly limits for lead effluent discharge.

In addition, the Safe Drinking Water Act (SDWA) regulates mine drainage that contains lead. Lead is defined under the SDWA in the primary drinking water standards (Note 27, Appendix A13). These regulations are designed to protect drinking water sources (surface and groundwater) from potential contamination by lead mine drainage.

But a large portion of waters that are contaminated with lead are the result of nonpoint source mine wastewaters. These wastewaters may be tailings, or simply surface drainage that contacts the lead ores. These nonpoint sources are regulated under S319 of the CWA as amended in 1987. This section requires that states submit a report that identifies nonpoint sources and proposes management actions to control these sources (Note 28, Appendix A13). This method should work if states develop regional and watershed scale plans that take into account the differences in ecosystems from one scale to the next.

In addition to runoff waters, stored mine wastewaters are regulated under the Migratory Bird Treaty Act (MBTA) (Note 29, Appendix A13) and the Endangered Species Act (ESA) (Note 30, Appendix A13). Any water that is stored has the potential of being inhabited by any migratory or local fauna. These statutes serve to protect animals from the potential harm created by the storage of these mine waters. The effect this statute has is to prevent mining companies from storing water in lieu of discharging it, where regulated by the CWA and the SDWA.

The application of these statutes may need to be reexamined. If water is held on the site, especially in wetlands, and allowed to slowly reenter the hydrologic system, the wetland may serve as a filter for potentially harmful materials. This becomes especially important when consideration is given to the high cost of some treatment methods. The shift in thinking must move from removing the ecosystem from the industrial cycle to safely utilizing the ecosystem within that cycle. To make this shift new environmental laws must be developed.

THE INDUSTRIAL–ECOLOGICAL SYSTEM

The mining laws were established to encourage metals development without regard to associated costs. The cost and benefit issues of the late 1800s encouraged immediate development. It was generally believed that concentrated resources should be extracted and utilized. The economic benefits of mining and selling lead and lead products were large, and the cost of damaged ecosystems seemed minor. The regulations governing the mining and processing of lead were few, as seen in Table 13.1; and because the resources were so concentrated, the cost of reuse or recycling would

Table 13.1   Laws That Regulate Lead Industrial Ecology in Three Economic Periods

Period	Law
Growth	Mining Law of 1872
Sustaining	Resource Conservation and Recovery Act
	Clear Air Act
	Clean Water Act
	Comprehensive Environmental Response and Liability Act
	Safe Drinking Water Act
	National Environmental Policy Act
	Emergency Planning and Community Right to Know Act
	Migratory Bird Treaty
	Endangered Species Act
Decrease	Environmental Wetland Filtering
(possible new laws)	Emergy Evaluation of Choices between Mining, Reuse, and Environmental Recycle

Note: For listing of statutes, see Appendix A13.

have been much greater. The lack of government regulation was clearly the result of a public attitude that growth and development were important at any cost.

Figure 13.1 is an energy systems diagram of how the government controls resource extraction. The left side of the diagram shows an energy source and natural production of a resource, in this case lead. The storage of economic resource rights is fed by the laws the government creates to regulate resource extraction. The action of law is shown with shaded pathways. The rights controlled by laws, among others, include property rights, selling rights, and the right to produce wasteful by-products. This control is shown in Figure 13.1 with the switch symbol. The rights in this storage tank control a switch which determines the economic use of the resource.

Once the economic use begins, the resource is fed into the economy and stimulates the circulation of money. The price is determined in our current system by the amount of human service used to extract the lead. Determined in this way, the price does not take into account the value of the environmental work.

The Mining Law of 1872 served, for all practical matters, to give resources to anyone who could develop them into goods. If we look at the systems diagram, during this initial period the government has very little control placed on the rights of economic use and development (Figure 13.1), and thus the Economic Resource Rights storage tank is full. With this tank full, individuals and companies had almost unlimited rights to exploit resources. The switch was constantly being turned on, and resources were continually being used.

Also during this period the storage of waste begins to build up and place a drain on the economy. As Figure 13.1 indicates, when the storage of waste increases, the amount of economic cost to deal with this increasing waste draws money from useful purposes in the economy. But, because the economic drain from waste is small in the beginning and the price of obtaining new resources is low, recycling is slow to develop and the benefits of recycling are not fully realized by the system.

As the concentration of rich lead deposits decreases, the price for their extraction increases as does the amount of stored wastes. At this point we enter the second economic period, sustainability. With it we can see in Table 13.1 the laws that govern the rights of economic use. The way these laws work to regulate resource extraction and use is by limiting the amount of associated waste production.

In the diagram in Figure 13.1 the flow of money is based on price, and as the price of lead becomes artificially inflated, the amount of real value received from the lead resource decreases. At the same time the storage of wastes reaches a point where its drain on the economy is significant enough to inflate the price. The controls that the laws place on rights during this period serve to stimulate reuse of materials rather than wasteful use.

The laws and regulations such as the Clean Air Act and Clean Water Act limit the amount of wasteful discharge an industry can create. The darkened pathways in Figure 13.2 show the action of these laws restricting outflow of wastes ("end-of-pipe regulation"). These laws either increase the cost of production by complicating production methods, or simplify production methods in favor of better reuse technologies. Reuse is the ideal goal of most of the statutes, but the former, also known as end-of-the-pipe regulation, seems to be the outcome. In most cases the law does not describe how industrial processes must reduce their wasteful production, rather the law states that reduction must be achieved. Terms like 'best available technology' are used to describe process adaptations. But in the long run the industry still has to deal with the wastes.

This leads us into a possible third stage, a prosperous decline (Table 13.1). The third stage requires some policy changes that should reflect a decrease in use and an increase in recycle. But the recycle to people (reuse) and recycle to nature are important. Recycle to nature returns waste products back to the environment in a dilute form and an appropriate place. See recycle on Figure 13.1.

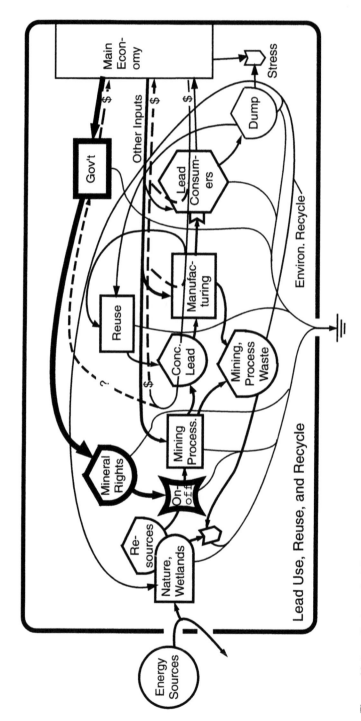

Figure 13.1   System of mining, reuse, and recycle. Darkened pathways are laws giving rights to mine.

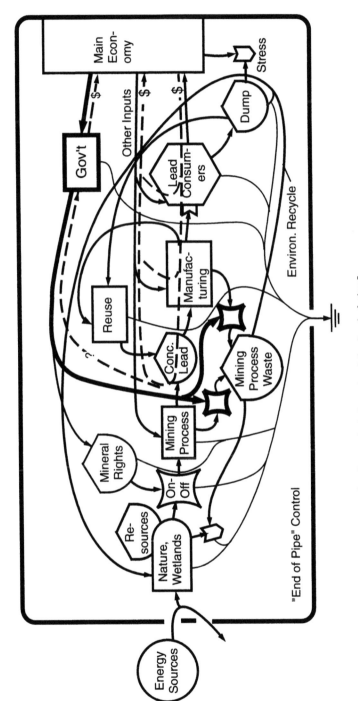

Figure 13.2 System of mining and use showing control of waste at the "end of pipe."

The laws during this period need to recognize the work the environment can essentially do for free. The recycle of materials back into the environment would have to be based on scientific principles and new thinking about values. These scientific principles and values systems are well established in the emergy theory. The evidence that wetlands may be a valuable option in recycling lead back into the environment is laid out in this book. Therefore, the policy for the prosperous decline should incorporate this new value system in its determination of resource rights.

For the upcoming period of prosperous decline, policymakers will need to recognize the responsibility of industry. New laws need to allow a greater role for industry in processing, recycle, and reuse. Laws will need to move away from regulating polluting actions, toward actions that will encourage industries to make the whole material cycle beneficial. The examination of laws in relation to the whole environmental system in this chapter suggests *systema operandi* rather than *modus operandi*.

Research for this chapter was carried out under the supervision of Professor A.C. Flournoy, College of Law, University of Florida, Gainesville.

CHAPTER 14

Summary, Policy for Heavy Metals and Environment

CONTENTS

SUMMARY

- This book contains the scientific studies of lead and wetlands made on the D.T. Sendzimir project, a joint investigation by the Centers for Wetlands and Environmental Policy of the University of Florida, Gainesville and the University of Mining and Metallurgy, Krakow, Poland, 1990 to 1998.
- Two wetlands with high concentrations of lead from wastewaters were studied, one in Florida and one in Poland (which also included zinc). The results of these studies plus work with published data emerging elsewhere were used to generalize about lead in wetlands, in society, and the global biogeochemical cycle of lead.
- From evidences of many kinds we now realize that wetlands and their humic peat, as they have evolved over geologic time, are a gaia mechanism for making the biosphere safe for life and, conversely, evolving the kind of life that makes the geologic processes compatible.
- Because several kinds of processes were found binding lead in wetlands, the lead not bound by one may be captured by another (physical filtration of particles, binding of soluble lead by negative surfaces of clays in sediment, binding by humic organic matter and peat, binding as insoluble sulfide crystals, binding in wood, substitution in shells and skeletons, precipitation as oxides and carbonates, etc.). Different processes are found in different degrees of prominence in different kinds of wetlands. Very little lead from rain and runoff gets past a wetland.
- Understanding of the distribution and movements of lead was found by developing simulation models of moderate complexity on the scale of time and space of whole wetlands. Important rates included were the inflows of lead, movements of water, percolation, binding capacities, growth of plants, toxicity to plants, sediment disturbance, and organic matter respiration. The much faster rates of chemical reaction that are part of these larger-scale processes were not included separately, but were aggregated as part of the main components and pathways.

This "top down" methodology models at the correct scale where the problems, time, and space scale are that of the environment and society. Aggregate models are simple enough to understand and calibrate accurately. They are on the same scale as policy thinking. This methodology contrasts with models that start with the ideal of including all the chemical and microphysical processes, combining them into a complex model that is cumbersome, calibrated with difficulty, hard to test, rarely finished, and subject to verification difficulties.

- Ecological Microcosms were studied in a greenhouse with one tree seedling in each and the normal soils and microbial processes transplanted from the field. As with other toxicity and biogeochemical studies in the past, microcosms helped understand the smaller-scale processes of the wetland that were also evaluated in the field, such as transpiration and rate of uptake of lead from waters. Using microcosms allowed replications and controls.
- Energy and economic-based evaluation of wetland filtration was made that showed great monetary savings to society from wetland filtration. By fitting civilization into the water–wetland systems, human society gains free benefit of earth life support and more competitive economy.
- All the processes of the earth can be arranged on a scale of "energy hierarchy" according to the amounts of energy that have been transformed in series. For the biogeosphere the scale ranges from low values in the fast processes of the atmosphere to large values required for the slow processes of building continental land.

In this study the position in the energy hierarchy of the states of lead in stages in its cycle was measured by calculating the emergy per mass of each. (Emergy is the available energy of one kind previously used to make a transformation to another kind.) Emergy per unit mass of lead increased with the concentration of lead, as you might expect since it requires more work (emergy use) to generate a higher concentration.

During the self-organization of the environment and society, each chemical cycle is observed circulating in a limited range of the universal energy hierarchy. For lead, the normal and appropriate place for most of the lead to circulate is in land processes and wetlands, not atmosphere and open waters where it is toxic to life.

- The studies of the wetlands with high lead concentrations showed that these areas had been of great value to public safety for many years and should be continued. Suggestions were made to vary hydroperiod to sustain vegetation and filtration capacity.
- Controversies remain in environmental lead management. There may be a limit to sewage sludge with heavy metals that should be applied to land used in agriculture (Chapter 3).
- Allowing sediments and peats with concentrations of heavy metals to be covered over with normal sediments by action of natural processes uses natural restoration work at low cost. Knutson et al. (1987) studied heavy metals in an embayment of the Hudson River 12 years after waste release from nickel–cadmium battery operations ceased. High level deposits of nickel and cadmium had been covered over by several centimeters of new sediments. Wetlands catch and bind these elements, too. Controversy exists on which areas have the risk of being disrupted, releasing toxic elements again. The concentration level and emergy evaluation determine when reprocessing environmental deposits to reconcentrate metals for use is a beneficial option.
- For future management of the lead-containing Sapp Swamp (Chapter 11), emergy evaluation of alternatives by Ton et al. (1998) found the highest net benefit ($2,870,000 emdollars over 20-year period) from leaving sediments undisturbed except for planting trees.
- Arguments continued on whether electric cars and increased battery use will be a new threat (Chapter 2). Emergy evaluations show other transportation with more net yield. Even if nickel–cadmium or nickel–metal hydride batteries replace lead batteries, there would still be releases of toxic heavy metals of environmental concern.
- Controversy remains as to the global condition of lead pollution. Socolow and Thomas (1997) summarized lead use with graphs of U.S. lead production, gasoline lead, lead in human blood, and the global lead cycle update. They suggest lead can be the first hazard to be appropriately managed. But Nriagu (1994, 1998) concludes that lead emission rates are decreasing in developed countries but increasing elsewhere. Global metal pollution is still increasing, although less in the atmosphere.
- There are still hazards to human health from the low levels of lead in urban soils and old houses with lead paint. Natural levels of lead in human blood are already close to that considered a toxicological limit, which leaves little margin for lead exposure.

- Understanding the lead processes, cycle of lead, and the role of wetlands is the basis for recommending policies for management of lead using wetlands. Extrapolating values from the study sites showed abundant capacity for existing and/or constructed wetlands to process the entire budgets of low concentration lead emissions in the U.S. and Poland. Runoffs need to be routed through the wetlands.
- Reviews cited in Chapters 3 and 4 showed many similarities among heavy metals in energy hierarchical position and the way the self-organization of earth cycles processes them in the geobiosphere. These elements are scarce, with high transformity and physiological impact when concentrated. To use these elements well means developing patterns of human civilization, industrial ecology, and ecological engineering that keep their processes and uses in separate pathways from humans and their ecosystems.
- Many more research papers and summaries show the ability of humic substances and peat to form binding complexes with heavy metals and many organic toxicants as well (Fuhr, 1987; Senesi and Miano, 1994; Hessen and Tranvik, 1998). It is now possible to conclude that humic substances are generated in all ecosystems including the sea in a wide range of molecular sizes. Brown humic substances accompany all life and help keep the biogeochemical cycles and global ecosystems compatible.
- Summaries show that the concepts for use of wetlands can be extended to other heavy metals. Many papers document the ability of wetlands to absorb and bind arsenic, cadmium, nickel, mercury, uranium, silver, copper, and many others. The following suggestions can help make a better partnership between mining and manufacturing, society, and environment.

SUGGESTED POLICIES

Based on new understanding about the role of wetlands and the behavior of lead in the biogeosphere, the following guidelines are suggested.

1. Where possible, restore the original biogeochemical cycle of lead. This means minimizing lead passage through the atmosphere and open waters. Wetlands are a means of keeping lead on land.
2. Where possible, where lead has been concentrated in sedimentary depositions in estuaries and lakes, allow these levels to be buried by normal sedimentation to become part of the geological cycle.
3. In order to correct the excess lead in global atmosphere and ocean waters, develop international treaties to further eliminate lead additives in transportation fuels, as already accomplished in the U.S.
4. Avoid processing lead-containing materials through incinerators in order to prevent lead release to the atmosphere.
5. Where possible, restore the pattern of water filtration by wetlands that originally existed. This means restoring water flows through wetlands. In many places it means adding constructed wetlands as necessary to stop lead from reaching open lakes, groundwater supplies, or the open ocean.
6. Wetland interfaces are needed between all runoffs and waters. This means restoring variable water levels to streams and lakes so that they will develop wider wetlands (longer hydroperiod). Remove weirs and unnecessary dams. The stormwater ponds arranged to catch urban runoff need to be managed as wetlands rather than as bare reservoirs. Very small wetlands need to be restored or constructed within city parks, housing developments, and road ditches to help filter the still high lead washing off streets and out of urban soils.
7. In order to catch heavy metals in sediments and organic substances, as well as for other reasons, flooding should be restored to floodplains and deltas by removal of levees and channelization. Housing within wetlands can be protected with permanent elevated foundations or by surrounding areas with a local encircling levee.
8. Where pore waters of upland soils and former solid waste sites have high content of mobile lead, drainage arrangements should be arranged for fringing wetlands or downstream filtration through constructed wetlands.
9. Point sources with concentrations of lead, runoffs around battery operations and smelting, and acid mine drainages need to run through a series of constructed wetlands as an economical and efficient means of filtration and holding of lead. Avoid excess discharge of sulfates to freshwater wetlands

so as to prevent excess hydrogen sulfide. pH neutralization may be required to prevent tree mortality and maintain biodiversity.

10. Sludge from treatment processing that contains high concentrations of lead should be dispersed in permanent wetlands, rather than in uplands where lead mobility is greater.

11. Wherever a series of wetlands has functioned to absorb large concentrations of lead in its sediments and peat, the lead is not readily released so long as the wetlands receive their normal water regimes. Regulation, tax incentive, or land purchase for protective purpose may be necessary to keep these areas operating as wetlands. It may be good public relations policy to fence these areas, advertise their history, and use them for educational purposes. Do not dig up the wetland and transfer it to upland. Lead levels reaching wildlife are small.

12. Wetlands should be managed at their natural pH levels, not exposed to very low or high pH that may cause bound lead to become more mobile; nor should wetlands be drained and dried out causing oxidation and leaching.

13. It may be a good policy for industries processing lead for useful commercial purposes to make and operate wetlands concurrently, thus taking more of the responsibility for global biogeochemical management. For this purpose tax incentives are suggested. After operations cease, the wetlands should be placed in public endowment (item #11).

14. Further implement the trend toward almost 100% reprocessing and reuse (anthropogenic recycle) of lead as well as other scarce metals.

APPENDICES

Symbols Used in Systems Diagrams

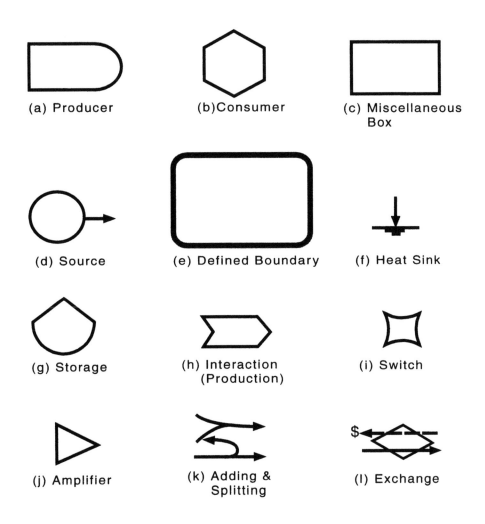

(a) Producer

(b)Consumer

(c) Miscellaneous Box

(d) Source

(e) Defined Boundary

(f) Heat Sink

(g) Storage

(h) Interaction (Production)

(i) Switch

(j) Amplifier

(k) Adding & Splitting

(l) Exchange

Biogeochemical Cycle of Lead and the Energy Hierarchy

Howard T. Odum

Table A4.1 Global Storages of Lead

Note	Item	E10 Grams
1	Seawater recently	2.74 E3
2	Seawater originally	21.9
3	Atmosphere now	1.84
4	Atmosphere originally	0.28
5	Soil, glaciers	4.8 E5
6	Deltas, wetlands, sediments	4.8 E9
7	Land	3.2 E9
8	Ore	1.4 E4
9	Civilization	3.47 E4

Notes:

1, 3, 5, 6. Nriagu (1978, page 10).

2. Use concentration in the deep sea as representative: 0.02 μg/kg (from Chow and Patterson, 1966).

   (2 E-5 g/m^3)(1.37 E9 m^3 seawater) = 2.7 E4 g.

4. In original air 5.3 E-10 g/m^3 surface air (from Patterson, 1965).

   (5.3 E-10 g/m^3)(1 E-3 m^3/g surface air)(5.2 E21 g atmosphere) = 0.28 E10 g.

7. Lead is 16 ppm in rock (Bowen, 1979) 2.2 E24 g.

   Uplifted continental sediments: (16 E-6 g/g rock)(2.2 E24 g) = 3.15 E19 g.

8. Lead reserves 141 E6 short tons (Kesler, 1978).

   (141 E6)(0.907) = 128 E6 tonne = 1.28 E14 g.

9. Civilization storage, order of magnitude estimate: Net production rate in Figure 4.6 (4000–534 E9 g/year)(100 years) = 3.47 E14 g.

Table A4.2 Global Flows of Lead in Figure 4.6

Note	Item	E9 Grams/Year
1	Seawater to sediments	2.5
2	Seawater to atmosphere	2 E-5
3	Atmosphere to seawater	210
4	Atmosphere to ecosystems	320
5	Ecosystems runoff to deltas	720
6	Predevelopment runoff	180
7	Deltas to open seawaters	34
8	Deltas and sediments to deep earth	?
9	Deltas, sediments to land	94
10	Sedimentary land to ecosystems	400
11	Normal weathering to ecosystems	18
12	Deep earth to land	?
13	Continental land to ores	?
14	Ores to economy	4000
15	Deep earth to ores	?
16	Economy to atmosphere	440
17	Economy via rivers to sediments	60
18	Economy solids to land	34
19	Volcanoes to atmosphere	0.4
20	Land to atmosphere	5.9
21	Land dust and organics to atmosphere	32

Notes:

1–5, 8, 10, 12–19, 21. Nriagu (1978).

6.   River runoff before wastes: 180 E9 g/year (from Bowen, 1966).

Natural denudation 1.1 E10 g/year (Tatsumoto and Patterson, 1963).

7.   To open sea 34 E9 g/year (after Tatsumoto and Patterson, 1968 quoted by Chow, 1978).

9.   Land cycle: (2.4 cm/1000 year)(2.6 g/cm^3)(1.5 E18 cm^3) = 9.36 E15 g/year.

(9.36 E15 g/year)(10 E-6 g lead/g land) = 93.6 E9.

11.   180,000 tons/year from natural weathering of rocks (Volesky, 1990).

20.   Land to atmosphere: 5.9 E9 g/year (Lantzy and Mackenzie, 1979).

Table A4.3 Emergy per Mass and Concentration of Lead Graphed in Figure 4.7

Note	Item	Concentration (g/m3)	Emergy per Mass (sej/g)
1	Ocean	3 E-5	0
2	Inland cycle	16	1 E9
3	Recycled in wetland	100	1.7 E9
4	Lead core	6.55 E4	4.5 E9
5	Refined lead	11.3 E6	7.3 E10

Notes: g = gram; m^3 = cubic meter; sej = solar emjoules; cm^3 = cubic centimeter.

1.   Zero emergy when no further dispersion is possible (no available energy in the concentration); ocean has lowest lead concentration of the main phases of the geobiosphere (Garrels, Mackenzie, and Hunt, 1975).

2.   Assigned a share of the global emergy contribution to the continental earth cycle (Kuroda, 1982; Drever et al., 1988). Transformity is that of the land cycle = (9.44 E24 sej/year)/(9.36 E15 g/year) = 1.0 E9 sej/g.

3.   Evaluated from wetland system containing lead (Figure 4.7). Emergy in support, 6.3 E10 $sej/m^2/year$ from Table 5.4 divided by 0.1 $g/m^2/day$ from Figure 8.2 times 365 days/year = 1.73 E9 sej/g. See Figure 4.6. Concentration in wetland = 100 ppm.

4.   Assigned a share of the global emergy contribution (9.44 E24 sej/year) to the earth cycle in maintaining orographic uplift (2.15 E15 g/year). Lead in ore, 6.55% (Kesler, 1978).

5.   Emergy per gram from Pritchard, Appendix A11, Figure A11.7 and Table A11.6; lead per volume = (11.3 g/cm^3)(1 E6 cm^3/m^3) = 11.3 E6 g/m^3.

Field Measurement Methods

Lowell Pritchard, Jr.

This appendix provides details for the diurnal oxygen method for primary production under water, for using leaf area index for production by emergent plants, for tests of toxicity by planting seedlings, and for the study of small invertebrate animals and their biodiversity.

DIURNAL OXYGEN MEASUREMENTS

Gross primary productivity and community respiration were estimated for underwater components (submerged and algae) with diurnal oxygen measurements every 3 h over a 24-h period. Dissolved oxygen measurements were made with an oxygen meter (YSI model 54A) with the probe attached to a 2-m rod. the probe was gently moved about in the water at a depth of about 20 cm to provide current necessary for electrode operation. Temperature was recorded at the time of measurements.

Several standard Winkler titrations were performed to confirm D.O. meter readings, using the azide modification (American Public Health Association, 1985). Sampling was done with a BOD bottle (300 ml) in a sampler designed to minimize aeration during collection. The sampler was extended on a 3-m pole held forward so that gas bubbles released from the sediments by footsteps would not influence the composition of the sample. Reagents were added, and titrations were carried out within a few hours.

CALCULATION OF DIFFUSION CONSTANT

Based on movement of dye and water depth, a diffusion coefficient was calculated each time according to the Bansal equation (Bansal, 1973). The very slow rates of water movement in open water areas of Steele City Bay were estimated using a drop of dye, stopwatch, and a meter stick. Rates of dye front movement were calculated. Surface water turnover as shown by dye movement was related to wind movement and proximity to windbreaks, and the fact that the winter measurements gave a higher diffusion coefficient was due to greater windiness.

The summer 1990 conditions were similar for all sites, but the winter 1991 wind and water movement conditions were variable, so the sites were broken into three groups and a different average value for water velocity was applied to each group.

Rates of flow and water depth are the two parameters that determine the overall reaeration coefficient in the empirical work of Churchill et al. (1962) and the analysis of Bansal (1973). Bansal gives the reaeration coefficient relationship as the general formula

$$K_{2(\text{base } 10,\, 20°C)} = \frac{cV^a}{D^b}$$

where K_2 = reaeration coefficient in reciprocal seconds.

To convert $K_{2(\text{base } e)}$, $K_{2(\text{base } 10)}$ was multiplied by 2302. The effect of temperature on the reaeration coefficient is given by the following empirical relationship (Bansal, 1973):

$$K_{2(T°)} = K_{2(20°)} * (1.016)^{T-20}$$

When this form of the reaeration coefficient is used, results are given as concentration change per unit time (dC/dt) rather than mass flux per area per unit time (dm/dt), because the depth of water is included in the calculation of the reaeration coefficient. A separate calculation is required for every depth and velocity condition. The formula for change in gas concentration due to diffusion comes from the two-film theory of gas transfer (Metcalf and Eddy Inc., 1979) and can be expressed as

$$\frac{dC}{dt} = K_{2(\text{base } e)} * (C_s - C)$$

where dC/dt = change in concentration, ppm h^{-1}
            K_2 = diffusion coefficient h^{-1}
            C_s = saturation concentration of oxygen in solution, ppm
            C = concentration of oxygen in solution, ppm

The diffusion coefficient is sometimes expressed as total oxygen flux across the water surface per hour (g O_2 m^{-2} h^{-1}) for a 100% oxygen deficit (i.e., 0% saturation). At a 100% oxygen deficit, C above = 0, and the change in the concentration equation reduces to

$$\frac{dC}{dt} = K_2 C_s'$$

Units of ppm-h^{-1} are equivalent to g m^{-3} h^{-1}, so multiplying by depth in meters yields the desired units of g O_2 m^{-2} h^{-1} at 100% oxygen deficit. These values are given in Table A5A.1. They are comparable to values for the diffusion coefficient given by Odum (1956), which range from 0.03 to 0.08 g O_2 m^{-2} h^{-1} for still water.

The diurnal curve method of calculating metabolism was used (Odum, 1985). From the raw data for dissolved oxygen concentration and temperature, rates of change were calculated. Oxygen deficit or excess was determined from a table of solubilities (American Public Health Association, 1985). From the diffusion-corrected rate-of-change curve for oxygen, gross production, net production, and community respiration were determined graphically using a compensating polar planimeter (see Figure 5.4). As a simplification, respiration was calculated to increase linearly throughout the day.

WATER LILY LEAF AREA INDEX

The leaf area index of floating vegetation was measured using a line-intercept transect 5 m in length. The intercept lengths were recorded for individual leaves of *Nymphaea*. Totals of intercept lengths divided by transect length gave a value for the leaf area index.

Table A5ᴀ.1 Calculation of Reaeration Coefficient from Measured Water Velocity and Depth According to Bansal Equation

Site	Velocity (ft s⁻¹)	±S.E.	Depth (ft)	$K_{2(base 10,20°C)}$ (s⁻¹)	$K_{2(base e,20°C)}$ (h⁻¹)	Oxygen flux at 100% deficit (g O_2 m⁻²h⁻¹)
			Winter			
B	0.017	0.007	2.62	0.000001	0.010	0.064
C	0.017	0.007	2.46	0.000001	0.011	0.066
D	0.017	0.007	2.95	0.000001	0.009	0.061
F	0.017	0.007	1.15	0.000003	0.032	0.089
G	0.017	0.007	1.64	0.000002	0.019	0.077
			Summer			
A	0.106	0.031	1.80	0.000006	0.051	0.299
B	0.037	0.005	2.95	0.000001	0.014	0.131
C	0.037	0.005	2.79	0.000001	0.015	0.134
D	0.016	0.001	3.28	0.000000	0.007	0.075
F	0.016	0.001	2.13	0.000001	0.013	0.089
G	0.037	0.005	2.30	0.000002	0.019	0.145
RP	0.106	0.031	2.30	0.000004	0.036	0.271

Note: D = depth of water column.

V = water velocity.

c = 0.000054 s⁻¹ at 20°C.

a = 0.6 a constant.

b = −1.4, a constant.

s = seconds; ft = feet; h = hours; g = grams.

K^2 = aeration coefficient using logarithm to the base 10.

From Bansal, 1973. $K_{2(base1\,0,20°C)}$ = cVᵃ/Dᵇ, where a = 0.6, b = 1.4, and c = 0.000054.

CANOPY LEAF AREA INDEX

Leaf area index for (*Nyssa*) trees in areas of very low canopy coverage (locations F and G) was estimated by grouping branches into size classes, visually estimating (from the ground) leaves per branch for each class, and branches per trunk, for five trunks of known diameter in each sample area. The length and width of 100 leaves from each sample area were measured. The area of each leaf was calculated assuming elliptic proportions (area = πab, where a and b are the lengths of the semiaxes). Average leaf area was multiplied by leaves per trunk to obtain total leaf area for five trunks. The total leaf area per unit basal trunk area was calculated and multiplied by the total basal area of trees in marked plots (see below under Woody plant sampling). This number, divided by the area of the plots, gave a leaf area index.

The leaf area index for areas of much higher canopy coverage was estimated using a vertical line-intercept method. Using a bow and arrow, a string was shot vertically into the canopy. Leaves touching the string were counted. This procedure was repeated at 20 different points. If the area overhead was open sky, a value of zero was recorded.

Leaf area index was also calculated from collected litterfall. Litterfall baskets were attached to trees in all plots in forested areas (F, G, H, and the reference forest; see Figures 1.3 and 5.2). Litter was collected on each field trip, separated into leaves and other material, dried, and weighed.

An area/mass ratio was determined for dried leaves by weighing uniformly punched circles of known area. Petiolar mass was subtracted from the collective leaf mass, and the area/mass ratio was used to convert the corrected leaf mass to leaf area. The leaf areas were summed over the year and divided by the area of the trip to determine the leaf area index for the leaf fall shadow of the tree (assuming on average a vertical drop). For locations without closed canopies, the calculated leaf area index was multiplied by the fraction of the area actually canopied to calculate the overall leaf area index for the location. For the 20 × 20-m plots in locations F and G, the canopied fraction of the plot was estimated by counting *Nyssa* greater than 10 cm dbh and multiplying by their estimated individual canopy area.

PRODUCTION SUMMARY

The leaf area index for the reference forest canopy (control pond, station RF) was converted to a gross primary productivity value using an LAI/gross primary productivity linear regression from data on wetland forests given by Brown et al. (1984). Gross primary productivities for the other areas were assigned in proportion to their relative leaf area indices.

Gross primary production was calculated for *Nymphaea* by setting the highest leaf area index equal to two times a conservative estimate of freshwater marsh net primary production (= 1000 g dry weight/m^2; Mitsch and Gosselink, 1986). Productivities for other locations were assigned based on the ratios of leaf area indices. For both trees and water lilies the energy conversion value of 4.5 kcal/g dry weight was used (E.P. Odum, 1983).

Aquatic gross primary production reported above in terms of g O_2 $m^{-2}day^{-1}$ was converted to energy terms using the conversion of 3.5 kcal/g O_2 from the simple formula for photosynthesis (Cole, 1975).

MACROINVERTEBRATES

At each sample location, five cores 7.7 cm in diameter and 10 cm in depth were taken with a cylindrical mini-Wilding-type sampler designed to isolate a portion of the water column above the sediment. The sampled material was transferred to a U.S. Standard No. 30 sieve bucket (Weber, 1973). Fine particulates were removed in the field by partially submerging and agitating the bucket, taking care not to allow exchange of materials except through the sieve bottom. Remaining material was drained, placed in screw-top 1-gal plastic containers, labeled, and then saturated with rose bengal stain solution. After a few hours the stain solution was drained. Since the peaty material retained a significant amount of water, 90% ethanol was added as a preservative, rather than the recommended 70%.

PROCESSING, IDENTIFICATION, AND ANALYSIS

From each sample, small aliquots were removed, washed under water in a No. 30 sieve, and placed in a water-filled pan (Weber, 1973). Macroinvertebrates visible to the naked eye were removed with forceps and stored in vials in 70% ethanol. This was repeated for the entire sample.

Whole specimens were identified to the family level, with the exceptions of crustaceans, gastropods, and oligochaetes. Where specimens were damaged, only portions with heads were counted in the analysis. Early instars and pupae were identified to the lowest reliable level. Chironomidae were separated into feeding guilds, and Culicidae were identified to genus or to species where possible. References for identification included Pennak (1978), McCafferty (1981), and Merritt and Cummins (1984). Data collected were summarized for taxonomic groups. Densities (individuals/area), family richness (number of families/sample), and diversity indices were calculated for each sample (Tables A5B.2 and A5B.3).

DIVERSITY INDICES

Diversity was calculated using three indices. The Shannon diversity index is given by

$$H' = \sum_{i=1}^{n} p_i \ln_2(p_i)$$

where H' is the information in bits per individual, p_i is the proportion of individuals in a sample belonging to taxon i, and n is the total number of taxa in the sample. Sample variance of H' is given by

$$s^2 = \frac{\Sigma f_i(\ln_2 f_i)^2 - (\Sigma f_i \ln_2 f_i)^2/N}{N^2}$$

(Zar, 1984), where N is the total number of individuals in the sample, f_i is the frequency of observation of each taxon, and the degrees of freedom are

$$DF = \frac{(s_1^2 + s_2^2)^2}{\dfrac{(s_1^2)^2}{N_1} + \dfrac{(s_2^2)^2}{N_2}}$$

Simpson diversity was calculated using the dominance measure

$$L = \frac{\Sigma n_1(n_1 - 1)}{N(N - 1)}$$

Simpson diversity is then simply

$$D_s = 1 - L$$

with variance

$$s^2 = 4[\Sigma p_i^3 - (\Sigma p_i^2)^2]/N$$

Margalef's (1968) diversity index was calculated as

$$Ma = \frac{(S - 1)}{\ln_2 N}$$

where S is the total number of taxa in the sample.

SEEDLING SURVIVAL

To determine whether regeneration by seedling had been hindered either by the toxicity of metals in the sediments or by flooding, seedlings of bald cypress (*Taxodium distichum*), pond cypress (*T. ascendens*), and blackgum (*Nyssa sylvatica* var. *biflora*) were planted on recently exposed sediments in Steele City Bay (locations F, G, and H) and in the reference forest area (see Figures 1.3 and 5.2). The hydrologic conditions were recorded along with the heights of the individuals planted. On each successive field trip that water levels permitted, the height and condition of each individual were recorded.

Data on Biota in Sapp Swamp

Lowell Pritchard, Jr.

Table A5B.2 Benthic Macroinvertebrate Sampling

Taxon	Lowest Level	Stage	Location									
			A1	A2	B	C	D	F	G	H	RF	RP
Raw Data, August 21, 1990												
Annelida												
Oligochaeta	Oligochaeta	?					5		4			
Arachnida												
Acarina	Hydracarina	?					2	10				
Isopoda		?										
Ostracods/Clodocerans		?	1									
Copepoda		?			5	3	3		1	1		
Coleoptera												
Dytiscidae		A										
Dytiscidae	Peltodytes	L						3				
Haliplidae		A								1		
Haliplidae		L										
Hydrophilidae		A							2	1		
Hydrophilidae		L		2								
Noteridae		A			1	2	1		2	1		
Chrysomelidae		L			4	1	2	1	3	6		
Diptera												
Culicidae	Unknown pupae	P	5									
Culicidae	Aedes	L		1	2							
Ceratopogonidae	Bezzia complex	L			2	2		31	10	9		
Dolichopodidae		L							1	2		
Chironomidae	Tanypodinae	L	8	12	2	14			2	5		
Chironomidae	Chironomini	L	236	46	17	6		3	3	124		
Chaoboridae		L			2	8	10					
Hemiptera												
Notonectidae		A	1									
Mesoveliidae		N			1				1			
Dipsochoridae		N							1			
Odonata												
Anisoptera		N										
Zygoptera		N		1								
Libellulidae		N						1				
Collembola		C		1	1					4		
Raw Data, February 3, 1991												
Annelida												
Oligochaeta	Oligochaeta	?									3	
Nematoda											72	

Taxon		Stage[a]									
Arachnida											
Acarina	Hydracarina	?					1	2			2
Isopoda		?					1	2	2		1
Ostracods/Cladocerans		?	10			3	1		2	3	
Copepoda		?	3		3	1		1	2	8	
Coleoptera											
Dytiscidae		A	1								10
Dytiscidae		L									3
Haliplidae	Peltodytes	A	1								
Haliplidae		L									
Hydrophilidae		A									5
Hydrophilidae		L									5
Noteridae		A	1		1	1	1				1
Chrysomelidae		L			1				1		1
Curculionidae									1		
Diptera											
Unknown pupa		P						1			
Culicidae	Aedes	L	2								39
Ceratopogonidae	Bezzia complex	L	2	6	6	5	12	19	11	1	10
Dolichopodidae		L						1	2		1
Chironomidae	Tanypodinae	L	3		21	15	2	2	2		75
Chironomidae	Chironomini	L	53	1	1	34	54	94	21	12	2
Chaoboridae		L			1	4		1	1		
Empididae							1			1	
Tabanidae								1	1		2
Tipulidae											2
Unknown				1		3					2
Hemiptera											
Notonectidae		A									
Mesoveliidae		N									
Dipsochoridae		N				1					
Unknown Homoptera											
Odonata											
Anisoptera		N		1							
Zygoptera		N									
Libellulidae		N	2		1	1	1	10	1	1	2
Unknown 1			1			1	1	1	3	1	3
Unknown 2			1								
Unknown 3											
Collembola							1				

[a] A = adult, L = larva, N = nymph, P = pupa.

Table A5ʙ.3 Observed Vascular Plant Species Over 2 Years

	Occurrence								
	A	B	C	D	F	G	H	RF	RP
Woody species						X	?	?	
Acer rubrum								X	
Styrax americana					X	X	X	X	
Cephalanthus occidentalis				X	X	X	X	X	X
Nyssa sylvatica var. *biflora*						X	X	X	X
Taxodium ascendens							?	X	
Ilex cassine var. *myrtifolia*								X	
Quercus laurifolia									
Herbaceous species									X
Nymphoides aquatica	X	X	X	X	X	X	X		X
Nymphaea odorata	X								
Eleocharis equisetoides	X								
E. baldwinii		X	X	X	X	X			X
Utricularia sp.						X	?	X	
Myrica cerifera		X							
Juncus effusus					X	X			
Scirpus cyperinus					?	X			
Physalis sp.					?	X			
Ludwigia leptocarpa					X	X			
L. decurrens					?	X		X	
Cyperus erythrorhizos					?	X			
Polygonum hydropiperoides						X			
Panicum sp.						X			
Eupatorium capillifolium						X			
Saururus cernuus						X			
Number of species	3	3	2	3	10	16	7	9	5

Methods Used for Chemical Analysis of Waters and Sediments

Shanshin Ton and Joseph J. Delfino

The following are the chemical analysis methods used to obtain data reported in Chapter 6.

CHEMICAL ANALYSIS OF WATER SAMPLES

Surface water samples were analyzed for total phosphorus, total nitrogen (Kjeldahl nitrogen, ammonium nitrogen, and nitrite + nitrate nitrogen), and total lead. The analytical methods used to determine the water quality parameters in this study followed those of the U.S. Environmental Protection Agency (EPA, 1979) and/or *Standard Methods* (American Public Health Association, 1985).

ANALYSIS OF SEDIMENT AND VEGETATION TISSUE

Following the procedures which were described by Delfino and Enderson (1978), all samples were digested with nitric acid and hydrogen peroxide and filtered before analysis by atomic absorption spectrophotometry.

Approximately 0.1 g dry sample was accurately weighed into a 250-ml Erlenmeyer flask and 1 ml of DDI water was added to moisten the sample. Concentrated nitric acid (10 ml) was added slowly and the flask was swirled to insure good mixing. The flasks were placed on hot plates and the mixture evaporated slowly to dryness. After cooling, another 5 ml of concentrated nitric acid was added. This step was repeated until all visible organic matter was destroyed (indicated by light-colored residue). Boiling was continued until the reddish-brown fumes ceased. The containers were removed from the hot plates and cooled to room temperature. DDI water (1 ml), 2 ml of concentrated nitric acid, and 5 ml of H_2O_2 (30%) were added to the flasks. The containers were returned to hot plates and warmed gently. The containers alternatively were removed from the hot plates to allow any effervescence to subside and then rewarmed. This process was continued until subsequent warming did not produce any further effervescence. Another 5 ml 30% H_2O_2 was added and the previous process was repeated. All containers then were removed from the hot plate and cooled down for the filtration.

The digestate and residue were separated by filtering through a Whatman No. 42 filter paper. The container and filter paper were twice rinsed with *circa* 5 ml of DDI water. The filtrate and rinses were collected in a 125-ml Erlenmeyer flask which was placed on a hot plate to bring the final volume to 10 ml. The concentrated samples were stored in 32-ml glass vials until AAS analysis was performed.

DETERMINING SEQUENTIAL CHEMICAL EXTRACTION

Chemical reagents were used to extract sediment samples to fractionate the metals into different forms, including soluble, exchangeable, organically bound, inorganic precipitation, sulfide, and residual categories. The extraction procedure followed those reported by Stover et al. (1976) and Rudd et al. (1988a and 1988b).

The sequence of chemical reagents in this study is listed in Table A6a.1 with extraction times and ratios of solid-to-reagent volume.

Table A6A.1 Conditions for Sequential Chemical Extraction of Sediment Samples

Designated Form Extracted	Reagent (A.R. Grade)	Extraction Time (h)	Solids: Solution/ Volume Ratio
Exchangeable	1.0 M KNO$_3$	16	1:50
Adsorbed	0.5 M KF (pH 6.5)	16	1:80
Organically bound	0.1 M Na$_4$P$_2$O$_7$	16	1:80
Carbonate	0.1 M EDTA (pH 6.5)	16	1:80
Sulfide	6.0 M HNO$_3$	16	1:50

Adapted from Stover et al. (1976) and Rudd et al. (1988a and 1988b).

Approximately 0.4 g of dried samples was measured and extracted directly in 50-ml polypropylene centrifuge tubes. Those tubes were agitated in an autoshaker for 16 h, after which the mixtures were centrifuged for 10 min at 4000 rpm. The supernatants were retained for metal analyses and the solids were washed by DDI water before proceeding to the next reagent. The samples were extracted and washed repeatedly through the sequence of reagents. The sample residues were treated by using a HNO$_3$-H$_2$O$_2$ digestion.

INSTRUMENTATION ANALYSES

To ensure validity of the data, certain actions were taken to meet acceptable quality assurance and quality control (QA/QC) requirements. The QA/QC program in this study included the analysis of EPA-known evaluation samples in every analysis series, the running of 10% duplicate samples, and checking the recoveries of standard-solution spikes.

All of the metal *analyses* were performed using flame atomic absorption spectrophotometry (AAS). A Perkin-Elmer model 5000 AAS with double-beam system and air-acetylene fuel was employed with an instrumental detection limit for Pb equal to 0.01 mg/l or 10.0 µg/g for solid samples.

Dissolved oxygen (DO) was measured using a YSI model 54A oxygen meter with an instrumental detection limit of 0.1 mg O$_2$ per liter. Winkler titrations were performed to confirm DO meter readings both before and after each sampling trip.

Redox potential and pH measurements were carried out using a Fisher model 900 pH meter. Instrumental calibration for the pH meter was achieved by using two standard buffer solutions at pH 4.0 and 7.0. The manufacturer's calibration procedures for redox potential measurements were followed as well.

Electrical conductivity was measured using a Fisher digital conductivity meter with an output range of 0 to 200 S/cm. Water depth was measured using a meter stick, with every datum representing an average of three measurements within a square meter area. A standard thermometer with a range from –10 to 110°C was used to obtain the temperature readings.

Total organic carbon (TOC) in the surface water samples was measured using an Ionics model 555 organic carbon analyzer. Dissolved CO$_2$ in the water samples used for TOC analysis was removed by acidifying the sample with concentrated phosphoric acid and stripping with a nitrogen purge. Potassium acid phthalate solutions were prepared as standards and used to calibrate this instrument in the range 1.0 to 10.0 mg TOC per liter.

METHOD OF EXTRACTING HUMIC SUBSTANCES FROM WATER

Aquatic humic substances were isolated from water samples using the procedure described in *Standard Method* 5510 C (Clesceri et al., 1989). The basic procedure includes acidification of the 0.45-μm filtered water sample to pH 2, concentration on a macroporous resin base (NaOH), and substitution of cations on an H-saturated cation exchange column. This procedure is similar to that used for isolation of the hydrophobic acid fraction of dissolved organic carbon from natural waters, as reported by Leenheer (1981).

The aforementioned method is an analytical isolation method. It was modified for preparative-scale isolation simply by scaling up the size of the resin column (Leenheer, 1981). It has been used to reproducibly isolate aquatic humic substances from a variety of water sources on a preparative scale (Davis, 1993). The modified schematic extraction procedure for aquatic humic substances is shown in Figure A6A.1. Details of the materials and methods used for the isolation and characterization of humic substances were described by Davis (1993).

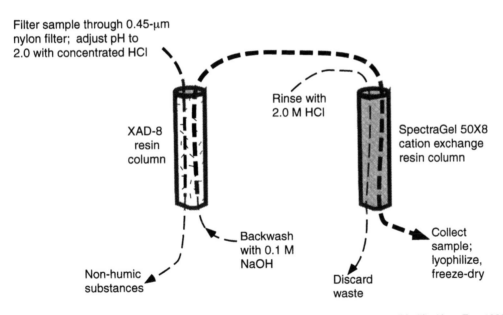

Figure A6A.1   Sketch of procedure for extracting humic substances from water sample. (Modified from Ton, 1993.)

DIALYSIS PROCEDURES FOR MEASURING LEAD BINDING TO HUMIC SUBSTANCES

Equilibrium dialysis was used to determine the binding capacity of humic substances and conditional stability constants (Saar and Weber, 1980). A continuous-flow system (Figure 6.6) was selected to perform the equilibrium dialysis, as is popular in pharmaceutical processes (New, 1990). A Spectrum Molecular/Por® Polysulfone, hollow fiber cartridge (HFC) was used to perform the dialysis analysis. The HFC is a sturdy bundle composed of 90 hollow fibers with inside diameters ranging from 0.5 to 0.7 mm and a molecular-weight cutoff at 2000 Da. The specifications of the HFC feature a wide range of chemical compatibility and pH values, from 1 to 13.

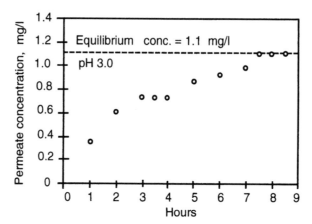

Figure A6A.2   Example of the lead binding by humic substances in the dialysis apparatus (Figure A6A.1). Equilibrium level is used to evaluate coefficients of dissociation (Ton, 1993).

The bundle was securely placed into a cap assembly and housed in a 1000-ml polymethylpentene (PMP) Fleaker®. A small 500-ml PMP Fleaker® was connected by Tygon® tubing to a Cole-Parmer Masterflex® pump to circulate the process solution through fibers. The solution in the 1000-ml Fleaker® also was circulated by pump. Two on-line sample outlets were installed to collect samples during analysis, and two Fisher stirring plates were used to blend the samples. A diagram of the apparatus used for dialysis analysis is shown in Figure 6.6. A test with lead on one side but without organic matter reached an equilibrium with lead concentrations the same on both sides of the membrane in 11 h or less (Figure A6A.2).

Two humic substances, Aldrich humic acid (AHA) and aquatic humic substances (SAPP 1), were used to determine the Pb binding capacity of humic substances and the conditional stability constants of Pb-organo complexes. Solutions of humic substances, AHA and SAPP 1 (approximately 20 mg humic substances per liter), were prepared in 0.1 M KNO$_3$ (Saar and Weber, 1980). Nitrogen gas was used to purge dissolved oxygen from the solutions before pH adjustment. The pH of the solution was adjusted as needed using dilute HNO$_3$ and KOH solutions (in 0.1 M KNO$_3$).

One liter of organic solution was used for dialysis against 0.1 l of 0.1 M KNO$_3$ for 72 h. Initially, 0.5 ml of 1000-mg/l Pb^{2+} standard solution was added to the electrolyte to perform the dialysis analysis, and 1.0 ml of 1000-mg/l Pb^{2+} standard solution was added for overnight dialysis analysis. Solutions were equilibrated for 6 h and 12 h in daytime overnight analyses, respectively, to assure sufficient equilibration (Figure A6a.2). Before the next metal addition, equilibrated solutions were subsampled (*circa* 2 ml) through on-line sample outlets from both Fleakers. The Pb concentrations, M_f for free metal concentration and M_t for total metal concentration in the equilibrated solutions, were measured using flame atomic absorption spectrophotometry.

COMPLEXING CAPACITY DETERMINATION AND SCATCHARD PLOT METHODOLOGY

When natural organic matter was dialyzed against a metal-ion solution, metal ions permeated through the membrane of the fibers and formed complexed compounds in the Fleaker containing organic matter (Alberts and Giesy, 1983; Saar and Weber, 1980; Stevenson, 1982; Truitt and Weber, 1981a, 1981b; Tuschall, 1981; Weber, 1983). At equilibrium, the free-metal concentration M_f was measured in the Fleaker containing electrolyte. Total metal concentration M_t, free plus complexed, was measured in the Fleaker containing organic matter. The complexed metal M_c then was calculated by a simple mass-balance equation:

$$M_c = M_t - M_f$$

The Pb complexing capacity was obtained from a plot of freely dissolved Pb concentration vs. total Pb concentration. The curve was extrapolated to the abscissa in order to obtain the Pb complexing capacity (Davis, 1993; Truitt and Weber, 1981a).

The conditional stability constant (β) can be estimated using the Scatchard method (Stevenson, 1982; Tuschall, 1981). It was assumed that

$$\beta = \frac{M_c}{(M_f)(nA_T - M_c)}$$

where A_T is the total ligand concentration, in terms of humic substances, and n is the number of binding sites per ligand molecule.

The equation above can be rearranged to

$$\frac{M_c}{(M_f)(A_T)} = \beta \left[n - \frac{M_c}{A_T} \right]$$

By substituting V for M_c/A_T, the final form of the equation becomes

$$\frac{V}{M_f} = \beta(n - V)$$

Thus, a plot of M_c/M_f vs. V should produce a curve with slope $-\beta$. This data analysis has been attributed to Scatchard, and a plot of V/M_f vs. V is termed a Scatchard plot (Stevenson, 1982; Tuschall, 1981). A theoretical Scatchard plot for titration of organic matter with metal is given in Figure A6A.3. The illustrated approach suggests two categories of β, with one "strong" site and

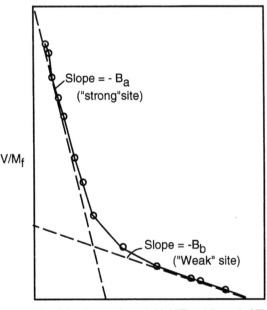

Figure A6A.3   Theoretical Scatchard plot used to evaluate stability constants (Ton, 1993).

one "weak" site. Changes of b in aqueous humic-substance samples (SAPP 1) with different pH values were examined.

A potential problem with the equilibrium dialysis technique is the leakage of humic substances across the membrane. This would lead to an underestimation of the Pb binding capacity and conditional stability constants (Truitt and Weber, 1981a). Lowered accuracy and reproducibility of analytical measurements further increased the uncertainty for metal binding capacities and determinations of stability constants (Haworth et al., 1987).

Chemical Data on the Cypress-Gum Swamps of Steele City Bay, Jackson County, Florida

Shanshin Ton and Joseph J. Delfino

This appendix contains data tables on the area studied for lead uptake. The analyses were first made at six stations (A, B, C, D, F, G) shown in Figure 6.1, where A is closest to the battery washing site. Later, sampling sites were expanded further downstream in a series of sites (A, B, C, F, OF1, OF2). Descriptions are given by Ton (1993). Sampling sites OF1 and OF2 were designated as checkout points for the Steele Bay Swamps and the boundary of the study area, respectively.

1-56670-401-4/00/$0.00+$.50
© 2000 by CRC Press LLC

Table A6B.2 Concentrations of Pb in a Wetland that Received Discharges from the Sapp Battery Superfund Site and in a Nearby Freshwater Wetland

	Background Data[a]	Livingston (1983–1985)		E & E (1986) Report		E & E (1989) Report	
		Station 1[b]	Station 2	E. Swamp	Steele City Bay	E. Swamp	Steele City Bay
Surface water (mg/l)	0.01	0.41–4.55	0.25–0.80	0.03	0.06	0.06	0.02
Sediments (μg/g)	9.0	184–1939	20.5–999.5	78.6	24–820	NA	33–940
Vegetation (μg/g)	8.00[c]	NA	NA	NA	NA	NA	NA

Note: NA = not available.

[a] Data cited from E & E (1985) study.
[b] The locations in this table do not refer to sampling site.
[c] Data cited from Casagrande and Erchull (1977), Okefenokee Swamp, Georgia.

Table A6B.3 Water Quality of Sampling Sites that Received Discharge from the Sapp Battery Superfund Site, Jackson County, Florida

	A[a]	B	C	D	F	G	Background Data[b]
Distance (m)[c]	0	40	244	387	259	>600	—
pH	3.4	3.6	3.8	4.4	4.5	3.9	4.3–5.0
Temperature (°C)	32	31	32	28	29.5	—	—
Water depth (cm)	51	52	63–89	>90	38–51	31–51	—
Conductivity (μmho/cm–25°C)	22	106	76	31	47.5	76.5	60–70
Nitrite + nitrate (mg-N/l)	0.08	0.01	0.02	0.04	0.04	0.02	0.005–0.11
Ammonia (mg-N/l)	3.76	0.18	0.08	0.16	0.35	0.16	0.01–0.56
Total Kjeldahl N (mg/l)	5.08	1.43	2.06	3.60	2.31	1.08	1.04–1.71
Total P (mg/l)	0.03	0.07	0.09	0.20	0.08	0.02	0.05–0.16
Total Pb (mg/l)	0.28	0.03	<0.01	<0.01	0.01	0.01	—

Note: — indicates no data; mg/l = ppm.

[a] See Figure 5.2 for the locations of the sampling sites.
[b] Background data are cited from Ewel and Odum, 1984.
[c] Distance measured from County Road 280. 0 m indicates inside the source area.

Table A6B.4 pH and Electrical Conductivity, 1989–1992

Site	Apr 1989	Feb 1990	Aug 1990	Sept 1990	June 1991	Jan 1992	May 1992	Avg	SD	Max	Min
				pH in Surface Water							
A	3.4	—	—	—	3.6	4.2	3.5	3.7	0.37	4.2	3.4
A0	—	—	—	—	—	—	3.9	3.9	—	3.9	3.9
B	3.6	3.9	4.1	3.8	4.2	4.9	4.1	4.1	0.42	4.9	3.6
C	3.8	—	3.7	4.0	4.2	4.8	4.0	4.1	0.37	4.8	3.7
D	4.5	—	4.2	—	4.3	5.0	4.6	4.5	0.32	5.0	4.2
E	—	3.9	—	—	4.8	5.2	4.4	4.6	0.56	5.2	3.9
F	4.5	—	3.8	—	4.3	5.5	4.6	4.5	0.62	5.5	3.8
G	3.9	—	4.6	—	4.4	4.9	4.8	4.5	0.39	4.9	3.9
H	—	—	5.0	—	5.0	5.8	4.6	5.1	0.50	5.8	4.6
OF1	—	—	—	4.0	4.2	5.9	5.0	4.8	0.87	5.9	4.0
OF2	—	—	—	3.8	4.5	4.6	4.5	4.3	0.36	4.6	3.8
PC	—	—	—	—	4.9	7.0	5.0	5.6	1.20	7.0	4.9
				Electrical Conductivity in Surface Water (S/cm)							
A	322.0	—	—	—	42.3	180.0	139.7	171.0	116.1	322.0	42.3
A0	—	—	—	—	—	—	61.0	61.0	—	61.0	61.0
B	106.0	—	38.3	4.9	34.0	49.6	40.5	55.5	27.05	106.0	34.0
C	76.4	—	46.9	44.6	27.0	37.0	36	44.6	17.06	76.4	27.0
D	30.8	—	28.6	—	23.2	28.0	30.7	28.3	3.09	30.8	23.2
E	—	—	—	—	29.2	65.2	43.4	45.9	18.13	65.2	29.2
F	47.4	—	61.3	—	30.5	48.0	32	43.8	12.76	61.3	30.5
G	76.5	—	100.1	—	27.0	46.6	28	55.6	31.92	100.1	27.0
H	—	—	72.7	—	31.6	45.0	42	47.8	17.55	72.7	31.6
OF1	—	—	—	39.7	25.0	32.0	34	32.7	6.05	39.7	25.0
OF2	—	—	—	56.8	25.5	67.4	26.8	44.1	21.21	67.4	25.5
PC	—	—	—	—	29.5	46.7	40	38.7	8.67	46.7	29.5

Table A6ʙ.5 Lead in Surface Water, Sediment, and Animals (Ton, 1993)

Site	Apr 1989	Feb 1990	Aug 1990	Jan 1992	May 1992	Avg	SD	Max	Min
			Lead in Surface Water (mg/l)						
A	0.28	—	0.01	0.14	0.09	0.13	0.11	0.3	0.0
A0	—	—	—	<0.01	0.01	0.01	0.00	0.0	0.0
B	0.03	0.01	0.02	0.01	0.01	0.02	0.01	0.0	0.0
C	<0.01	0.01	<0.01	<0.01	0.01	0.01	0.00	0.0	0.0
D	<0.01	<0.01	0.01	<0.01	<0.01	0.01	0.00	0.0	0.0
E	—	—	0.01	0.20	0.06	0.09	0.10	0.2	0.0
F	0.01	0.01	<0.01	0.01	<0.01	0.01	0.00	0.0	0.0
G	<0.01	0.01	<0.01	<0.01	<0.01	0.01	0.00	0.0	0.0
H	—	0.03	0.01	<0.01	<0.01	0.02	0.01	0.0	0.0
OF1	—	—	<0.01	<0.01	<0.01	<0.01	0.00	0.0	0.0
OF2	—	—	<0.01	<0.01	<0.01	<0.01	0.00	0.0	0.0
PC	—	—	<0.01	<0.01	<0.01	<0.01	0.00	0.0	0.0
			Lead in Sediments (μg/g)						
A	385.7	—	393.9	27.7	23.4	207.7	210.3	393.9	23.4
A0	—	—	—	—	36.2	36.2	0.0	36.2	36.2
B	210.9	—	763.7	113.0	74.5	290.5	320.6	763.7	74.5
C	234.2	—	278.9	23.5	48.8	146.3	129.0	278.9	23.5
D	59.4	—	303.6	10.6	49.2	105.7	133.6	303.6	10.6
E	—	477.8	—	342.1	512.0	444.0	89.8	512.0	342.1
F	88.5	—	472.2	99.9	125.7	196.6	184.4	472.2	88.5
G	94.3	—	99.6	87.4	62.0	85.8	16.7	99.6	62.0
H	—	—	215.7	36.3	112.9	121.6	90.0	215.7	36.3
OF1	—	133.3	—	23.5	10.6	55.8	67.4	133.3	10.6
OF2	—	—	—	10.6	10.6	10.6	0.1	10.6	10.6
PC	—	—	—	23.3	23.4	23.4	0.0	23.4	23.3

Lead in Animals, October 1989 (μg/g)

Beaver (*Castor fiber*)	Site L	13.3	(liver)
Sunfish(*Centrarchus*)	Site F	66.7	
Sunfish (*Centrarchus*)	Site F	50.0	
Pickeral (*Esox lucius*)	Site G	16.7	

Table A6ʙ.6 Lead Content in the Soil/Sediment Profile of Steele City Bay, Jackson County, Florida

Depth (cm)	Location[a]											
	A		B		C		D		F		G	
	1[b]	2	1	2	1	2	1	2	1	2	1	2
0–15	286.6	484.7	344.9	76.8	228.3	240.0	53.5	65.2	146.7	30.2	88.5	100.1
15–30	18.6	18.6	566.3	100.1	158.4	41.9	41.9	30.2	123.4	18.6	18.6	18.6
30–45	6.9	18.6	438.1	18.6	100.1	30.2	6.9	18.6	41.9	18.6	30.2	18.6

Note: All data in this table are in the units of μg/g (= ppm, dry weight).

[a] See Figure 5.2 for the locations of the sampling sites.
[b] Numbers indicate different sampling locations, 15 m apart from each other.

Table A6ʙ.7 Lead Content in Vegetation Samples in Steele City Bay, Jackson County, Florida (Ton, 1990)

Vegetation	Item	A	B	C	D	F	G
Cypress	Bark	*112.5	*56.3	*75.0	<5.0	<5.0	*12.5
Taxodium ascendens	Stem (0–3 cm)	* 37.5	*12.5	*175.0	12.5	<5.0	<5.0
	Leaves	—	<5.0	<5.0	<5.0	<5.0	—
Black gum	Bark	—	—	—	18.8	—	25.0
Nyssa sylvatica	Stem (0–3 cm)	—	—	—	<5.0	—	37.5
	Leaves	—	12.5	—	—	—	12.5
Water lily	Stem and leaves	—	75.0	12.5	12.5	12.5	—
Nymphaea odorate	Root	—	62.5	12.5	—	50.0	—
Eleocharis baldwinii		—	487.5	—	—	—	—
Algae		—	—	—	—	150.0	—

Notes: * Indicates dead tree; — indicates no data, due to lack of specimen in the sampling area.

The leaf samples (B, C, D) were collected from live trees on the edge of the swamp. All data are in the units of μg/g (= ppm, dry weight).

[a] See Figure 5.2 for the location of the sampling sites. Detection limit: 5.0.

Table A6B.8 Lead in Vegetation Samples (μg/g)

Apr 1989		Aug 1990			Jan 1992			May 1992	
A-EL	487.5	A-EL	eq.AG	61.2	A-EL	eq.AG(N)	35.9	AI-AG	35.8
A-TAa	112.5	A-EL	eq.BG	736.7	A-EL	eq.BG(N)	86.0	AI-BG	125.6
A-TAb	37.5	A-EL	sp.AG	235.7	A-EL	eq.(O)	789.2	AII-AG	10.5
B-TAa	56.3	A-EL	sp.BG	620.2	A-EL	sp.AG	241.3	AII-BG	151.3
B-TAb	12.5	B1		23.3	A-EL	sp.BG	600.1	AIII-AG	10.6
B-TAc	<5.0	B2		10.6	B1(N)		36.0	AIII-BG	959.0
B-NSc	12.5	B3		61.6	B1		73.7	AIV-AG	164.6
B1&2	75.0	C1		23.0	B2		74.5	AIV-BG	762.9
B3	62.5	C2		10.5	B3		23.5	AV-AG	87.6
C-TAa	75.0	C3		10.6	C1		23.5	AV-BG	444.1
C-TAb	175.0	D1		10.6	C2		23.4	AVI-AG	267.5
C-TAc	<5.0	D2		0.5	C3		10.5	AVI-BG	585.0
C1&2	12.5	D3		0.4	D1		10.5	AVII-AG	<10.0
C3	12.5	F1		48.7	D2		10.6	AVII-BG	176.5
D-TAa	<5.0	F2		49.0	D3		10.5	A1	369.5
D-TAb	12.5	F3		23.2	E1		18.4	A2	341.8
D-TAc	<5.0	G1		10.6	E2		252.6	A3	23.3
D-NSa	18.8	G2		23.1	E3		183.5	A01	74.7
D-NSb	<5.0	G3		10.5	E-JL		201.4	A02	61.7
D1&2	12.5				E-JR		1241.2	A03	<10.0
D3	—				E-JL(OLD)		418.6	A04	10.6
F-TAa	<5.0				F1		10.6	B1	23.5
F-TAb	<5.0				F2		10.6	B2	10.6
F-TAc	<5.0				F3		10.6	B3	10.5
F1&2	12.5				G1		36.1	B4	<10.0
F3	50.0				G2		23.4	C1	23.4
F-ALG	150.0				G3		10.6	C2	0.6
G-TAa	12.5				OF1-1		23.5	C3	<10.0
G-TAb	<5.0				OF1-2		10.6	D1	10.6
G-TAc	—				OF1-3		10.6	D2	10.6
G-NSa	25.0				PC1		10.6	D3	<10.0
G-NSb	37.5				PC2		10.5	E1	203.1
G-NSc	12.5				PC3		10.5	E2	51.3

Feb 1990

E-JL	850.0
E-JR	1716.7
F-TAc	<10.0

E3	317.6
EJL	95.1
EJR	2068.2
F1	36.1
F2	10.5
FF4	<10.0
G1	10.6
G2	<10.0
G3	10.6
H1	<10.0
H2	10.6
H3	<10.0
OF1-1	10.6
OF1-2	10.6
OF1-3	10.6
OF2-2	10.6
OF2-3	<10.0
P1	<10.0
P2	<10.0
P3	<10.0
P4	<10.0

Abbreviations:

TA, *Taxodium ascendens* and NS, *Nyssaa sylvatica*: a, bark; b, stem; c, leaves
A[B..]X, *Nymphaea odorata*: 1, leaves; 2, stem; 3, root; 4, flower
J, Juncus: L, leaves; R, root
AG, aboveground biomass; BG, belowground biomass
El sp., *Eleocharis* sp.; El eq., *Eleocharis equisetoides*; ALG, algae
AI, *Amphicarpum muhlenbergianum*; AII, *Eleocharis equisetoides*; AIII, *Hypericum fasciculatus*; AIV, *Eleocharis* sp.; AV, *Ludwigia* sp.; AVI, *Xyris* sp.; AVII, *Cyperns* sp.

Table A6B.9 The Distribution of Chemical Forms of Lead in Steele City Bay, Jackson County, Florida (Ton, 1990)

Chemical Form	A[a]	B	C	D	F	G	Avg
Exchangeable	24.4	17.5	11.9	11.9	20.3	14.9	16.8
Adsorbed	1.4	1.1	1.6	0.1	0.0	0.1	0.7
Organically bound	45.2	32.8	31.0	38.0	36.3	20.6	34.0
Carbonate	20.5	34.9	37.4	27.1	29.0	29.2	29.7
Sulfide	6.0	7.1	9.8	11.9	6.3	22.4	10.6
Residual	2.6	6.6	8.3	11.0	8.1	12.9	8.2

Note: All numbers in this table are in % of total lead in the soil/sediment.

[a] See Figure 5.2 for the locations of the sampling sites.

Details and Statistics on Microcosm Studies

Shanshin Ton

Details are included here on setting up the microcosms. The experimental design matrix for the microcosm is shown in Table A7.1. Six replicates of cypress and black gum were used for each treatment. Six additional seedlings served as the experimental control for each species. Nine units without seedlings were used as blanks for each treatment. In total, 129 test units were used for the microcosm study.

Experimental microcosms were set up in a greenhouse near the Center for Wetlands, University of Florida, in June 1991. The microcosms were confined to two regions, each with an area of 1.8 × 1.8 m. Each region was divided into eight rows and eight columns, with the columns designated by letters from A to H. The rows were designated by numbers from 1 to 16. The combination of a number and a letter thus indicated each individual test unit. Totally, there were 128 sections created for the test. To compensate for the difference among test units (129) and created sections (128), a section, termed 16I, was added to the last row.

To satisfy statistical requirements, the seedlings were set in random order. They were planted individually in plastic double plots, using a mixture of potting soil, peat, pine bark, sand, clay, etc. from a commercial supplier (Traxler Peat Co., Orange Heights, FL) as the bedding substrata. One teaspoon of fertilizer (Osmocote 17-6-10) was spread under the root zone of each plant before replanting to provide the nutrients required for growth. In this study, reagent-grade sulfuric acid and standard solution of Pb from Fishers Scientific, Inc. were used to adjust the pH and provide the source of Pb, respectively.

A distribution system with automatic irrigation programming was constructed to perform the chemical treatments. Each replicate sample was treated as an individual to avoid interference from all others. To standardize treatments, the test solution was delivered through a polyvinyl chloride (PVC) pipe and small vinyl tube to each pot.

SAMPLING AND MEASUREMENTS

Initial Pb concentrations in the water, sediment, and vegetative tissue were sampled. Nominal Pb concentrations and pH of the microcosms were monitored during each time of preparation. Actual values of these chemical reagents were regularly measured in the plant pots (Table A7.2).

The heights of the seedlings were measured with a Stanley Powerlock® II tape approximately once each month from August 1991 through December 1992. Growth rates of the seedlings were calculated using the net growth value for each month.

1-56670-401-4/00/$0.00+$.50
© 2000 by CRC Press LLC

Table A7.1   Matrix of the Experimental Design for the Microcosm Study — Chemical Treatments for the Microcosms

pH [Pb]	O(W)	0.5(A)	5.0(B)
7.0(W)	1*	2	3
2.0(C)	4	6	8
4.0(D)	5	7	9

Note: * Numbers indicate the treatment combinations:

1. Three different pH values 7.0(W), 4.0(D), 2.0(C), and three Pb concentrations 0(W), 0.5(A), 5.0(B) mg/l were used for the chemical treatment test.

2. Six replicate seedlings of each species (cypress, black gum) were tested in each treatment combination.

3. Six extra seedlings served as an experimental control for each species, using the same conditions as treatment 1.

Table A7.2   Effects of Chemical Treatment on the Growth of Pond Cypress (*Taxodium ascendens*) and Black Gum (*Nyssa sylvatica*) Seedlings

		Chemical Reagent			Test Species			
		Pb (mg/l)		Acidity (pH)		*Nyssa sylvatica* Mean Growth (cm/month)		*Taxodium ascendens* Mean Growth (cm/month)	
Treatment	(n)	Nominal	Actual	Nominal	Actual				
3	8	4.88	1.52	6.4	5.4	8.13	$r^2 = 0.997$	4.70	$r^2 = 0.984$
8	8	4.88	2.35	2.4	4.3	7.33	$r^2 = 0.984$	6.22	$r^2 = 0.993$
9	8	4.88	2.32	4.5	5.6	8.19	$r^2 = 0.996$	6.73	$r^2 = 0.984$
2	8	0.44	0.13	6.4	5.6	6.99	$r^2 = 0.996$	6.64	$r^2 = 0.986$
6	8	0.44	0.30	2.4	4.7	7.59	$r^2 = 0.999$	5.71	$r^2 = 0.983$
7	8	0.44	0.23	4.5	5.7	7.43	$r^2 = 0.995$	4.22	$r^2 = 0.991$
1	8	0	0	6.4	5.7	8.16	$r^2 = 0.996$	7.37	$r^2 = 0.985$
4	8	0	0	2.4	4.4	7.91	$r^2 = 0.994$	6.89	$r^2 = 0.981$
5	8	0	0	4.5	5.4	7.81	$r^2 = 0.999$	7.11	$r^2 = 0.987$
Control	8	0	0	6.4	6.4	7.89		6.99	

Note: (n) indicates numbers of collected data.

STATISTICAL ANALYSIS

A statistical analysis for multifactorial experiments was used to analyze the growth rate (Montgomery, 1984). The analysis used was a 3^2 factorial design; that is, a factorial arrangement with two factors at each of three levels. The treatment combinations for this design were shown earlier in Table A7.1.

Since there were $3^2 = 9$ treatment combinations, there were eight degrees of freedom among the treatment combinations. The main effects of pH (as X) and [Pb] (as Y) each have two degrees of freedom, and the XY interaction has four degrees of freedom. If there are n replicates, i.e., eight in this analysis, there are n $3^2 - 1$ total degrees of freedom and $3^2 (n - 1)$ degrees of freedom for error.

A statistical software package, MINITAB® (Minitab, Inc., 1991), available for Apple Macintosh® computers, was used for the statistical analysis. The ANOVA (analysis of variance) tables were generated, and the arithmetic means of the growth rates also were calculated (Tables A7.3 to A7.5).

Variance (MS) is the mean of the sum of the square (SS) of the deviations from the mean. The ratio between the variance between treatments and the residual variance ("error" term, the variance among duplicates) is the F ratio, which can be used to estimate the probability that this much difference could occur by chance. Customarily, a difference is not regarded as proven unless the probability (P) obtained with F ratio tables is less than 0.05 (occurring by chance less than 5% of the time).

Table A7.3   The Anova (Analysis of Variance) for Growth of *Nyssa Sylvatica* with Chemical Treatments

Factor	Type	Levels		Values	
Acid	Fixed	3	1	2	3
Pb	Fixed	3	1	2	3

Analysis of Variance for Growth

Source	DF	SS	MS	F	P
Acid	2	0.52	0.26	0.00	0.996
Pb	2	5.59	2.79	0.05	0.955
Acid × Pb	4	5.22	1.30	0.02	0.999
Error	63	3785.41	60.09		
Total	71	3796.73			

Note: DF, degrees of freedom; SS, sum of squares; MS, mean square; F-test with denominator: error.

Denominator MS = 60.09 with 63 degrees of freedom (a = 0.05).

Table A7.4   The Anova (Analysis of Variance) for Growth of *Taxodium ascendens* with Chemical Treatments

Factor	Type	Levels		Values	
Acid	Fixed	3	1	2	3
Pb	Fixed	3	1	2	3

Analysis of Variance for growth

Source	DF	SS	MS	F	P
Acid	2	0.89	0.44	0.02	0.983
Pb	2	33.73	16.87	0.65	0.524
Acid × Pb	4	41.64	10.41	0.40	0.806
Error	63	1627.97	25.84		
Total	71	1704.23			

Note: F-test with denominator: error.

Denominator MS = 25.84 with 63 degrees of freedom (a = 0.05).

Table A7.5   The ANOVA (Analysis of Variance) for Growth of *Nyssa sylvatica* and *Taxodium ascendens* with Chemical Treatments

Factor	Type	Levels		Values	
Species	Fixed	2	1	2	
Acid	Fixed	3	1	2	3
Pb	Fixed	3	1	2	3

Analysis of Variance for Growth

Source	DF	SS	MS	F	P
Species	1	86.34	86.34	2.01	0.159
Acid	2	0.16	0.08	0.00	0.998
Pb	2	29.96	14.98	0.35	0.706
Species × acid	2	1.25	0.63	0.01	0.986
Species × Pb	2	9.36	4.68	0.11	0.987
Acid × Pb	4	19.05	4.76	0.11	0.979
Species × acid × Pb	4	27.81	6.95	0.16	0.957
Error	63	5413.38	42.96		
Total	71	5587.31			

Note: F-test with denominator: error.

Denominator MS = 42.96 with 126 degrees of freedom (a = 0.05).

Equations, Programs, and Calibration Table for Simulation Models

Shanshin Ton and Howard T. Odum

1-56670-401-4/00/$0.00+$.50

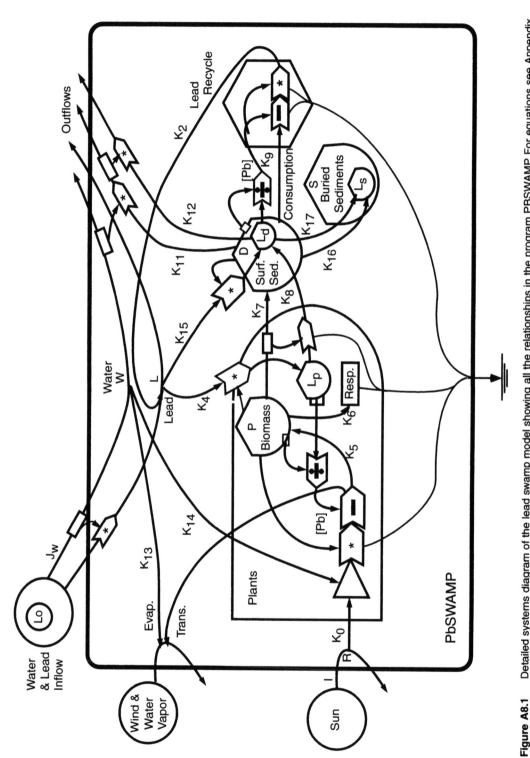

Figure A8.1   Detailed systems diagram of the lead swamp model showing all the relationships in the program PBSWAMP. For equations see Appendix Table A8.1, spreadsheet calibration in Appendix Table A8.2, and program listing in Appendix Table A8.3.

Table A8.1   Equations for Simulation Model PBSWAMP (Appendix Figure A8.1)

Explanation of letters

D	= surface organic sediments (detritus)
I	= flow of incident light
Jw	= water inflow (rain plus streams)

"K"s are coefficients evaluated in Appendix Table A8.2

L	= outflow of lead remaining in water
L0	= lead concentration in inflowing waters
Ld	= lead in sediments
Lp	= lead in plants
Ls	= lead in buried sediments
P	= plant biomass
R	= outflow of light remaining (albedo)
S	= buried sediments
Tox1	= K1*Lp/P where Lp/P is lead concentration in plants
Tox2	= K3*Ld/D where Ld/D is lead concentration in sediments
W	= outflow of remaining water

Remaining light which flows out equals incident light I minus that used by plant production as diminished by lead toxicity

$$R = I - K0*R*P*(1 - Tox1), \text{ therefore } R = I/(1 + K_0*P*(1 - Tox1))$$

Water outflow equals inflow Jw minus evaporation and transpiration

$$W = Jw - K13*W - K14*R*P*(1 - Tox1); \text{ therefore } W = [Jw - K14*R*P*(1 - Tox1)]/(1 + K13)$$

Remaining lead L which flows out equals lead inflow plus recycle minus uptake by plants and uptake by sediments

$$L = L0*Jw + K10*D*(Ld/D)*(1 - Tox2) - K4*L*P - K15*L*D; \text{ therefore } L = [L0*Jw + K10*D*(Ld/D)* (1 - Tox2)]/(1 + K4*P + K15*D)$$

Rate of change of plant biomass P equals production reduced by lead toxicity minus respiration and biomass passed to sediments

$$DP = K5*R*P*(1 - Tox1) - K6*P - K7*P$$

Rate of change of organic sediments D equals input from plants minus that used in consumption and flowing out in stream waters

$$DD = K7*P - K9*D*(1 - Tox2) - K11*W*D - K16*D$$

Rate of change of lead in plants Lp equals that captured by plants minus that carried with biomass to the sediments

$$DLp = K4*L*P - K8*P*Lp$$

Rate of change of lead in surface organic sediments Ld equals that in biomass from plants plus that absorbed from waters minus that consumed and that outflowing in streams

$$DLd = K8*P*Lp + K15*D*L - K2*D*(Ld/D)*(1 - Tox2) - K12*W*D*Ld - K17*Ld$$

Rate of change of buried sediments S equals the small flow from surface sediments

$$DS = K16*D$$

Rate of change of lead in buried sediments Ls equals the small flow from lead in surface sediments

$$DLs = K17*Ld$$

Table A8.2   Calibration Table for Model PbSWAMP (Figure A8.1)

Calibration Values

Jw	2.00E + 00
P	1.00E + 04 g/m²
D	5.00E + 04 g/m²
Lp	1.00E + 01 g/m²
Ld	1.00E + 01 g/m²
L	5.00E − 03
R	2.00E + 02 kcal/m²/day
W	5.00E − 01
S	0.00E + 00
Ls	0.00E + 00

Calibration Values

At T = 0	TOX1 = 0.9	TOX2 = 0.05	
k0*R*P*(1 − TOX1) = 3800		k0 =	1.90E − 02
k1*(Lp/P) = 0.9		k1 =	9.0E + 02
k2*LD*D*(1 − TOX2) = 0.1		k2 =	2.11E − 08
k3*(LD/D) = 0.5		k3 =	2.50E + 02
k4*L*P = 0.045		k4 =	9.00E − 04
k5*R*P*(1 − TOX1) = 500		k5 =	2.50E − 03
k6*P = 100		k6 =	1.00E − 02
k7*P = 400		k7 =	4.00E − 02
k8*LP*P = 0.1		k8 =	1.00E − 06
k9*D*(1 − TOX2) = 390		k9 =	8.21E − 03
k10*D*LD*(1 − TOX2) = 0.1		k10 =	2.11E − 08
k11*W*D = 10		k11 =	3.20E − 04
k12*W*D*LD = 0.01		k12 =	4.00E − 08
k13*W = 0.5		k13 =	1.00E + 00
k14*R*P*(1 − TOX1) = 1		k14 =	5.00E − 06
K15*L*D = 0.05		k15 =	2.00E − 04
K16*D = 2		K16 =	4.00E − 05
K17*Ld = 0.002		K17 =	0.0002

Table A8.3 Simulation Program PBSWAMP in BASIC

```
10 REM Macintosh
20 REM PbSWAMP-Steel City Bay
30 CLS
40 LINE (0,0)-(360,250),,B
50 LINE (0,50)-(360,50)
56 LINE (0,100)-(360,100)
60 LINE (0,140)-(360,140)
62 LINE (0,180)-(360,180)
70 REM Input sources:
80 DIM A(12)
90 DATA 2200,2000,2500,3300,3900,4000,4200,3600,3600,3500,2800,3000
100 FOR M = 0 TO 11
110 READ A(M)
120 NEXT
130 JW=2
140 L0 = 0
150 REM Initial conditions of storages:
160 P=1000
170 D=5000!
180 Lp= 0
190 Ld= 0
200 REM Coefficients:
210 K0=.0019
220 K1=900
230 K2=2E-08
240 K3=250
250 K4=.0009
260 K5=.00025
270 K6=.01
280 K7=.04
290 K8=.000001
300 K9=.0078
310 K10=2E-08
320 K11=.00032
330 K12=4E-8
340 K13=1
350 K14=.0000005
360 K15=.0002
365 K16=.00004
368 K17=.0002
370 REM Scaling factors:
380 I0=.005
390 LP0=2
400 LD0=.4
410 D0=.0006
420 P0=.0045
430 TX=(60*365)
440 T0=365/TX
450 DT=10
460 REM Equations:
465 IF T >(365*5) THEN L0 = .04
467 IF T >(20*365) THEN L0 = .0005
470 TOX1=K1*(Lp/P)
480 IF TOX1>1 THEN TOX1=1
490 TOX2=K3*(Ld/D)
500 IF TOX2>1 THEN TOX2=1
510 I=A(M)
```

continued

Table A8.3 (continued)   Simulation Program PBSWAMP in BASIC

```
520 R= I/( 1+K0*P*(1-TOX1))
530 W= (JW-K14*R*P*(1-TOX1))/(1+K13)
540 L= (JW*L0+K10*Ld*D*(1-TOX2))/(1+K4*P+K15*D)
550 DP = K5*R*P*(1-TOX1)-K6*P-K7*P
560 DD =K7*P-K9*D*(1-TOX2)-K11*W*D – K16*D
570 DLP = K4*L*P-K8*P*Lp
580 DLD = K8*P*Lp +K15*L*D-K2*D*Ld*(1-TOX2)-K12*W*D*Ld -K17*Ld
585 DS = K16*D
587 DLs = K17*Ld
590 Lp= Lp+DLP*DT
600 IF Lp < .0000001 THEN Lp = .0000001
610 Ld=Ld+DLD*DT
620 IF Ld < .00000001# THEN Ld = .00000001#
630 D=D+DD*DT
631 IF D < 5000 THEN D = 5000
640 P=P+DP*DT
641 IF P <1000 THEN P = 1000
642 S = S +DS*DT
646 Ls = Ls + DLs*DT
650 T=T+DT
660 Y= INT (T/365)
670 Z= 12*Y
680 M= INT (T/30.4) -Z
700 PSET (T0*T, 50-P*P0)
710 PSET (T0*T, 100-D*D0)
715 PSET (T0*T, 100-S*D0)
718 PSET (T0*T, 140-L*2000)
720 PSET (T0*T, 180-Lp*LP0)
725 PSET (T0*T, 250-Ld*LD0)
730 PSET (T0*T, 250 – Ls*LD0)
760 IF T*T0 < 360 GOTO 460
770 END
```

APPENDIX **A9**

Data on the Biala River Wetland and the Results of the Field Experiments

Wlodzimierz Wójcik and Malgorzata Wójcik

Details given in Appendix A9 use the following abbreviations: min = minute; km = kilometer; km² = square kilometer; m = meter; m³ = cubic meter; cm = centimeter; cm³ = cubic centimeter; mm = millimeter; TSS = total suspended solids; µg = microgram.

SITE LOCATION AND CONFIGURATION OF LAND

The investigated area is the wetland of the Biala River at its end section where the river valley forms a floodplain between rather low hills covered with forest. The Biala River is a left-bank tributary of the Biala Przemsza River which enters 29.22 km from the confluence of the Biala Przemsza with the Vistula River (Chapter 9, Figure 9.1). The distance between the river source and its outlet is about 11 km. The mean longitudinal slope is about 0.24%.

In its upper section, the river valley is rather narrow and shallowly incised in the sandy bottom. Near the village of Laski the valley turns a little northwest. The investigations included the Biala River valley from the village of Laski to its mouth where it meets the Biala Przemsza River. The wetland is about 3.5 km long, and 70 to 300 m wide, usually between 100 to 150 m. The walls of the valley are 4 to 8 m high, steeply undercut along almost their entire length, with a slope of 16 to 25%. The slope is less on the northern side. The right bank has the shape of a small ridge diversified by a series of low dunes. On the left bank there is a flat area 3 to 8 m high, with a small ridge up to 60 m high running parallel to the valley. Numerous tributary valleys cut the side of this ridge with winding courses.

The longitudinal slope of the wetland varies from 0.01 to 0.6% with three zones. From the Dabrowka Channel to the ruins of the old mill in Reczkowe village the slope is about 0.22%, oscillating from 0.2 to 0.5% (less than 0.1% only along short sections). From the old mill to the bridge in Kuzniczka village, the mean slope is 0.11%, with the slopes in some sections below 0.05%. Along the section from the bridge to the mouth the slope increases to about 0.29%, but immediately after passing the bridge it is 0.6%. Along the wetland, the Biala River meanders. In the central part ponds of some tens of meters wide occur locally. At many places the water of the main river bifurcates, forming one or more additional beds which return afterward to the main stream. Starting from about 800 m upstream from the bridge in Kuzniczka, water flow is spread through the entire valley. The total area of the Biala River wetland is 70 ha.

Access to the wetland is best by the road in the village of Laski and near the bridge in Kuzniczka. From this bridge an unpaved road runs along the northern side up to the ruins of the mill, but along some sections it is not passable at times because of deep sand or water-logged soil. The area has access to electric power transmission lines.

HYDROLOGY

The Biala River had its source in an area of springs. Discharge of two main springs was 3.5 and 2.0 l/s. With a total area of 53.54 km², the Biala River includes the following subbasins:

1. The basin above a conjunction of Biala River with the Ponikowska Adit is about 6 km long and 17.32 km² in area.
2. The Ponikowska Adit basin is about 6 km long and 15.24 km² in area.
3. The Dabrowka Channel basin is about 6.23 km long and 13.37 km² in area.
4. The basin between the outlet of Dabrowka Channel and outlet of Ponikowska Adit is 0.38 km long and 0.39 km² in area.
5. The basin from the Biala River mouth to the outlet of Dabrowka Channel (location of the wetland study) is about 4.5 km long and 7.22 km² in area.

The wetland floodplain formed in the terrain of the U-shaped prevalley of the Biala River on Quaternary sediments of fine- and medium-grained sand. Much later, mine waters from zinc and lead mines, some of which carried great quantities of suspended solids, deposited in the wetland. Currently, four zones can be distinguished (Chapter 9, Figure 9.1):

1. Upper part: from 4.72 to 2.6 km from the confluence of the Biala River with Biala Przemsza. The water level of the main river bed is below the ground level. Only one side branch (from 4.7 to 2.45 km from the Biala River mouth) exists beside the main river bed. The main stream is strongly tortuous, and meandering is prevented only by trees growing at the sides in the lower section. Here, processes of secondary erosion dominate, causing more flow in the main stream and gradual disappearance of the branches.
2. Central part: from 2.6 to 2.2 km above the mouth of the Biala River. Water table and the ground are approximately on the same level. Water released from the mill raises the water table, causing overflow from the central river bed toward the right side among the trees and tufts of sedge. This area is very soft and reached with difficulty. Accumulation and erosion processes may be in balance here.
3. Lower part: from 2.2 to 0.73 km from the river mouth. The water level is above the ground level and this area is practically inaccessible in summer. There is no recognizable central stream channel. Accumulation processes dominate. Near a bridge in Kuzniczka village (about 1 km from the river mouth) two streams appear, converging into one.
4. Last part: from 0.73 km to the river mouth. Phenomena and processes are like those in the upper part, except where a row of old wooden piles causes water dissipation among the tufts of sedges. The mouth of the Biala River joins the Biala Przemsza in one concentrated bed.

The average time for water to flow through the wetland was determined by tracing salt (NaCl) to be about 330 min. Mean velocity varied between 0.2 and 0.7 m/s.

As the result of groundwater extraction by the "Boleslaw" Mining and Metallurgical Works, the groundwater table has been lowered to more than 100 m below the ground surface in the Pomorzany mine area. Consequently, the natural flow in the Biala River and the Ponikowska Adit disappeared. The Dabrowka Channel receives waterwater from the Pomorzany mine and effluents from the municipal wastewater treatment plant and water treatment plant in Olkusz City. An appearance of natural runoff only occurs after heavy rains or periods of extensive thaws.

HISTORY OF THE BIALA RIVER WETLAND

The story of the effect of human activities on this area is a long one. The Ponikowska Adit canal was built in the 16th century to discharge mine water from the ore deposits in the Olkusz region into the Biala River at Laski. The varying flows were usually equal to or greater than the natural flow of the Biala River. The water diversion decreased the groundwater table to 313 m above sea level.

Historical research located data on the flow in the Biala River since the year 1880 (Chapter 9, Figure 9.2). Increased flow reached a maximum about the year 1910, as much as four times greater than the natural flow of the river. (Natural flow was estimated to be about 20 cm^3/min — approximately equal to the quantity of two main springs.) Starting in 1880 flow was greater than the river bed could hold and flooded the valley. High flow continued for about 30 years, from 1880 to 1920. In the period of 1920 to 1960 the total runoff through the Biala River bed

Table A9.1   Water Flow in the Biala River Wetland Since 1955 (in m^3/s)

Year	Natural Flow Biala River	Natural Flow Ponikowska Adit	Mine Water	Tailing Water	Municipal Wastewater	Total
1955	0.36	0.12	0.12			0.60
1956	0.30	0.12	0.12			0.54
1957	0.30	0.12	0.08			0.50
1958	0.30	0.12	0.02	0.06		0.50
1959	0.30	0.12	0.08	0.06		0.56
1960	0.30	0.05	0.09	0.06		0.50
1961	0.30	0.05	0.33	0.06		0.74
1962	0.30	0.05	0.35	0.05		0.75
1963	0.24	0.05	0.88	0.05		1.22
1964	0.20	0.05	1.17	0.06		1.48
1965	0.16	0.05	1.10	0.06		1.37
1966	0.16	0.05	0.18	0.05		0.44
1967	0.16	0.05	0.09	0.07		0.37
1968	0.16	0.04	0.15	0.04		0.39
1969	0.16	0.04	0.09	0.05		0.34
1970	0.16	0.04	0.08	0.08	0.05	0.41
1971	0.11	0.02	0.07	0.12	0.06	0.38
1972	0.19	0.02	0.20	0.11	0.06	0.58
1973	0.18		0.43	0.13	0.01	0.75
1974	0.05		1.91	0.13		2.09
1975	0.02		2.70	0.14	0.04	2.90
1976	0.01		2.69	0.15	0.09	2.94
1977			2.22	0.10	0.08	2.40
1978			2.23	0.09	0.09	2.41
1979			2.43	0.11	0.10	2.64
1980			2.08	0.03	0.12	2.23
1981			2.13	0.04	0.13	2.30
1982			2.10	0.13	0.13	2.36
1983			2.36	0.12	0.13	2.61
1984			2.17	0.13	0.17	2.47
1985			2.06	0.15	0.15	2.36
1986			2.19	0.08	0.19	2.46
1987			2.00	0.01	0.20	2.21
1988			1.84	0.11	0.18	2.13
1989			1.98	0.09	0.16	2.23
1990			1.86	0.12	0.16	2.14
1991			1.85	0.14	0.22	2.21
1992			1.80	0.10	0.29	2.19

(natural runoff and waters disposed into the river) was reduced to about one third to one half of the earlier flow. Flow of mine wastewater from the Olkusz mine increased through the Roznos Channel 1961 to 1965 (Table A9.1 and Chapter 9, Figure 9.3).

After World War II, waters were contributed from the Boleslaw mine through the Channel Boleslaw, from the Olkusz mine through the Roznos Channel, and from sedimentation basins of flotation tailings. In the 1970s waters from the Olkusz sewage treatment plant, after mechanical and biological stages of treatment, and wastewaters from the Pomorzany mine were added. At present, all these waters and sewage are carried together into the Biala River through the Dabrowka Channel. Periodically, in periods of drought, small amounts of water are supplied through the Ponikowska Adit to private fish ponds in Laski. Due to intensive pumping at the Pomorzany mine, the sources and flows of the Biala River and its confluents disappeared in 1975 along the section to Laski, except after exceptionally heavy rains and spring thaw. Thus, the waters running through the wetland are chiefly those from the Dabrowka Channel together with several small seepages on the left-hand side of the river and flat alluvial cones within 50 to 100 m of the talweg of the Biala River. Because of the geological formation with loess over permeable sand and Triassic formations, neither runoff from the left nor right side is great.

A hydrometric current meter was used to measure water velocity, and the Cullman method was applied to calculate the water flow. From 1989 to 1993, little difference was observed between flow at the upper and lower ends of the wetland, whereas measurements in 1978 showed greater difference (Table A9.2). In 1989 the temporary flow at the beginning of the wetland was 2.3 to 2.19 cm³/s, and at the end of the wetland 2.0 to 2.15 cm³/s. These results showed that infiltration and evapotranspiration from the wetland slightly exceeded the runoff in normal atmospheric conditions. Field inspections confirmed the lack of water draining from the southern bank. On the other hand, the left northern bank of the valley was supplying the wetland with water from numerous springs continuously. Also, small valleys and ravines carried water occasionally. An inventory was made of 16 small springs and water courses that are permanent or potential sources of the runoff. In normal conditions the runoff from the northern side is slight and dispersed. However, with extended or heavy rains, an increased runoff can be expected because of the morphology of the left bank and the groundwater at the surface. Especially in the upper part of the valley, the wetland formed at the left bank is sologenic (classified as low spring peat marsh), while all the rest is fluviogenic (classified as a low flooded peat marsh).

Table A9.2   Flow of the Water on the Upper and Lower End of the Biala River Wetland in 1978, 1989, and 1993

Year	Beginning		End	
	Flow (m³/s)	Average velocity (m/s)	Flow (m³/s)	Average velocity (m/s)
1978[a]	2.80	1.20	4.10	0.30
1989 April 26	2.30	0.65	2.00	0.49
1989 May 31	2.25	0.70	2.15	0.55
1989 December 19	2.14	0.54	2.12	0.53
1993 November 15	2.10	0.51	2.25	0.59

[a] Historical data.

RIVER LEVEL

The water table in the main stream of the river was measured from the bridge in Kuzniczka. Variation in level at this station did not exceed 3 cm, i.e., about 5% of the depth. Apparently, few variations in water flows due to weather were reaching the river.

TIME FOR WATERS TO FLOW THROUGH THE WETLAND

The time for water to flow through the wetland was studied by tracing salt (NaCl was added upstream). Stations for measuring the tracer were the same stations used for studying water quality along the wetland (Chapter 9, Figure 9.4c). The times of maximum salt concentration were located on graphs of salt with time at each station. Rate of flow was estimated using the differences in time between station maxima (Table A9.3). Time of flow through the entire wetland was 336 min, with an average velocity of 0.25 m/s. Highest velocities were observed between the points P1-P2, P3-P4, and P6-P7 (Chapter 9, Figure 9.4c).

Table A9.3   Times for Tracer Flows across the Wetland

Section Beginning	Section End	Length of Section (m)	Time of Flow (min)	Average Velocity (m/s)
P1	P2	1240	46	0.45
km 3+420	km 3+900			
P2	P3	750	42	0.30
km 3+900	km 3+150			
P3	P4	750	29	0.43
km 3+150	km 2+400			
P4	P5	1040	152	0.11
km 2+400	km 1+360			
P5	P6	630	38	0.28
km 1+360	km 0+730			
P6	P7	730	29	0.42
km 0+730	km 0+000			
Entire wetland		5140	336	0.25

Note: See Figure 9.1 in Chapter 9 for sampling point location.

FUTURE FLOW

Whereas the amount of mine waters starting in the year 2000 is expected to be considerably smaller, the total flow may be greater because of an increased amount of municipal wastewater (Chapter 9, Figure 9.17). However, the amount of municipal wastewater from Olkusz may not increase. The project to expand the wastewater treatment plant and channel from the plant to the Dabrowka Channel has not yet been approved. It is possible that the wastewater will be sent to another catchment area.

QUALITY OF WASTEWATERS FLOWING INTO THE WETLAND

To study quality of discharged water, water samples were collected two or three times a month at the outlet of the Dabrowka Channel since 1977. In the years 1978 and 1979 sampling was made by the Polish Academy of Sciences and in 1989 and 1990 as a part of this project. The highest concentrations of total solids, zinc, and lead were observed in 1979: up to 3543, 37.8, and 31.6 g/m^3, respectively (Table A9.4 and Figures 9.13 to 9.15). Later, the concentration of substances gradually decreased.

The hourly variation of discharge substances was examined in the Dabrowka shaft from which mine water flows (Table A9.5). Periodic increase of concentrations of zinc and lead occurred (hours 4, 7 to 10, 12 to 14, and 19 to 23). Because changes in total suspended solids (TSS) coincided with changes in concentrations of zinc and lead, it appeared that these metals were in suspended solids and colloidal forms (see Chapter 9, Figure 9.13). To prove it, wastewater was filtered in the first stage through laboratory paper and in the second stage through a ceramic filter (Table A9.6). These experiments showed that the ionic forms of zinc and lead are small.

Table A9.4 Quality of Wastewater Discharged to the Biala River Wetland 1977–1988 — Historical Data

Year		pH	TSS (mg/dm³)	Zn (mg/dm³)	Pb (mg/dm³)	SO₄ (mg/dm³)	Fe (mg/dm³)	Mn (mg/dm³)	Hardness Total (mval/dm³)
1977	Average	7.55	540	1.75	1.26	116.2	19.95	1.88	5.48
	range	7.2–8.0	tr–2278.0	nd–7.95	nd–4.80	98.80–134.0	0.86–111.0	nd–8.00	3.25–6.13
1978	Average	7.55	540	1.75	1.26	116.2	19.95	1.88	5.48
	range								
1979	Average	7.88	951	11.87	8.71	116.61	12.48	—	5.47
	range	7.4–8.5	0.0–3543.0	0.0–37.80	0.00–31.6	65.00–193.0	1.25–58.00	0.00–0.34	4.07–6.60
1984	Average	8.20	68.39	2.30	0.94	135.20	4.12	0.22	6.02
	range	7.80–8.50	6.00–240.4	0.85–7.45	0.44–3.12	9.13–212.8	0.95–13.36	0.09–0.54	4.92–9.49
1985	Average	8.22	33.89	2.38	0.72	132.99	3.45	0.34	7.36
	range	8.10–8.80	0.50–129.8	1.31–6.42	0.31–6.42	2.01–220.0	0.64–16.70	0.21–0.41	5.85–9.45
1986	Average	8.06	77.07	2.32	0.87	153.54	5.13	0.36	7.2
	range	7.40–8.65	6.00–245.0	0.81–8.01	0.32–72.20	2.67–261.2	1.56–112.4	0.20–1.07	5.35–8.88
1987	Average	8.25	59.45	1.78	0.68	125.29	3.30	0.45	7.76
	range	7.40–8.50	9.80–462.5	0.72–11.86	0.32–72.20	2.20–226.8	1.17–30.89	0.16–0.75	0.08–9.66
1988	Average	8.23	145.10	2.38	7.49	111.87	12.87	0.44	7.67
	range	8.20–8.30	5.20–337.8	1.00–3.37	0.80–20.50	4.0–117.0	2.87–29.74	0.35–0.61	6.95–8.52
1989	Average	8.18	58.00	1.95	0.51	190.20	4.78	0.27	7.66
	range	7.70–8.40	0.00–532.0	0.79–5.95	0.20–1.78	141.0–235.0	0.47–27.7	0.14–0.55	6.56–8.88
1990	Average	8.25	48.00	1.65	0.61	207.30	4.63	0.32	8.20
	range	7.95–8.50	5.00–110.0	0.59–4.61	0.16–1.56	154.0–276.0	2.02–11.50	0.40–0.50	6.85–9.56

Hardness Carbonate (mval/dm³)	Alkalinity (mval/dm³)	TDS (mval/dm³)	Chlorides (mg/dm³)	Lignin Sulfonate (mg/dm³)	COD (mg/dm³)	COD KMnO₄ (mg/dm³)
	3.30		21.25	2.22	201.90	
	3.00–3.70		9.0–30.0	1.20–4.00	22.90–810.3	
	3.30		21.25	2.22	201.90	
	3.48		23.25	5.29	—	
	3.10–3.90		16.0–36.0	2.30–7.60	—	
4.22	3.99	471	21.18	6.22	29.03	34.78
2.99–4.42	3.00–4.25	386–598	17.73–20.0	2.00–12.50	12.90–41.56	31.28–44.24
4.17	4.18	481	20.76	2.55	22.79	37.60
3.49–4.41	3.50–4.45	373–686	17.73–26.5	1.00–4.00	11.29–29.67	30.97–48.66
4.19	4.20	513	21.62	2.76	26.09	41.67
3.87–4.59	3.90–4.50	392–645	17.73–24.8	trace–4.50		24.02–56.65
4.45	4.46	503	19.53	2.40	23.04	40.61
3.79–4.89	3.80–4.90	361–670	0.96–26.59	1.50–4.00		33.18–50.56
4.56	4.56	498	17.43	1.73		37.49
4.49–4.63	4.50–4.70	495–502	15.96–18.6	1.50–2.00		32.23–40.76
4.19	4.32	561	19.41	3.00		39.54
4.10–4.89	4.10–4.50	428–688	12.40–24.8	1.50–4.10		27.80–45.50
4.33	4.33	625	16.50	2.20	26.62	38.57
4.00–4.79	4.10–4.50	574–692	11.50–2.22	0.80–3.70	19.62–33.0	28.44–44.87

Table A9.5   Hourly Changes in Mine Wastewater Quality — October 19, 1989

Sampling Time	pH	Zn (g/m³)	Pb (g/m³)
6:15	8.05	1.42	0.36
7:15	7.90	4.20	3.71
8:15	7.90	3.20	2.84
9:15	7.90	2.17	0.71
10:15	7.85	3.03	1.32
11:15	7.90	1.97	0.75
12:15	7.85	1.92	0.64
13:15	7.90	2.29	0.89
14:15	7.85	1.47	0.50
15:15	7.85	1.81	0.18
16:15	7.85	1.15	0.28
17:15	8.00	1.32	1.32
18:15	7.95	1.29	0.32
19:15	7.90	2.32	0.96
20:15	7.95	1.22	0.53
21:15	7.90	1.82	0.50
22:15	7.95	1.44	0.53
23:15	8.00	6.69	0.93
24:15	8.00	2.70	0.28
1:15	8.00	2.17	0.39
2:15	8.00	1.73	0.21
3:15	7.80	1.44	0.39
4:15	7.85	1.64	2.07
5:15	7.90	1.29	0.32

Table A9.6   Filtration of Wastewater from Dabrowka Channel

	Date	Raw Wastewater	After Filtration through 3 Layers of Laboratory Paper	After Filtration through Ceramic Filter
Zn	6/01/1989	1.67	0.21	0.16
	6/22/1989	1.23	0.35	0.10
	6/23/1989	42.72	0.57	—
Pb	6/01/1989	0.34	0.04	0.04
	6/22/1989	0.27	0.12	0.04
	6/23/1989	21.18	0.46	—

METEOROLOGICAL CONDITIONS

Precipitation

Table A9.7 shows the amounts of precipitation in the neighboring stations in 1974 to 1987 ranging from 716 to 800 mm. The lowest annual precipitation was 500 to 600 mm; the highest from 972 to 1191 mm. The highest monthly precipitation varied from 162 to 201 mm.

Table A9.7   Rainfall for Stations Located Near the Biala Wetland 1974–1987 (in mm)

Station	Average Annually	Lowest Annual/Year	Highest Annual/Year	Lowest Monthly/Date	Highest Monthly/Date
Boleslaw	800.0	547/1984	1191/1974	6/Mar/1974	201/June/1974
Chechlo	733.0	500/1984	1000/1974	3/Mar/1974	180/July/1985
Golczowice	773.0	600/1982	986/1974	1/Mar/1974	162/July/1985
Ryczow	732.8	509/1984	1040/1974	1/Mar/1974	177/June/1974

Air Pollution Wastes

Air pollution wastes come to the Biala River area from the industrial plants of the Silesia situated in the west (mostly from the Katowice Still Mill) and partly from the south, from the mining and metallurgical complex in Bukowno, and from the industrial plants of Trzebinia. The wind rose for the town of Olkusz, situated at a distance of about 8 km from the wetland, shows winds prevailing from the west (Figure A9.1).

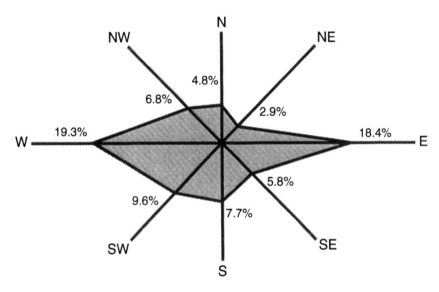

Figure A9.1   Wind rose for Olkusz City.

The air samples were collected by the SANEPID in three stations, each located at a distance of about 1 km from the wetland. The mean fallout of dust in the period 1985 to 1987 for these three localities was 125 tons/km²/year (Table A9.8). The mean concentration of dust suspended in the air was 186 µg/m³ of air. The mean fallout of lead was 116 mg/m³/year, that of cadmium 8.6 mg/m³/year, and that of zinc as much as 1838 mg/m³/year. These values are rather high, but they are small compared with the amounts of pollutants flowing into the wetland from the mine waters.

Table A9.8 Average Fallout of Dust, Lead, Zinc, and Cadmium on the Biala River Wetland, 1985–1987

Station	Dust (t/km²/year)	Suspended Dust (µg/m³)	Pb (mg/m²/year)	Zn (mg/m²/year)	Cd (mg/m²/year)
Laski	153	198	151	3589	10.6
Krzykawa	134	190	140	1305	13.4
Rudy	87	170	56	620	1.9
Average from stations	125	186	116	1838	8.6

GEOLOGICAL CONDITIONS

Quaternary, Triassic, and Permian geological formations are found at the surface in this area.
The Quaternary period formation mainly contains a few meters of medium- and fine-grained sands and gravels of fluvioglacial origin of various thickness with insertions of silt, clays, mire,

and organic soil. From place-to-place in the bottom part, dolomite-sandstone rock waste and, sporadically, dust and mud formations and quartz gravel occur.

The Quaternary formation in the study area was surveyed in detail with 29 bore holes 6 to 15 m deep by the Hydrogeological Enterprise at Krakow May 4 to June 12, 1971 in connection with the flood control project of the Biala River. Cores showed the surface soil 0.2 to 0.6 m thick, and peaty sediment 0.6 to 3.2 m thick. The loose ground below has high permeability.

Below, the Upper Triassic-Keuper formation has low permeability mud and argillites of characteristic red and cherry-red coloring with insertions of sandstone, limestone, and dolomite. The mean thickness of these formations together is about 30 m (maximum of 100 m).

Middle Triassic-Muschelkalk formation is represented by the dolomites, limestones, and marls. In the top part of the Muschelkalk there are diplopore dolomites up to 30 m thick with a substratum of the ore-bearing dolomites up to 90 m thick. The bottom part of Muschelkalk is represented by the Gogolin Bed containing pellitic limestones, fine-crystalline marls, or conglomerates up to 20 m thick.

In the top part of the Lower Triassic-Mottled sandstone, marine formations of the Roethian period occur in the form of fine-grained or zoolithic dolomites and dolomite marls about 30 m thick. Below are mottled, various grained sandstones of gray color up to 3 m thick. Underneath the mottled sandstone, or immediately under the Roethian, are the Rottigendes, formations of the Lower Permian period, 100 to 200 m thick. In the top part they developed as mottled siltstones or sandstones with underlying conglomerates.

The ore-bearing deposits in the Muschelkalk and Roethian formations, 1.5 to 30 m thick (4 m average), were originally crystaline limestone transformed into dolomite. The deposits contain 2.1 to 13.7% zinc and 0.1 to 11.6% lead.

About 60% of the ores contain zinc and lead in the form of sulfides (for example, galena), and about 30% in the form of zinc and lead oxides. About 10% of the ores contain minerals of zinc and lead.

In the examined area one can distinguish Quaternary and Triassic water-bearing horizons and a paleozoic water-bearing complex connected with a poorly water-bearing Permian formation of the Rotliegendes series. Where Quaternary sands are deposited directly on impermeable Keuper silts of considerable thickness, the Quaternary and Triassic water-bearing horizons can be distinguished. There is one common Quaternary-Triassic water-bearing horizon in the areas where the Keuper formations underwent erosion, and the Quaternary packet rests directly on the Muschelkalk or the Roethenian. The Quaternary sands are easily permeable, with a filtration coefficient determined in the laboratory to be about $k = 0.0285$ cm/s. The Keuper formations are poorly permeable and isolate waters of the Triassic horizon in the region about 500 m upstream from the Kuzniczka bridge. The Quaternary formations at this place are lying on a Keuper layer only 3 m thick.

ANALYSES OF SOIL AND ACCUMULATED SEDIMENTS

With samples collected from wetland stations (Chapter 9, Figure 9.4a), physicochemical properties of the soil and sediments were analyzed as follows:

1. The mechanical composition was determined by the Bouyoucosa-Casagrande areometric method.
2. The content of hydroscopic water was determined according to Polish Standard PN-55/B-04487.
3. pH reaction was measured by means of an electronic pH-meter.
4. The content of organic matter was determined by baking a dried sample at 550°C until the weight stabilized, in this way oxidizing the organic matter.
5. Specific weight was determined according to Polish Standard PN-55/B-04486.
6. Bulk density was determined according to Polish Standard PN-55/B-04488.
7. Porosity was determined according to Polish Standard PN-55/B-04488.
8. The content of elements in the samples was determined using an atomic adsorption spectrophotometer, Perkin-Elmer Models 403 and 2380. Samples were dried and mineralized by digesting with $HClO_4$ and HNO_3.

Table A9.9a   Physical and Chemical Characteristics of Soil from the Biala River Wetland

Sample Number	Depth (cm)	Fraction Contents (in %) of Diameter (in mm)						Weight Density (g/cm³)	Bulk Density (g/cm³)	pH H₂O	pH KCl	Organic Matter (%)
		1.0–0.1	0.1–0.05	0.05–0.02	0.02–0.006	0.006–0.002	>0.002					
31a	0–30	48	20	18	8	6	0	2.11	0.52	5.9	5.8	
31b	30–70	75	7	8	2	5	3	2.26	0.87	6.0	5.7	
31c	70–80	51	10	12	10	5	12	2.62	1.53	6.4	6.3	
32a	0–30	14	6	6	37	25	12	2.87	1.21	7.5	7.5	4.21
32b	30–50	66	14	10	6	2	2	2.32	0.79	7.2	7.2	19.34
32c	50–80	75	7	5	5	3	5	2.62	1.65	6.0	5.6	2.31
32c	0–30	20	6	22	42	9	1	2.97	1.28	7.7	7.7	2.62
33a	30–60	13	4	7	36	26	14	2.93	1.28	7.4	7.4	2.36
33b	60–90	15	7	24	31	15	8	2.79	1.18	7.4	7.2	5.73
33c	0–10	0	14	33	39	10	4	2.93	1.36	7.5	7.5	4.33
34a	10–20	4	10	36	34	10	6	3.02	1.54	7.8	7.8	0.76
34b	20–60	3	7	11	42	23	14	2.82	1.21	7.6	7.5	2.96
34c	60–100	13	12	26	28	17	4	2.75	1.13	7.5	7.3	7.87
34d	0–25	59	14	19	4	2	2	2.28	0.91	6.4	5.9	23.86
35a	25–75	14	22	40	10	5	9	2.64	1.46	6.8	6.5	1.38
35b	75–150	88	6	3	0	1	2	2.65	1.76	6.9	6.9	0.20
35c	0–15	13	7	8	38	25	9	2.85	1.20	7.5	7.5	2.78
36a	15–55	11	2	8	34	31	14	2.89	1.28	7.6	7.6	0.69
36b	55–80	28	10	16	16	18	12	2.54	0.84	7.4	7.2	11.79
36c	80–130	33	14	30	11	0	12	2.37	0.99	6.5	6.2	20.48
36d	0–40	51	31	12	3	2	1	3.09	1.75	7.5	7.3	1.91
37a	40–75	29	4	19	25	15	8	2.61	0.33	6.8	6.8	7.00
37b	75–100	90	3	2	0	1	4	2.61	1.44	7.0	6.6	2.06
37c	0–30	14	33	30	17	5	1	3.02	1.72	7.5	7.5	1.46
40a	30–70	10	6	19	42	15	8	3.01	1.43	7.5	7.4	0.57
40b	70–100	40	18	26	10	0	6	1.94		6.7	6.5	46.61
40c	0–30	0	15	47	30	5	3	2.98	1.62	7.8	7.8	0.62
41a	30–60	8	5	21	40	17	9	3.06	1.45	7.7	7.7	0.43
41b	60–120	38	20	24	10	0	8	2.17	0.68	6.5	6.2	30.78
41c	0–15	22	28	27	16	6	1	3.01	1.63	7.5	7.5	2.65
42a	15–70	36	16	20	14	12	2	2.54	0.98	7.0	6.8	13.66
42b	70–150	50	45	2	2	1	0	2.63	1.44	7.5	6.2	1.50

continued

Table A9.9a (continued) Physical and Chemical Characteristics of Soil from the Biala River Wetland

Sample Number	Depth (cm)	Fraction Contents (in %) of Diameter (in mm)						Weight Density (g/cm³)	Bulk Density (g/cm³)	pH H₂O	pH KCl	Organic Matter (%)
43a	0–10	15	18	38	14	6	9	2.57	1.32	5.6	5.1	4.69
43b	10–50	11	13	47	14	4	11	2.63	1.47	6.1	5.8	1.12
43c	50–70	9	12	44	19	7	9	2.67	1.54	5.1	4.1	0.59
43d	70–110	8	4	25	28	7	28	2.67	0.57	5.0	3.8	1.25
45a	0–15	77	8	5	2	5	3	2.48	1.17	7.0	6.7	10.50
45b	15–25	95	2	2	1	0	0	2.64	1.71	7.6	6.7	0.50
45c	25–50	97	1	1	0	1	0	2.65	1.74	7.5	6.8	0.20
46a	0–5	88	3	4	4	1	0	2.12	0.72	6.7	6.2	33.86
46b	5–15	33	8	3	25	20	11	2.90	1.14	7.2	7.0	4.07
46c	15–60	88	3	4	5	0	0	2.43	1.25	7.4	7.1	10.93
47a	0–25	72	9	8	6	4	1	2.29	0.75	6.0	5.6	2.27
47b	25–50	73	8	8	5	3	3	2.59	1.32	5.9	5.9	4.43
47c	50–80	83	4	4	1	4	4	2.64	1.59	6.8	6.1	1.19
48a	0–10	74	11	9	4	0	2	2.41	1.13	6.0	5.4	15.90
48b	10–60	93	4	1	2	0	0	2.59	1.54	6.3	6.3	1.09
48c	60–80	93	2	1	1	0	3	2.64	1.61	6.2	6.2	0.30
49a	0–15	81	11	3	2	1	2	2.57	1.30	5.6	5.0	4.10
49b	15–30	83	9	3	1	2	2	2.65	1.41	5.5	4.8	1.96
49c	30–50	95	4	1	0	0	0	2.64	1.62	6.4	5.8	1.32

Table A9.9b   Physical and Chemical Characteristics of Soils from the Biala River Wetland

Sample Number	Depth (cm)	Weight Density (g/cm³)	Bulk Density (g/cm³)	Porosity (%)	Organic Matter(%)	Hygroscopic Water	
						Value (%)	Maximum (%)
47a	0–25	2.29	0.75	67.2	28.27	5.9	13.5
47b	25–50	2.59	1.32	49.0	4.43	1.0	2.4
47c	50–80	2.64	1.59	39.8	1.19	0.3	0.9
48a	0–10	2.41	1.13	53.1	15.90	3.3	7.2
48b	10–60	2.59	1.54	40.5	1.09	0.2	0.5
48c	60–80	2.64	1.61	39.0	0.30	0.1	0.3
49a	0–15	2.57	1.30	49.4	4.10	0.7	1.6
49b	15–30	2.65	1.41	46.8	1.96	0.5	1.1
49c	30–50	2.64	1.62	38.6	1.32	0.1	0.2
50a	50–100	2.82	1.29	54.2	5.50	2.4	4.2
50b	100–150	2.57	1.31	49.0	6.00	2.2	3.8
50c	150–200	2.65	1.50	43.4	1.60	0.8	1.0
51a	50–100	2.72	1.56	42.6	2.40	1.0	1.8
51b	100–150	2.65	1.76	33.6	0.30	0.1	0.2
51c	150–200	2.62	1.62	38.2	0.90	0.3	0.6
52a	50–100	2.82	1.41	50.0	1.90	1.3	2.8
52b	100–150	2.53			9.70	2.8	4.7
52c	150–200	2.61	1.47	43.7	2.10	0.8	1.5
53a	50–100	2.67	1.63	38.9	1.40	0.4	0.7
53b	100–150	2.66	1.72	35.3	0.10	0.1	0.2
53c	150–200	2.65	1.84	30.6	0.20	0.2	0.3
54a	50–100	2.69	1.23	54.3	6.00	1.8	3.9
54b	100–150	2.72	1.54	43.4	1.80	0.9	1.7
54c	150–200	2.66	1.73	35.0	0.10	0.1	0.1
55a	50–100	2.66	1.51	43.2	1.30	1.5	2.8
55b	100–150	2.64	1.70	35.6	0.30	0.1	0.3
55c	150–200	2.60	1.67	35.8	0.40	0.2	0.4
56a	50–100	2.87	1.69	41.1	1.00	0.3	0.5
56b	100–150	2.64	1.54	41.7	2.90	0.7	1.3
56c	150–200	2.62	1.43	45.4	2.60	0.6	1.2
57a	5–15	2.90	1.35		1.50	1.3	2.7
57b	40–50	2.86	1.53		0.90	0.5	1.8
57c	60–80	2.86	1.51		0.90	0.6	1.8
58a	10–20	2.69		68.0	12.10	3.6	8.7
58b	30–50	2.51	1.10	60.5	11.50	3.7	7.9
59	10–20	2.50	1.30		9.00	2.4	5.3
60a	10–20	2.92	1.45	50.3	2.70	1.0	2.4
60b	30–50	2.65		35.8	0.20	0.0	0.1
61a	15–30	2.77	1.32	52.3	7.50	1.4	3.3
61b	50–60	2.63	1.52	42.2	2.50	0.9	2.1
62a	10–20	2.82	1.76	35.1	0.8	0.2	0.6
62b	20–30	2.75		28.7	0.5	0.1	0.1
62c	30–50	2.90		42.4	0.5	0.3	0.8
63a	50–100	2.63			6.1	1.8	4.0
63b	100–150	2.66			3.8	1.5	3.6
63c	150–200	2.56			5.8	1.4	3.2
64a	50–100	2.60			5	0.9	2.3
64b	100–150	2.64			2.1	0.4	1.2
64c	150–200	2.64			0.7	0.2	0.4
65a	50–100	2.66	1.70		0.2	0.1	0.2
65b	100–150	2.66	1.77		0.2	0.0	0.2
65c	150–200	2.67	1.63		0.4	0.0	0.2
66a	50–100	2.61			4.0	0.8	1.8
66b	100–150	2.64	1.57		1.7	0.5	1.0

continued

Table A9.9b (continued) **Physical and Chemical Characteristics of Soils from the Biala River**

Sample Number	Depth (cm)	Weight Density (g/cm³)	Bulk Density (g/cm³)	Porosity (%)	Organic Matter(%)	Hygroscopic Water	
						Value (%)	Maximum (%)
66c	150–200	2.67	1.64		1.3	0.4	1.0
67a	50–100	2.62	1.55		2.5	0.4	1.1
67b	100–150	2.63	1.60		1.7	0.4	1.0
67c	150–200	2.62	1.62		0.9	0.2	0.5
68	0–20	2.84	1.34		4.2	1.3	2.9
69	0–20	2.77	1.11		5.7	1.3	3.7
70	0–20	2.80	1.13		5.9	1.2	3.7
71	0–20	2.78	1.16		5.6	1.5	4.0
72	0–20	2.80	1.17		6.2	1.4	3.9
73	0–20	2.87	1.39		3.0	1.0	2.6
74	0–20	2.82	1.06		4.8	1.6	4.4
75	0–15	2.95	1.54		2.0	0.6	1.5
76	0–15	2.93	1.46		3.7	0.7	1.8
77	0–20	2.91	1.27		1.9	0.9	2.3
78	0–20	1.86			46.6	7.2	20.8
79	0–20	2.31	1.59		24.5	5.1	14.1
80	0–20	2.41	0.86		15.8	2.7	7.2
81	0–20	2.71	1.00		7.3	2.4	7.0
82	0–20	2.60	1.01		10.2	2.3	7.0

Table A9.9 contains analyses of the physical and chemical properties of sediment samples taken from various depths. The shallower layers contained high density sediments (up to 3.09 g/cm³) where the percentage of organic matter was small. Sediments were up to 7.8 pH in KCl. The grain composition was uniform silty sand to silt (Figure A9.2). These results indicate that sediments from the Dabrowka Channel were deposited on the wetland. The thickness of accumulated sediments ranges from 0 to 150 cm. Their accumulation is mostly evident along the section from the outlet of the Dabrowka Channel to about 1.5 to 2.0 km down the valley. There was very little deposit in the water-logged lower parts of the valley, and particularly in stations situated close to the valley banks.

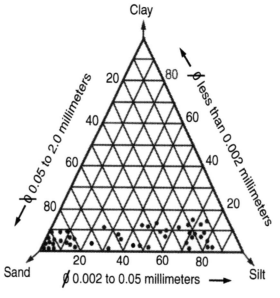

Figure A9.2   Texture of soil samples shown by plotting data for the percents of sand, silt, and clay on triangular coordinates.

X-RAY DIFFRACTION STUDIES

Mineral particles were studied with X-ray diffraction. The measurements were carried out using an ORON-2 X-ray diffractometer (Table A9.10). Ground material was pressed in a metal frame, and orientation improved by removing the outside layer. The measurements were performed within angular variation of 3 to 60° or 3 to 50°.

From the X-ray photographs the absorption coefficient was estimated, although quantitative results from this method are only approximate. The relative error ranged from 10 to 40% with greater error for smaller quantities. For contents less than 0.5% the determinations should be regarded as uncertain.

Accurate analysis of sphalerite (zinc sulfide) and galenite (lead sulfate) was difficult due to the superposition of many X-ray reflections of these minerals. The clay minerals present in most samples were small in quantity and were not examined in detail. These included illite, mixed-pack minerals, chlorite, and montmorillonite.

High dolomite content was found in samples collected from the upper part of the wetland (samples no. 33 to 41). Not much galenite was recorded, since the percent content of lead in the sediments was small, as also shown by chemical investigations. However, X-ray analysis can be used only to detect crystals and does not reveal amorphous forms of elements.

The following characteristic groups of sediments were found:

1. Most common are clastic components (quarts, feldspar) without carbonates and sulfides or containing only their traces. These samples contain a little more clay minerals. Examples here are samples no. 48b, 48c, 50c, 54c, and 55a.
2. Samples containing mainly carbonates (first of all dolomite — also calcite as a secondary component) were accompanied by small amounts of sulfides (pyrite, marcasite, sphalerite, and galenite) and usually a small amount of quartz. Two samples contained gypsum (Stations 33 and 50).
3. Samples containing both clastics and carbonate-sulfide components were found in the top, most recent deposits at Stations 51, 52, 54, and 56.
4. Samples with considerable quantities of undetermined amorphous materials probably contained more organic substance (observed microscopically in samples from Stations 36 and 41).

ZINC AND LEAD DISTRIBUTION IN SEDIMENTS

High concentrations of heavy metals were found in the sediments with zinc, lead, and cadmium — as much as 4.46, 1.34, and 0.02%, respectively (Table A9.11), particularly in the upper part of the wetland (Stations 32 to 34; see Chapter 9, Figure 9.6a). High concentrations were found to a depth of 1.5 m (Stations 33, 34, 36, 37, 40, 41, 42, 50, 52, 54, 56, 57, 61, 62, 63, 68, 69, 70, 71, 72, 73, 74, 76; see Figures A9.3 (a to g) and Chapter 9, Figure 9.6). Concentrations of the heavy metals generally decreased with depth in sediment. High concentrations of heavy metals in the deeper layer in some stations can be explained by sedimentation of the suspended solids in local sinks holes (Stations 58 and 81; see Figures A9.3a and A9.3d).

At some stations concentrations in the upper 10 to 15 cm were less than those deeper (see Stations 34, 36, 41, 46, 48, 57; Figures A9.3a, b, c, and d and Figure 9.6 in Chapter 9). In the 1970s, wastewaters deposited high quantities of suspended solids, zinc, and lead on the wetland, which were covered over later by deposits with less contaminants, followed by soil-building processes.

The total amount of deposited zinc and lead was calculated with the krigging method and the SURFER computer program, which showed that the weight of zinc and lead was 3927 and 1887 tons, respectively.

Table A9.10   Results of X-ray Analysis of Soil Samples

Number of Sample	Depth (cm)	Minerals (in %)								
		QUA	ORT	ALB	CAL	DOL	PIR	MAR	SPH	GAL
33a	0–30	6.1	0	0	4.6	61.5	2.8	5.6	0.9	0.6
33b	30–60	5.8	0	0	11.4	66.3	4.3	5.1	1.0	0.7
33c	60–90	6.6	0	0	2.9	61.1	4.3	4.2	0.4	0.5
36a	0–15	5.1	0	0	10.0	61.0	4.5	5.6	1.0	0.3
36b	15–55	6.0	0	0	12.0	67.2	4.3	5.1	0.5	0.6
36c	55–80	26.3	0	0	4.8	32.3	3.4	1.8	0.4	0.6
36d	80–130	47.7	9.8	3.0	0.7	3.4	0	0	0	0
41a	0–30	2.8	0	0	4.7	70.9	5.4	9.6	0.4	0.6
41b	30–60	5.2	0	0	9.0	66.0	6.3	9.2	0.9	0.7
41c	60–120	50.9	7.4	0	0	2.7	0.7	0.9	0	0
43a	0–10	62.0	7.6	3.2	0	0	0	0	0	0
43b	10–50	72.2	9.9	3.3	0	1.2	0	0	0	0
43c	50–70	67.3	12.1	4.3	0	0	0	0	0	0
43d	70–110	47.4	10.8	4.2	0	0	0	0	0	0
45a	0–15	75.4	3.1	0.8	0	3.1	0.5	0.6	0.2	0.3
45b	15–25	87.4	5.0	0.9	0	0	0.3	0.4	0	0
45c	25–50	86.4	4.9	0.7	0	0	0	0	0	0
48a	0–10	66.3	4.6	0.7	0	0	0.3	0.4	0	0
48b	10–60	84.7	8.3	1.1	0	0	0	0	0	0
48c	60–80	89.8	5.4	0.7	0	0	0	0	0	0
50a	50–100	7.4	0	0	73.3	0	5.1	5.6	0.7	0.6
50b	100–150	75.0	4.6	2.0	10.8	0	0.6	0.8	0	0

		QUA	ORT	ALB	CAL	DOL	PIR	MAR	SPH	GAL
50c	150–200	87.9	4.9	1.4	1.4	0	0.2	0.3	0	0
51a	50–100	60.1	0	0	24.7	6.6	2.4	2.9	0.5	0.4
51b	100–150	85.4	5.3	0.9	1.1	0	0.2	0.4	0	0.5
51c	150–200	88.5	5.8	0.9	1.2	0	0	0	0	0
52a	50–100	22.8	0	0	54.7	9.8	4.3	6.6	0.9	0.8
52b	100–150	65.2	6.7	1.6	20.0	2.1	1.0	1.3	0.4	0.4
52c	150–200	88.4	5.2	1.2	3.2	0	0.3	0.4	0	0
53a	50–100	77.1	4.4	1.1	9.0	1.4	0.9	1.0	0.3	0
53b	100–150	85.1	11.0	0.6	1.3	0	0.2	0.3	0	0
53c	150–200	87.2	5.7	1.5	1.0	0	0.2	0.3	0	0
54a	50–100	38.4	0	0	42.3	5.4	4.3	3.8	0.7	0.7
54b	100–150	55.3	4.1	1.0	30.0	5.0	1.9	1.9	0.6	0.5
54c	150–200	85.1	3.6	0.7	1.9	0	0.2	0.3	0	0
55a	50–100	83.0	9.6	3.2	0	0	0	0	0	0
55b	100–150	89.1	4.1	1.2	0	0	0	0	0	0
55c	150–200	85.2	8.4	0.7	0	0	0	0	0	0
56a	50–100	33.3	4.5	1.6	38.4	7.6	3.4	4.1	2.6	0.7
56b	100–150	73.9	5.4	2.0	9.5	1.2	0.7	0.7	0	0
56c	150–200	86.8	6.8	1.4	1.0	0	0.2	0.3	0	0

Note: QUA — quartz, ORT — orthoclase, ALB — albite, CAL — calcite, DOL — dolomite, PIR — pyrite, MAR — markasite, SPH — sphalerite, GAL — galena.

Table A9.11 Contents of Selected Elements in Sediments (in %)

Sample Number	Depth (cm)	Zn	Pb	Cd	Ca	Mn	Fe	S
32a	0–30	1.10	0.68	0.01	14.64	0.11	4.61	4.03
32b	30–50	0.17	0.10	0.003	1.23	0.02	0.42	0.59
32c	50–80	0.05	0.02	0.002	0.10	0.01	0.16	0.05
33a	0–30	1.04	0.7	0.010	14.11	0.11	5.35	5.18
33b	30–60	1.19	0.86	0.010	14.78	0.11	3.98	3.91
33c	60–90	1.04	0.62	0.005	13.36	0.27	5.09	3.41
34a	0–10	1.23	0.65	0.008	14.21	0.12	6.52	5.50
34b	0–20	0.83	0.54	0.007	16.14	0.13	6.39	6.64
34c	20–60	0.95	0.64	0.010	14.46	0.10	3.00	2.62
34d	60–100	0.96	0.54	0.004	13.00	0.16	4.56	2.91
35a	0–25	0.08	0.04	0.002	0.65	0.05	3.38	0.23
35b	25–75	0.01	0.02	0.002	0.09	0.01	0.58	0.07
35c	75–150	0.01	0.02	0.002	0.04	0.01	0.16	0.05
36a	0–15	1.23	0.76	0.010	13.86	0.11	5.35	3.90
36b	15–55	1.23	0.82	0.010	16.07	0.12	4.11	3.49
36c	55–80	1.5	0.72	0.010	6.53	0.07	4.04	2.63
36d	80–130	0.59	0.7	0.004	1.00	0.02	1.73	0.53
37a	0–40	4.46	0.67	0.002	9.68	0.08	8.09	9.73
37b	40–75	0.86	0.50	0.005	6.86	0.08	3.23	3.00
37c	75–100	0.19	0.04	0.002	0.64	0.01	0.58	0.64
40a	0–30	1.41	0.42	0.008	16.68	0.13	6.33	4.49
40b	30–70	1.39	0.88	0.001	15.82	0.13	6.24	5.40
40c	70–100	0.35	0.18	0.003	3.14	0.06	2.13	1.85
41a	0–30	0.65	0.32	0.005	17.00	0.13	4.33	6.45
41b	30–60	1.34	0.75	0.010	6.07	0.13	5.96	5.73
41c	60–120	0.34	0.20	0.002	1.34	0.02	1.45	0.81
42a	0–15	1.92	0.43	0.010	15.32	0.13	7.02	5.99
42b	15–70	1.20	0.64	0.010	6.96	0.10	3.55	1.76
42c	70–150	0.05	0.14	0.001	0.06	0.01	0.11	0.10
43a	0–10	0.03	0.02	0.001	0.10	0.02	0.65	0.10
43b	10–50	0.03	0.01	0.002	0.09	0.01	0.76	0.05
43c	50–70	0.01	0.03	0.001	0.01	0.02	0.95	0.05
43d	70–110	0.02	0.02	0.002	0.11	0.05	2.28	0.05
45a	0–15	0.28	0.07	0.002	1.19	0.03	0.98	0.49
45b	15–25	0.03	0.02	0.002	0.09	0.01	0.13	0.04
45c	25–50	0.02	0.02	0.001	0.04	0.01	0.12	0.04
46a	0–5	0.47	0.23	0.004	1.54	0.03	1.46	0.63
46b	5–15	1.31	1.12	0.010	13.78	0.13	6.22	3.95
46c	15–60	0.07	0.02	0.001	0.19	0.01	0.11	0.16
46d	60–80	0.02	0.02	0.001	0.06	0.01	0.05	0.06
47a	0–25	1.42	0.72	0.020	1.56	0.04	3.94	1.25
47b	25–50	0.21	0.06	0.002	0.30	0.01	0.48	0.17
47c	50–80	0.02	0.02	0.002	0.01	0.01	0.23	0.07
48a	0–15	0.10	0.03	0.001	0.56	0.07	1.83	0.08
48b	10–60	0.15	0.04	0.002	0.47	0.01	0.38	0.03
48c	60–80	0.02	0.02	0.002	0.06	0.01	0.12	0.04
49a	0–15	0.04	0.12	0.001	0.09	0.04	0.41	0.13
49b	15–30	0.02	0.02	0.002	0.01	0.01	0.27	0.20
49c	30–50	0.02	0.02	0.002	0.03	0.01	0.10	0.03
50a	50–100	0.91	0.52	0.008	1.94		4.99	4.03
50b	100–150	0.13	0.08	0.002	2.77		1.09	0.78
50c	150–200	0.04	0.02	0.001	0.36		0.22	0.28
51a	50–100	0.43	0.38	0.003	10.19		2.37	1.89
51b	100–150	0.04	0.03	0.003	0.63		0.30	0.30
51c	150–200	0.02	0.02	0.003	0.57		0.28	0.26
52a	50–100	0.92	0.64	0.009	17.21		4.83	4.02

Table A9.11 (continued)   Contents of Selected Elements in Sediments (in %)

Sample Number	Depth (cm)	Zn	Pb	Cd	Ca	Mn	Fe	S
52b	100–150	0.36	0.24	0.005	4.38		1.82	1.23
52c	150–200	0.08	0.06	0.002	1.06		0.43	0.57
53a	50–100	0.21	0.06	0.004	3.54		0.81	0.83
53b	100–150	0.03	0.02	0.003	0.59		0.26	0.41
53c	150–200	0.02	0.02	0.003	0.44		0.25	0.39
54a	50–100	0.93	0.46	0.008	10.93		3.58	2.63
54b	100–150	0.63	0.34	0.008	9.27		2.12	1.70
54c	150–200	0.09	0.03	0.004	0.96		0.29	0.34
55a	50–100	0.03	0.01	0.004	0.26		1.53	0.30
55b	100–150	0.02	0.01	0.004	0.11		0.23	0.20
55c	150–200	0.02	0.01	0.003	0.14		0.24	0.19
56a	50–100	2.63	0.36	0.015	14.71		3.41	4.32
56b	100–150	0.28	0.06	0.006	2.99		0.85	0.86
56c	150–200	0.09	0.01	0.006	0.50		0.27	0.37
57a	5–15	1.26	0.64	0.009	20.20		3.81	2.87
57b	40–50	1.49	0.42	0.011	19.80		3.23	2.30
57c	60–80	1.58	0.42	0.010	19.90		3.73	2.68
58a	10–20	1.64	1.34	0.014	11.40		7.29	2.69
58b	30–50	1.39	0.17	0.014	0.50		2.81	0.56
59	10–20	0.20	0.00	0.000	0.60		1.46	0.29
60a	10–20	1.37	0.74	0.015	16.80		6.51	5.22
60b	30–50	0.17	0.00	0.000	0.40		0.08	0.14
61a	15–20	1.25	0.82	0.013	16.00		3.27	2.14
61b	50–60	0.15	0.00	0.000	0.40		1.07	0.14
62a	10–20	1.57	0.30	0.007	13.20		4.81	5.16
62b	20–30	0.45	0.06	0.000	9.10		2.08	1.66
62c	30–50	0.68	0.36	0.003	18.60		3.73	3.34
63a	50–100	0.27	0.21	0.008	1.76	0.020	1.35	0.55
63b	100–150	0.20	0.08	0.005	1.04	0.019	1.43	0.32
63c	150–200	0.42	0.14	0.007	1.65	0.020	0.85	0.54
64a	50–100	0.24	0.11	0.003	1.15	0.015	0.81	0.66
64b	100–150	0.13	0.02	0.005	0.53	0.011	0.61	0.28
64c	150–200	0.05	0.02	0.004	0.18	0.002	0.24	0.19
65a	50–100	0.02	0.01	0.004	0.10	0.002	0.18	0.15
65b	100–150	0.03	0.01	0.002	0.09	0.002	0.19	0.15
65c	150–200	0.04	0.02	0.003	0.34	0.006	0.28	0.18
66a	50–100	0.10	0.05	0.004	0.66	0.009	0.45	0.35
66b	100–150	0.17	0.07	0.003	0.70	0.011	0.62	0.35
66c	150–200	0.08	0.02	0.004	0.38	0.009	0.49	0.21
67a	50–100	0.03	0.02	0.004	0.27	0.007	0.40	0.25
67b	100–150	0.06	0.02	0.002	0.37	0.010	0.41	0.27
67c	150–200	0.01	0.01	0.003	0.17	0.006	0.22	0.21
68	0–20	0.81	0.58	0.008	28.70		5.75	3.78
69	0–20	1.15	0.71	0.009	26.20		5.25	3.09
70	0–20	1.21	0.76	0.012	27.80		5.71	3.65
71	0–20	1.40	0.82	0.013	26.30		4.68	3.13
72	0–20	1.03	0.60	0.009	27.80		6.50	4.18
73	0–20	0.93	0.41	0.007	28.70		6.25	4.49
74	0–20	1.25	0.87	0.016	25.20		6.32	2.87
75	0–20	0.73	0.40	0.005	30.40		7.29	6.68
76	0–20	2.48	0.67	0.013	26.20		6.75	6.21
77	0–20	1.48	0.84	0.012	27.60		6.96	5.23
78	0–20	0.45	0.25	0.004	3.70		2.10	1.91
79	0–20	1.12	0.50	0.007	12.30		4.68	3.47
80	0–20	0.25	0.16	0.004	1.80		1.16	0.44
81	0–20	2.11	0.82	0.017	19.20		4.89	3.25
82	0–20	0.86	0.87	0.013	10.20		4.61	1.34

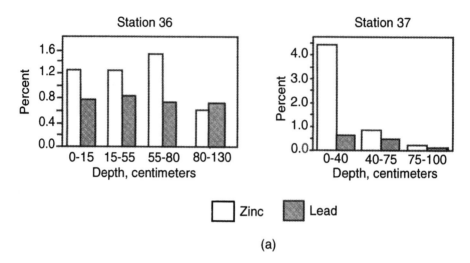

Figure A9.3a   Percent concentrations of lead (Pb) and zinc (Zn) with depth in soil in the Biala River wetland
(stations mapped in Chapter 9, Figure 9.4). (a) Stations 36 and 37; (b) stations 40 to 46;
(c) stations 49 to 52; (d) stations 53 to 58; (e) stations 59 to 65; (f) stations 68 to 75; (g) stations
76 to 82.

Zinc and lead in the soils in the wetland were compared with that in upland soils of the adjacent
region. Zinc content in the vicinity of the wetland is about 200 ppm, lead content 100 ppm (Chapter
9, Figure 9.7), and cadmium content about 3 ppm. Zinc and lead concentrations are very high
locally where there are outcrops of the ore-bearing geological formation.

VEGETATION CHARACTERISTICS

Extensive phytosociological research on the study of wetland and surrounding areas found 17
plant communities. For most, illustrative phytosociological pictures were made (occasionally drawing several pictures of one community to illustrate variation). From 55 pictures made, characteristics
are given in Tables A9.12a to A9.12g (a key to the pictures is in Table A9.12h). The prevailing
plant communities are the marshes, including extensive patches dominated by the Deschampsia
grass (*Deschampsia caespitosa*). Somewhat farther from the river bed are the remains of the old
wet meadows of the Caltion alliance.

VALLEY ZONES

Within the valley several vegetation zones were distinguished:

1. The communities of the aquatic and marshy plants dominate the Biala River above the outlet of
   the Dabrowka Channel in Laski where the area underwent swamping.
2. An unusual community dominated by the grass *Deschampsia caespitosa* can be observed from the
   outlet of the Dabrowka Channel to the narrowing of the valley above the mill ruins in the village
   Reczkowe. Other communities include marsh sedges and fragments of former wet meadows.
3. The alder (*Alnus gluticosa*) is entering into various communities in the vicinity of the mill ruins
   in the village Reczkowe — a section about 0.5 km long.

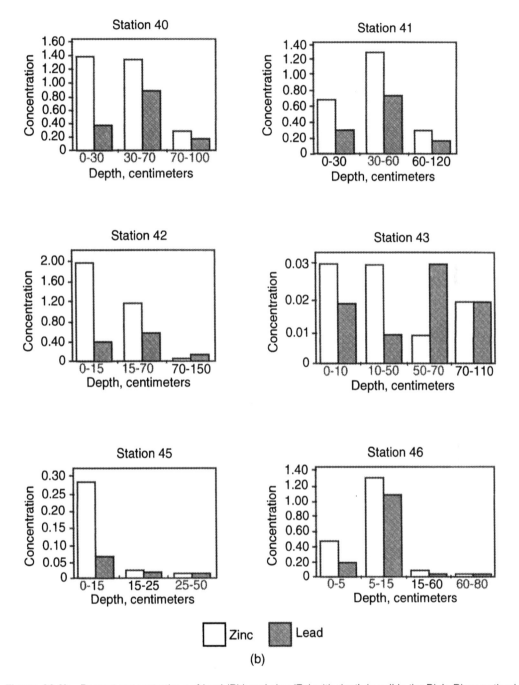

Figure A9.3b Percent concentrations of lead (Pb) and zinc (Zn) with depth in soil in the Biala River wetland (stations mapped in Chapter 9, Figure 9.4). (a) Stations 36 and 37; (b) stations 40 to 46; (c) stations 49 to 52; (d) stations 53 to 58; (e) stations 59 to 65; (f) stations 68 to 75; (g) stations 76 to 82.

4. The vegetation in the lower section below the ruins of the mill is mainly typical of the region, with marsh sedges and patches of aquatic plants.
5. The northeast and southwest sides of the valley are surrounded by pine-tree forest growing on sand dunes rising 1 to 10 m above the wetland.

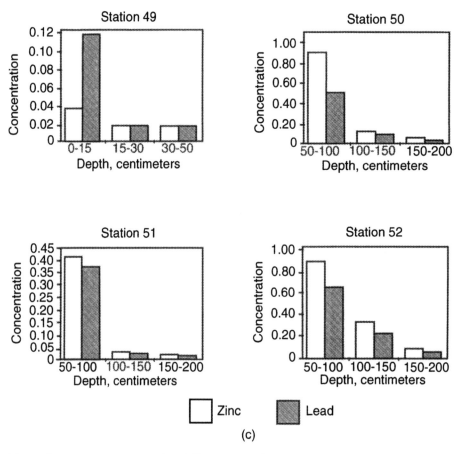

Figure A9.3c   Percent concentrations of lead (Pb) and zinc (Zn) with depth in soil in the Biala River wetland (stations mapped in Chapter 9, Figure 9.4). (a) Stations 36 and 37; (b) stations 40 to 46; (c) stations 49 to 52; (d) stations 53 to 58; (e) stations 59 to 65; (f) stations 68 to 75; (g) stations 76 to 82.

COMMUNITY TYPES

Below is a review of identified plant communities.

1. Early successional pine forest (*Vaccinio myrtilli-Pinetum*) surrounds almost the whole valley and is associated with the elevation formed by sand dunes. In the forest the pine trees of low age classes dominate, showing symptoms of damage by SO_2 air pollution. The predominant ground cover is redberry (*Vaccinium vitis-idaea*) or blueberry (*V. myrtillus*) or the grass Deschampsia (*Deschampsia flexuosa*) with a great number of pirolas (*Pirola secunda* and *P. rotundifolia*).

1a. Early successional pine forest, with atypical patches of heather (*Calluna vulgaris*) and birch (*Betula verrucosa*). Patches contain many individuals of similar, small size.

2. Swamp forest (*Vaccinio uliginiosi-Pinetum*). Several small areas occur under the valley sides, characterized by peat mosses in the ground cover under small pine trees.

3. Alder swamp (*Carici elongatae-Alnetum*) patches occurred in a narrow band in the wetland margin at the foot of land slopes, with black alder growing on small hillocks, and iris (*Iris pseudoacorus*) and marsh sedge (*Carex elongata*) numerous in-between hillocks at lower places, and sometimes the elm (*Ulmus scabra*). See Table A9.12f for floristic composition.

3a. Alder swamp in its initial stages is developing in the patches of the marsh sedge (*Caricetum gracilis*) in the middle part of the valley. In present conditions it may soon become a typical alder swamp.

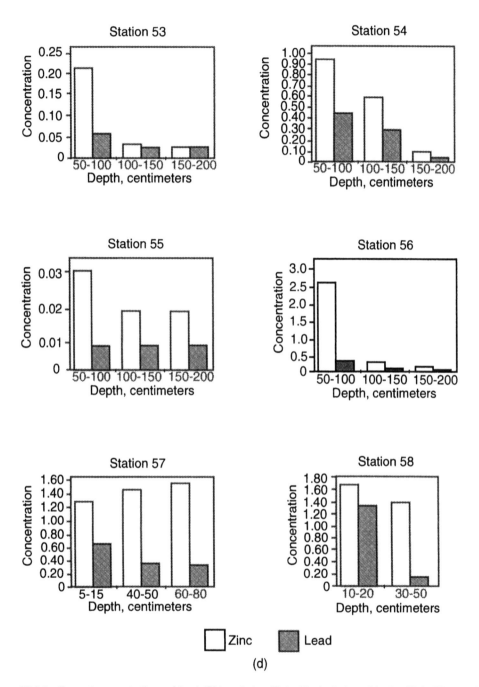

Figure A9.3d Percent concentrations of lead (Pb) and zinc (Zn) with depth in soil in the Biala River wetland (stations mapped in Chapter 9, Figure 9.4). (a) Stations 36 and 37; (b) stations 40 to 46; (c) stations 49 to 52; (d) stations 53 to 58; (e) stations 59 to 65; (f) stations 68 to 75; (g) stations 76 to 82.

4. Meadows and fresh pastures (order Arrhenatheretalia) occupy small areas on a higher terrace where they are used as grassland.
5. A community with the grass (*Deschampsia caespitosa*) dominates strongly silted places. The mass occurrence of this species cannot be easily explained, and it needs special investigation. The diversity is small (see Table A9.12a).
6. Wet meadows are observed on small areas situated closer to the village where they are mowed occasionally. The wet meadows are diversified, with rare, interesting species, such as the sundew

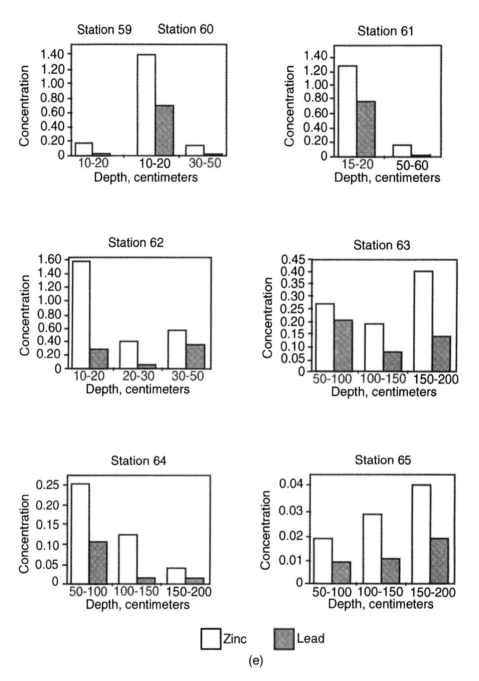

(e)

Figure A9.3e Percent concentrations of lead (Pb) and zinc (Zn) with depth in soil in the Biala River wetland (stations mapped in Chapter 9, Figure 9.4). (a) Stations 36 and 37; (b) stations 40 to 46; (c) stations 49 to 52; (d) stations 53 to 58; (e) stations 59 to 65; (f) stations 68 to 75; (g) stations 76 to 82.

(*Drosera rotundifolia*), the horsetail (*Equisetum variegatum*), and the orchids (*Orchis maculata, Epipactis palustris, Liparis loeselli*) (see Table A9.12d).

7. Communities of aquatic plants occur in the wetland pools and ponds. The main components are pondweed, *Potamogeon* sp., and duckweed, *Lemna minor*. The diversity is small.

8. A massive community of bog bean (*Menyanthes trifoliata*) occurs in the flooding areas up to 50 cm deep on the river near the bridge in Kuzniczka. Normally the bog bean occurs dispersed among

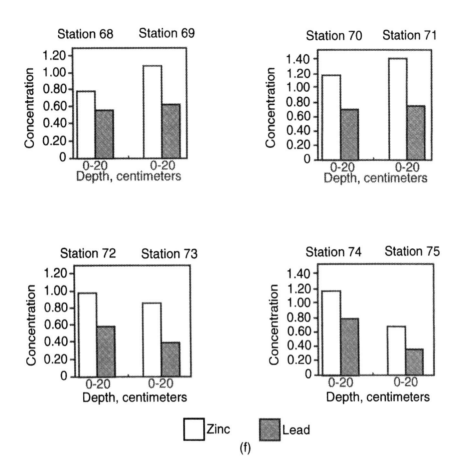

(f)

Figure A9.3f   Percent concentrations of lead (Pb) and zinc (Zn) with depth in soil in the Biala River wetland (stations mapped in Chapter 9, Figure 9.4). (a) Stations 36 and 37; (b) stations 40 to 46; (c) stations 49 to 52; (d) stations 53 to 58; (e) stations 59 to 65; (f) stations 68 to 75; (g) stations 76 to 82.

other plants on poor, wet meadows. Such mass occurrence of the bog bean has not been reported previously. Perhaps this single-species occurrence is related to the heavy metals.

9. Typical marsh (*Scirpo-Phragmitetum*) with cattail (*Typha latifolia*) occupies the water-logged areas with a long hydroperiod.

10. Reed marsh (*Scirpo-Phragmitetum*) with the reed (*Phragmites communis*) grows in less water-logged places with intermediate hydroperiod in the lower valley.

11. Horsetail marsh (*Scirpo-Phragmitetum* with *Equisetum limosum*) occupies small areas mainly in the lower part of the valley where the water depth is 20 cm or more continuously.

12. Sedge marsh (*Carex gracilis and Caricetum gracilis*) occurs frequently in low, wet places over the whole valley.

13. A sedge marsh (*Caricetum paniculatae*) community occurs in extensive areas in the upper parts of the valley with great hillocks formed by the sedge (*Carex paniculata*) and single clusters of the grass (*Deschampsia caespitosa*) on the banks.

14. Sedge marsh (*Caricetum rostratae*) occurs mainly in the upper part of the valley, on the passage between the marsh communities (*Scirpo-Phragmitetum* s.1.) and the communities of the wet meadows.

15. Sedge marsh (*Caricetum vesicariae*) appears on small areas without a major role in this territory.

16. Sedge marsh *Caricetum acutiformis* and *C. ripariae* occur mainly in the central part of the valley, at wet-logged places similar to sites for the sedge marsh *C. gracilis*.

17. Sedge marsh *Caricetum elatae* forms clusters in the lower part of the valley with, e.g., water mint (*Mentha aquatica*), bugleweed (*Lycopus europaeus*), and occasionally aquatic plants, e.g., duckweed (*Lemmna minor*) in-between.

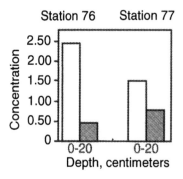

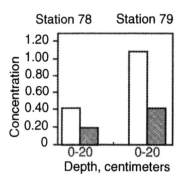

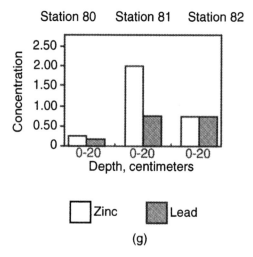

(g)

Figure A9.3g   Percent concentrations of lead (Pb) and zinc (Zn) with depth in soil in the Biala River wetland (stations mapped in Chapter 9, Figure 9.4). (a) Stations 36 and 37; (b) stations 40 to 46; (c) stations 49 to 52; (d) stations 53 to 58; (e) stations 59 to 65; (f) stations 68 to 75; (g) stations 76 to 82.

See Chapter 9, Figure 9.3 for the proportion of the particular communities. In order of the frequency of occurrence, the communities with the largest area are as follows:

1. Reed marsh (*Scirpo phragmitetum* with *Phragmites communis*)
2. Community with Deschampsia (*Deschampsia caespitosa*)
3. Typical marsh (*Scirpo phragmitetum* with *Typha latifolia*)
4. Sedge marsh (*Caricetum gracilis*)
5. Great stooled sedge marsh (*Caricetum paniculatae*)
6. Alder swamp (*Carici elongatae-Alnetum*)
7. Wet meadows of various types

MISSING SPECIES

The flora of the Biala River wetland area has some peculiar features. Although common elsewhere, willows (*Salix alba, S. fragillis, S. vincinalis*), poplars (*Populus alba* and *P. nigra*), and a meadows species *Bellins perennis* are absent.

Table A9.12a Phytosociological Characteristic of Plants in a Community of *Deschampsia caespitosa*

No. of Phytosociological Record in the Table	1	2	3	4	5	6	
No. of phytosociological record	21	9	30	7	31	13	S T
Date	1989 8/23	1989 8/17	1990 7/12	1990 6/21	1990 7/12	1989 8/22	A
Max height of plants (cm)	170	160	150	70	130	80	B I
Average height of plants (cm)	80	100	70	30	70	30	L I
Vegetation covering (%)	100	100	100	70	90	100	T Y
Area of record (m²)	30	50	50	50	50	6	
No. of species in record	10	8	8	12	7	4	
Deschampsia caespitosa	5.5	5.3	5.3	4.5	4.3	1.3	V
Agrostis stolonifera	+	2.2	+	1.2	3.3	5.5	V
Cirsium palustre	+1.2	+	1.2	1.1	+	.	V
Epilobium palustre	+.2	+	+	+	.	1.2	V
Calamagrostis epigeios	.	+	.	.	+	.	III
Equisetum limosum	.	1.2	1.1	+	.	.	III
Lythrum salicaria	+	.	.	+	+	.	III
Poa trivialis	+	+	.	1.1	.	.	III
Carex sp.	+	.	+.2	.	.	.	II
Equisetum palustre	.	.	+	1.1	.	.	II
Mentha aquatica	+	.	+	.	.	.	II
Angelica silvestris	.	.	.	+	.	.	I
Carex paniculata	.	.	.	.	+.3	.	I
C. rostrata	.	.	.	+	.	.	I
Galium palustre	.	.	.	+.2	.	.	I
Lycopus europaeus	.	.	.	.	.	1.3	I
Lysimachia vulgaris	.	.	.	.	+	.	I
Phragmites communis	+.2	+	.	.	.	.	I
Solanum dulcamara	.	.	.	.	.	.	I
Urtica dioica	.	.	.	+	.	.	I

Note: Field measurements were made by H. Tacik and S. Dubiel.

DIVERSITY

Another characteristic of the Biala River flora is the mass occurrence of a few species in low diversity. Starting with the most common, these are reed (*Phragmites communis*), Deschampsia (*Deschampsia caespitosa*), sedge (*Carex gracilis*), great stooled sedge (*C. paniculata*), water mint (*Mentha aquatica*), cattail (*Typha latifolia*), sedge (*C. rostrata*), and black alder (*Alnus glutinosa*).

Plant Diversity

Based on the phytosociological pictures and transformation of the Braun-Blanquet scale of the areas of covering, the Richness and Shannon indexes were calculated for 46 pictures (these located on the wetland), employing PCORD computer program (Table A9.13). The Shannon diversity indexes ranged from 0.526 to 3.048, with an average equal to 1.942. Richness ranged from 4 to 33 and average equal to 13.3.

Table A9.12b Phytosociological Characteristics of a Community of Reeds (*Scirpo-Phragmitetum*)

No. of Phytosociological Record in the Table	1	2	3	4	5	6	7	S
No. of phytosociological record	20	3	22	17	11	10	27	T
Date	1990 6/21	1990 6/21	1989 8/23	1989 8/22	1990 6/21	198 8/17	1990 7/12	A
Depth of water (cm)	30	5	20	5	—	15	20	B
Max height of plants (cm)	240	170	180	160	140	120	—	I
Average height of plants (cm)	70	90	70	—	100	70	—	L
Vegetation covering (%)	80	100	90	100	70	100	80	I
Area of record (m²)	50	25	75	100	50	100	100	T
No. of species in record	11	14	11	26	7	13	8	Y
Ch. Phragmition								
Typha latifolia	4.4	3.3	4.4	+.2	.	+	.	IV
Equisetum limosum	.	2.3	.	.	.	4.3	4.4	III
Phragmites communis	.	.	+.2	5.4	4.4	.	.	III
Sparganium ramosum	.	.	1.3	1.2	.	.	.	II
Typha Angustifolia	.	.	.	.	.	.	.	I
Ch. Phragmitetea								
Carex rostrata	2.2	1.1	2.2	.	.	.	2.2	III
Galium palustre	1.2	2.2	.	+	+	.	.	III
Lycopus europaeus	1.1	1.1	.	.	.	+	.	III
Lysimachia thyrsiflora	1.1	1.1	.	.	+	.	.	III
Peucedanum palustre	+	.	.	+.2	.	+	.	III
Scutellaria galericulata	+	.	.	+	.	.	.	II
Alisma plantago-aquatica	.	.	.	+	.	.	.	I
Carex gracilis	.	.	.	.	.	3.3	.	I
C. paniculata	.	.	.	+.2	.	.	.	I
Rumex hydrolapathum	.	.	.	+.2	.	.	.	I
Ch. Molinietalia								
Deschampsia caespitosa	.	.	1.3	+.2	1.2	+.2	+.2	IV
Cirsium palustre	.	.	+.2	1.2	+	+	.	III
Lythrum salicaria	+	1.1	.	1.2	.	2.2	.	III
Caltha palustris	.	1.2	.	.	.	+.2	3.2	III
Equisetum palustre	.	2.3	.	+	.	.	.	II
Lysimachia vulgaris	.	.	.	+	.	.	.	I
Myosotis palustris	.	+	.	.	.	.	.	I
Others								
Lemna minor	2.3	.	1.3	+	.	1.1	.	III
Comarum palustre	+.2	+	.	2.2	.	+.2	.	III
Epilobium palustre	.	.	1.1	1.2	.	2.1	+	III
Mentha aquatica	.	.	2.2	1.2	.	.	1.1	III
Epilobium hirsutum	.	.	.	+	.	.	+.2	II
Menyanthes trifoliata	.	3.4	+	.	.	.	.	II
Solanum dulcamara	.	.	+.2	.	.	+	.	II
Veronica scutellata	+	+	.	.	.	.	.	II
Agrostis stolonifera	.	.	.	+.2	.	.	.	I
Alnus glutinosa	.	.	.	+.3	.	.	.	I
Calamagrostis epigeios	.	.	.	+.2	.	.	.	I
Eupatorium cannabium	.	.	.	1.2	.	.	.	I
Galium mollugo	.	.	.	+	.	.	.	I
Mentha longifolia	.	.	.	.	.	.	1.2	I
Poa trivialis	.	.	.	+	.	.	.	I
Scrophularia nodosa	.	.	.	.	+	.	.	I
Stellaria uliginosa	.	+	.	.	.	.	.	I
Urtica dioica	.	.	.	.	1.1	.	.	I

Note: Field measurements measured by H. Tacik and S. Dubiel.

Table A9.12c Phytosociological Characteristic of the Community of Sedge
(*Caricetum gracilis*)

No. of Phytosociological Record in the Table	1	2	3	4
No. of phytosociological record	9	10	12	26
Date	1990	1990	1990	1990
	6/21	6/21	6/21	7/12
Max height of plants (cm)	80	50	70	—
Average height of plants (cm)	80	50	70	—
Vegetation covering (%)	90	70	85	100
Area of record (m²)	100	100	100	100
No. of species in record	4	7	9	5
Ch. Magnocaricion				
Carex gracilis	5.4	4.4	4.5	5.5
C. paniculata	+.2	1.2	+.2	.
C. acutiformis	.	.	.	1.2
Ch. Molinietalia				
Cirsium palustre	.	+	+	+
Deschampsia caespitosa	.	+.2	.	+.2
Lythrum salicaria	.	+	+	.
Equisetum palustre	.	.	1.1	.
Lysimachia vulgaris	.	.	+	.
Others				
Poa trivialis	.	+.2	1.1	.
Urtica dioica	+	.	.	+
Equisetum limosum	+	.	.	.
Ranunculus sceleratus	.	.	+	.
Salix cinerea	.	+.2	.	.
Solanum dulcamara	.	.	1.1	.

Note: Field measurements were made by H. Tacik and S. Dubiel.

Heavy Metal Concentration in the Plants

Atomic absorption analyses of the heavy metals concentration in plants growing on the Biala River wetland were carried out in 1989 and 1990 using sampling stations shown in Chapter 9, Figure 9.4. The samples included leaves, stems, underground parts (roots or rhizomes plus roots), aboveground parts, and in one case a whole plant (*Potamogeton* sp.).

In the laboratory the samples were dried and digested with perchloric and nitric acids ($HClO_4$ and HNO_3). A Perkin-Elmer atomic absorption spectrophotometer Model 403 and 2380 was employed for analysis. Heavy metals concentration was determined for nine plant species: *Phragmites communis*, *Carex* sp., *Typha latifolia*, *Mentha aquatica*, *Potamogeton natans*, *Scirpus lacustris*, *Sparganium* sp., *Menyanthes trifoliata*, and *Deschampsia caespitosa*.

The concentration of metals in dry matter of plants varied from 5 to 6500 mg/kg for zinc, from 3.5 to 1050 mg/kg for lead, and from 0.5 to 48 mg/kg for cadmium (Table A9.14 and Chapter 9, Figure 9.11). In most cases higher concentrations were found in the underground parts of plants than in leaves or stems. Average concentrations of zinc and cadmium in the rhizomes of *Phragmites communis* were 2 to 3 times greater than in leaves and stems, and 3 to 7 times greater for lead. In *Typha latifolia* the average concentration of zinc and lead was about 5 to 8 times greater in rhizomes. Cadmium values were the same for above- and underground parts. The difference in metals concentration between underground parts and aboveground parts of *Carex* sp. was even more significant, 10 to 20 times greater for zinc, lead, and cadmium.

The highest concentration of zinc, 6500 mg/kg, and lead, 1050 mg/kg, was found in underground parts of *Carex* sp. The highest concentration of cadmium, 48 mg/kg, was found in *Potamogeton natans* (whole plant). Highest concentrations of zinc, lead, and cadmium in Deschampsia grass were 2600, 480, and 12 mg/kg, respectively, for the roots (see Chapter 9, Figures 9.9 and 9.10).

Table A9.12d Phytosociological Characteristic of the Community of Sedge (*Caricetum paniculatae*)

No. of Phytosociological Record in the Table	1	2	3	4	5	6	S T
No. of phytosociological record	6	13	6	29	15	17	A B
Date	1989	1990	1990	1990	1990	1990	I
	6/22	6/21	6/21	7/12	6/21	6/21	L
Max height of plants in (cm)	120	100	80	150	80	170	I T
Average height of plants in (cm)	100	45	60	120	60	100	Y
Vegetation covering (%)	100	90	70	100	90	90	
Area of record (m²)	90	20	25	70	50	50	
No. of species in record	10	13	7	8	16	9	
Ch. Magnocaricion							
Carex paniculata	4.3	4.5	4.5	5.4	4.4	5.5	V
C. gracilis	3.3	+.2	+.2	.	.	.	III
C. vesicaria	1.2	.	.	.	.	.	I
Galium palustre	.	1.2	.	.	.	.	I
Ch. Phragmitetea							
Equisetum limosum	+	.	.	+	.	.	II
Lysimachia thyrsiflora	+	.	.	.	.+	.	II
Peucedanum palustre	.	.	.	.	2.1	.	I
Rumex hydrolapathum	.	+	.	.	.	.	I
Scutellaria galericulata	.	.	.	.	1.1	.	I
Ch. Calthion, Molinietalia							
Cirsium palustre	+	1.1	+.2	2.1	1.1	2.1	V
Deschampsia caespitosa	1.3	.	1.3	+.2	1.2	+.2	V
Equisetum palustre	.	1.1	.	.	.	+	II
Filipendula ulmaria	.	+	.	.	+	.	II
Lythrum salicaria	.	+	+	.	.	.	II
Caltha palustris	.	+.2	.	.	.	.	I
Lysimachia vulgaris	.	.	.	.	1.1	.	I
Others							
Urtica dioica	1.1	.	1.1	+	+.2	+.2	V
Agrostis stolonifera	.	.	+.2	1.2	1.2	+.2	IV
Poa trivialis	+	1.1	.	.	+	+	IV
Epilobium palustre	.	+	.	+	.	+	III
Mentha aquatica	.	1.1	.	+	.	.	II
Potentilla erecta	.	.	.	.	+	+.2	II
Carex fusca	.	.	.	.	1.2	.	I
Chaerophyllum sp.	+	.	.	.	.	.	I
Dactylis glomerata	.	+	.	.	.	.	I
Eriphorum latifolium	.	.	.	.	+	.	I
Frangula alnus	.	.	.	.	+	.	I
Salix cinerea	.	.	.	.	+	.	I

Note: Field measurements made by H. Tacik and S. Dubiel.

Highest concentrations of zinc, lead, and cadmium in *Scirpus lacustris* were found for roots as 1550, 800, and 11 mg/kg, respectively. Highest concentrations of zinc, lead, and cadmium in leaves of *Mentha aquatica* were 577.5, 370, and 5 mg/kg, respectively. Highest concentrations of zinc, lead, and cadmium in bog bean (*Menyanthes trifoliata*) were recorded as 180, 80, and 1 mg/kg, respectively. Highest concentrations of zinc, lead, and cadmium in *Phragmites communis* were 280 and 217 mg/kg in roots, and 40 mg/kg in leaves. Highest concentrations of zinc, lead, and cadmium in roots of cattail (*Typha*) were 220, 250, and 1.5 mg/kg, respectively.

Table A9.12e Phytosociological Characteristic of the Community of Sedge
(*Caricetum rostratae*)

No. of Phytosociological Record in Table	1	2	3	4
No. of phytosociological record	8	5	28	18
Date	1989	1990	1990	1990
	8/17	6/21	7/12	6/21
Depth of water (cm)	20	10	10	30
Max height of plants (cm)	80	60	80	50
Average height of plants (cm)	80	50	70	50
Vegetation covering (%)	100	95	90	95
Area of record (m²)	100	25	50	25
No. of species in record	7	19	12	10
Ch. Magnocaricion, Phragmitetea				
Carex rostrata	5.5	4.5	4.3	4.4
C. paniculata	.	+	+.3	+.2
Equisetum limosum	2.1	1.1	3.1	.
Galium palustre	.	2.2	.	+.2
Lysimachia thyrsiflora	.	.	+	2,1
Heleocharis palustris	.	2.2	.	.
Lycopus europaeus	.	.	.	+
Rumex hydrolapathum	.2	.	.	.
Sparganium simplex	.	.	.	+
Typha augustifolia	.	.	+.2	.
T. latifolia	+.2	.	.	.
Ch. Molinietalia				
Cirsium palustre	1.1	+	.	+
Equisetum palustre	.	3.3	1.1	.
Lythrum salicaria	1.2	1.1	.	.
Caltha palustris	.	2.2	.	.
Deschampsia caespitosa	.	.	.+.2	.
Myosotis palustris	.	1.1	.	.
Others				
Lemna minor	+	2.3	2.2	1.1
Mentha aquatica	.	1.1	2.2	+
M. longifolia	.	.	1.2	.
Calliergon sp. (d)	.	+.2	.	.
Cardamine pratensis	.	+	.	.
Comarum palustre	.	2.1	.	.
Epilobium hirsutum	.	.	+	.
E. palustre	.	.	2.1	.
Plagiomnium sp. (d)	.	+.2	.	.
Potamogeton natans	.	.	.	1.2
Ranunculus flammula	.	+	.	.
Valeriana simplicifolia	.	1.2	.	.
Veronica scutellata	.	1.1	.	.

Note: Field measurements were made by H. Tacik and S. Dubiel.

Table A9.12f Phytosociological Characteristics of a Swamp Community
Dominated by Alders (*Carici elongatae-Alnetum*)

No. of Phytosociological Record in Table	1	2
No. of phytosociological record	20	14
Date	8/23/1989	6/21/1990
Tree layer cover (%) A	40	70
Shrub layer cover (%) B	20	10
Herb layer cover (%) C	80	70
Average height of trees (cm)	22	12
Average diameter of trees (cm)	20	10

continued

Table A9.12f (continued)   Phytosociological Characteristics of a Swamp Community Dominated by Alders (*Carici elongatae-Alnetum*)

Area of record (sq. m)	160	100
No. of species in record	32	22
Trees		
Alnus glutinosa A	3.3	4.4
A. glutinosa B	+.3	+
Ulmus scabra A	1.3	.
U. scabra B	1.3	.
Shrubs		
Frangula alnus	1.3	1.1
Rubus idaeus	+.3	.
Solanum dulcamara	+.3	.
Herbs		
Ch. *Alnetea glutinosae*		
Carex elongata	2.3	+.2
Lycopus europaeus	2.2	1.1
Solanum dulcamara	3.3	2.2
Ch. *Phragmitetea*		
Peucedanum palustre	1.2	+
Carex gracilis	.	+
Iris pseudoacorus	+.3	.
Rumex hydrolapathum	+	.
Scutellaria galericulata	.	1.1
Ch. *Molinietalia*		
Cirsium palustre	+.2	3.2
Deschampsia caespitosa	2.3	3.3
Filipendula ulmaria	+.2	1.1
Angelica silvestris	+.2	.
Cirsium oleraceum	+.2	.
Crepis paludosa	.	1.2
Caltha palustris	2.2	.
Equisetum palustre	.	+
Galium palustre	1.3	.
Juncus effusus	1.2	.
Lychnis flos-cuculi	.	+
Lysimachia vulgaris	.	+
Lythrum salicaria	.	1.1
Scirpus silvaticus	3.3	.
Eupatorium cannabinum	+.2	+.2
Quercus robur + Q. sessilis	+	+
Agrostis alba	1.2	.
Athyrium filix-femina	+.2	.
Carex hirta	1.2	.
Carex sp.	+	.
Cardaminopsis halleri	.	2.3
Festuca gigantea	+.2	.
Geum urbanum	1.2	.
Galeopsis bifida	+	.
Heracleum sphondyl.	+	.
Holcus mollis	+	.
Mentha verticillata	+	.
Molinia arundinacea	.	+
Poa trivialis	.	1.1
Ranunculus repens	+.2	.
Rubus idaeus	+.2	.
Urtica dioica	1.2	.

Note: Field measurements were made by H. Tacik and S. Dubiel.

Table A9.12g Phytosociological Characteristics of a Forest Community *Vaccinio myrtilli-Pinetum*

No. of Phytosociological Record in the Table	1	2	3	4	5	
No. of phytosociological record	23	31	19	14	30	
Date	1989	1990	1989	1989	1990	
	8/23	8/30	8/12	8/22	8/30	
Exposition	W	N	SW	W	—	S
Slope	1	5	3	3	—	T
Tree layer cover (%) A	60	65	40	40	60	A
Shrub layer cover (%) B	20	2	30	60	45	B
Herb layer cover (%) C	80	85	90	90	70	I
Moss layer cover (%) D	30	5	5	40	7	L
Average height of trees (m)	20	20	20	—	25	I
Average diameter of trees (cm)	25	25	35	—	40	T
Area of record (m²)	100	200	100	250	200	Y
Ch. Dicrano-Pinion						
Chimaphila umbellata	1.1	.	+	.	.	II
Deschampsia flexuosa	4.4	4.4	.	.	.	II
Pirola chlorantha	.	+	.	.	.	I
Ch. Vaccinio-Picetea						
Vaccinium myrtillus	1.3	1.2	2.2	4.3	1.3	V
V. vitis-idaea	1.2	.	4.4	3.2	+.2	IV
Pirola secunda	1.2	2.3	1.1	+	.	IV
Trientalis europaea	+	.	+	+	1.1	IV
P. uniflora	+	1.1	+	.	.	III
Melampyrum vulgatum	1.1	1.1	.	.	.	II
P. cf rotundifolia	.	.	.	.	+	I
Others						
Agrostis vulgaris	+	1.2	1.1	2.2	+.2	V
Festuca ovina	1.3	+.2	1.2	2.2	1.2	V
Calluna vulgaris	1.3	1.2	1.2	+.2	.	IV
Potentilla erecta	.	1.2	1.2	2.2	+	IV
Anthoxanthum odoratum	.	+	+	+	.	III
Fragaria vesca	+.2	2.2	.	1.2	.	III
Chamaenerion angustifolium	.	1.1	.	1.2	.	II
Hieracium pilosella	1.2	.	+	.	.	II
H. umbellatum	.	.	+	.	+	II
Oxalis acetosella	.	.	.	.	4.3	I
Equisetum silvaticum	.	.	.	.	3.1	I
Trees						
Pinus silvestris A	4.4	4.3	3.3	3.3	3.3	V
P. silvestris B	2.2	.	2.2	.	.	II
P. silvestris C	.	.	+	.	.	I
Sorbus aucuparia A	.	.	.	.	2.2	I
S. aucuparia B	+.2	+	.	+.2	2.2	IV
S. aucuparia C	+	.	.	.	+	II
Shrubs						
Frangula alnus B	1.2	+	+	3.3	2.3	IV
F. alnus C	.	.	.	1.1	.	I
Juniperus communis B	.	.	1.2	2.2	.	II
J. communis C	.	.	+	+	.	II

Note: Field measurements were made by H. Tacik and S. Dubiel.

Table A9.12h Characteristics of Phytosociological Census

Phytosociological Indexes
Example: *Phragmites communis* 5.4
Surface covering index (first digit)

scale:	5	>75% of surface covering
	4	50–75%
	3	25–50%
	2	<25%, numerous plants
	1	5%, numerous plants
	+	<5%, not numerous plants

Sociability index (second digit)

scale:	5	whole fields
	4	large conglomerations
	3	intermittent conglomerations (large clusters)
	2	small clusters
	1	single plants
when marked only +		only few plants

Division of land vegetation:

	Layer A	trees
	Layer B	shrubs, small trees
	Layer C	herbs, small shrubby plants
	Layer D	bryophytes

For comparison, data on zinc, lead, and cadmium were collected from the literature (Table A9.15).
Results of comparing a small set of data on lead in selected plants and their soil (Figures A9.5 and Chapter 9, Figure 9.12) were complex without an obvious explanation. Perhaps lead in soil was high enough to cause erratic responses by plants.

To compare concentrations of heavy metals in soil with underground and aboveground parts of the plants, soils from the root zone of particular plants were excavated and analyzed (Figure A9.5 and Chapter 9, Figure 9.12). Concentrations of heavy metals in sediments were much higher than in roots.

Biomass Regrowth on Clipped Quadrats

For evaluation of yearly biomass growth, nine plots 1 m² each were cleaned in early spring, and then harvested in late autumn. The biomass in each plot is given in Table A9.16. The average biomass production studied by this method was 846 g/m². Regrowth data are an index of productivity, of the influence of pollutants, and the annual net deposition of sedimentary organic matter.

UPSTREAM-DOWNSTREAM ANALYSES OF PASSING WATERS

Upstream and downstream measurements of zinc and lead were carried out at time intervals to capture the same water as it passed. The sampling stations on the wetland are those in Figure 9.4c of Chapter 9. The results are in Table A9.17 and Figures 9.14 and 9.15 in Chapter 9.

MEASUREMENTS OF ZINC AND LEAD UPTAKE FROM WATER FLOW ON THE EXPERIMENTAL PLOTS

Lead and zinc were analyzed in waters passed through two plots of wetland vegetation.

Plot 1 was situated at the beginning of the wetland (Figure 9.5 in Chapter 9). A dike 4 m long and 0.8 m deep was constructed on the branch of the Biala River just behind the place where the Dabrowka Channel joins the river. The water retained by the dike was directed with three pipes through a channel (about 10 m long) into an old stream bed. The plot was 96 m long and 5 to 8 m wide, with a surface

Table A9.13   Diversity Indices for 46 Phytosociological Census Calculated with the PCORD Program

No. of picture	Mean	Standard Deviation	Total	Min	Max	S	E	H
1	1.279	6.994	229.000	.000	64.000	16	.797	2.211
2	1.458	7.385	261.000	.000	64.000	16	.819	2.272
3	1.229	6.469	220.000	.000	64.000	16	.806	2.235
4	1.726	7.928	309.000	.000	64.000	15	.855	2.314
5	1.447	7.531	259.000	.000	64.000	13	.818	2.099
6	1.575	6.969	282.000	.000	64.000	20	.841	2.520
7	2.235	7.442	400.000	.000	49.000	33	.872	3.048
8	1.737	7.614	311.000	.000	64.000	21	.832	2.535
9	1.631	5.661	292.000	.000	53.000	31	.879	3.019
10	1.525	6.062	273.000	.000	49.000	24	.865	2.748
11	.682	6.129	122.000	.000	81.000	10	.591	1.361
12	.754	6.380	135.000	.000	81.000	8	.635	1.321
13	.665	6.146	119.000	.000	81.000	8	.588	1.223
14	.715	4.995	128.000	.000	64.000	12	.745	1.851
15	.743	6.033	133.000	.000	64.000	7	.641	1.247
16	.603	6.153	108.000	.000	81.000	4	.604	.837
17	1.693	6.965	303.000	.000	64.000	25	.843	2.714
18	.888	6.272	159.000	.000	81.000	13	.731	1.875
19	2.067	8.012	370.000	.000	64.000	24	.859	2.731
20	2.123	7.873	380.000	.000	49.000	27	.858	2.828
21	.637	5.189	114.000	.000	64.000	7	.682	1.327
22	.682	6.129	122.000	.000	81.000	10	.591	1.361
23	.587	6.087	105.000	.000	81.000	7	.487	.947
24	.899	5.571	161.000	.000	64.000	11	.787	1.888
25	1.307	6.199	234.000	.000	49.000	14	.863	2.276
26	.927	5.599	166.000	.000	64.000	11	.804	1.929
27	1.318	6.576	236.000	.000	81.000	26	.811	2.642
28	.547	4.897	98.000	.000	64.000	7	.637	1.239
29	1.140	6.587	204.000	.000	64.000	13	.770	1.975
30	.939	6.349	168.000	.000	64.000	8	.765	1.591
31	.520	6.071	93.000	.000	81.000	4	.380	.526
32	.520	4.863	93.000	.000	64.000	7	.596	1.160
33	.598	4.937	107.000	.000	64.000	8	.685	1.424
34	.570	6.104	102.000	.000	81.000	5	.484	.778
35	.894	6.125	160.000	.000	64.000	10	.728	1.676
36	.765	5.033	137.000	.000	64.000	13	.769	1.972
37	.547	4.897	98.000	.000	64.000	7	.637	1.239
38	.754	6.380	135.000	.000	81.000	8	.635	1.321
39	.978	5.386	175.000	.000	64.000	16	.812	2.252
40	.749	6.359	134.000	.000	81.000	9	.615	1.351
41	.760	6.401	136.000	.000	81.000	7	.663	1.291
42	1.765	7.362	316.000	.000	64.000	19	.868	2.556
43	1.263	6.840	226.000	.000	64.000	12	.818	2.033
44	.732	5.238	131.000	.000	64.000	10	.727	1.673
45	1.006	7.325	180.000	.000	81.000	8	.710	1.476
46	1.151	6.627	206.000	.000	49.000	10	.808	1.860
Averages	1.072	6.307	191.957	.000	67.304	13.3	.731	1.842

Number of cells in main matrix = 8234

Percent of cells empty = 92.592

Matrix total = 8.8300 E+03

S = richness = number of nonzero elements in row

E = evenness = H/ln (richness)

H = Diversity = − sum [P$_i$ Ln (P$_i$)]

where Pi = importance probability in element i (element i relativized by row total)

Matrix mean = 1.0724

Variance of totals of species = 6.9188 E+03

Table A9.14 Concentration of Selected Heavy Metals in Plant Species from the Biala River Wetland (in mg/kg of dry matter)

Plant		Zn	Pb	Cd
Grass	Leaves	590	140	3
Deschampsia	Roots	2600	480	12
(*Deschampsia caespitosa*)	Leaves	185	40	2
	Leaves	35	15	1
	Roots	450	375	5
Sedge	Leaves	580	20	3
(*Carex* sp.)	Roots	6500	1050	16
	Leaves	610	240	4
	Leaves	140	45	2
	Rhizomes	550	490	5
	Leaves	23	3.5	1.7
	Leaves	87.5	16.5	<1
	Leaves	17	3.5	0.8
	Leaves	35	7.5	0.9
Reed	Leaves	120	30	2
(*Phragmites communis*)	Stems	66	25	<2
	Leaves	245	65	2
	Stems	160	25	<2
	Rhizomes	235	150	2
	Leaves	150	55	2
	Stems	74	40	<2
	Rhizomes	280	200	2
	Leaves	25	10	0.5
	Stems	45	5	1.5
	Rhizomes	235	130	2.5
	Leaves	65	10	0.5
	Stems	267.5	8.5	1
	Rhizomes	152.5	122.5	<1
	Leaves	87.5	35	1
	Stems	127.5	30	1.5
	Rhizomes	202.5	217.5	8.5
	Leaves	87.5	160	40
	Stems	77.5	9.5	<1
Cattail	Leaves	68	25	<2
(*Typha latifolia*)	Rhizomes	160	250	<2
	Clubs	57	20	<2
	Leaves	49	30	<2
	Rhizomes	265	110	<2
	Leaves	10	5	0.5
	Rhizomes	220	85	1.5
	Leaves	5	5	1
	Rhizomes	40	50	0.5
Bulrush	Leaves	470	210	4
(*Scirpus lacustris*)	Leaves	220	25	3
	Roots	1550	800	11
	Aboveground	5	5	1
Water mint	Leaves	255	20	1.5
(*Mentha aquatica*)	Stems	155	15	1.5
	Roots	550	55	3
	Leaves	142.5	26	0
	Stems	57.5	15	1
	Roots	152.5	65	<1
	Leaves	577.5	370	5
	Stems	192.5	119	1.5

continued

Table A9.14 (continued) Concentration of Selected Heavy Metals in Plant Species from the Biala River Wetland (in mg/kg of dry matter)

	Roots	167.5	203.5	2
	Leaves	224	22	2.5
	Stems	88.5	10	3
	Leaves	113.5	20	1.5
	Stems	58.5	10	2
Sparganium	Leaves	121	62	1.2
(*Sparganium* sp.)	Rhizomes	171	155	2.7
Pond weed	Whole plant	2037	797	48
(*Potamogeton natans*)				
Bog bean	Leaves	180	29	1
(*Menyanthes trifoliata*)	Stems	115	15	1
	Roots	150	80	1

Table A9.15 Concentration of Selected Heavy Metals in Different Plant Species from Literature Data (in mg/kg of dry matter)

Plant	Zn	Pb	Cd	Ref.
Typha domingensis	15,078	228	54	Dunababin (1988)
T. angustifolia				
Rhizomes			240–3000	Kneip and Hazen (1979)
Roots		0.1–0.5	0.1–0.20	Kufel (1989)
Aboveground parts		0.04–0.5	0.04–0.2	Kufel (1989)
		0.4	0.0	Kufel (1989)
	43			Seidel (1966)
		0.5–6.3		Seidel (1966)
Phragmites australis				
Leaves	12–33		>0.11	Lehtonen (1990)
Stems	27–170		<0.1	Lehtonen (1990)
Rhizomes	18–35		0.11–0.21	Lehtonen (1990)
Phragmites communis				
Aboveground parts	37			Seidel (1969)
Phragmites				
Aboveground parts		0.2–2.8	0.06–0.3	Kufel (1989)
Whole plants		0.1–5.9		Kufel (1978)
		0.6	0.00	Kufel (1989)
Parts not defined		0.5–3.4		Kufel (1989)
Rhizomes		2.70	0.10	Kufel (1989)
Roots		1.70	0.00	Kufel (1989)
Spartanium sp.				
Whole plant	35–260	0.5–31.33	0.15–1.57	Lehtonen et al. (1990)
Scirpus martimus				
Shoot	15–80		0.2–0.7	Lehtonen (1990)
Roots	30–440		0.4–6.0	
Scirpus lacustris				
Aboveground parts	50			Seidel (1966)
Carex stricta				
Aboveground parts	62			Seidel (1966)
Mentha aquatica				
Aboveground parts	78			Seidel (1966)

area of 590 m². The upper third of the plot was overgrown with cattails and the remainder with reeds (91 plants per square meter). In spite of the small plot size and short detention time, a good reduction of heavy metals was observed: up to 94% for zinc and 93% for lead (Table A9.18).

After plotting uptake as a function of time (Chapter 9, Figure 9.16), uptake of metals was correlated with time of flow using the least-square method and GRAPHER computer program. The resulting equations relate lead and zinc uptake to time:

$$YPb = 52.646 \, Lnt - 144.893$$

$$YZn = 53.482 \, Lnt - 146.282$$

where YPb and YZn = reduction of lead and Zn concentration, respectively, and t is time of flow.

Caution is advised on using the equations until there is a larger set of data involving more conditions.

Plot 2 was situated near the bridge in Kuzniczka and was overgrown bog bean. The natural configuration of the area made special construction unnecessary. The water depth was about 0.5 m. In the autumn months growth of the water mint was noticed, which accounted for about 20% of the vegetation on the plot. The flow time through the plot was 20 min, measured by tracing NaCl.

Four series of measurements were conducted that showed mean reduction of zinc and lead to be 46 and 43%, respectively (Table A9.19). In spite of the drop in air temperature on November 29 to –12°C, the heavy metals reduction was maintained at the same level.

DETERMINATION OF A COEFFICIENT "N" FOR THE MANNING EQUATION FOR FLOW-THROUGH VEGETATION

A coefficient "n" was determined for the Manning equation for the wetland overgrown by the reeds and cattails. These are plants most commonly used in wetlands constructed for wastewater treatment. The Manning equation is widely used in advance calculation of flows through wetlands.

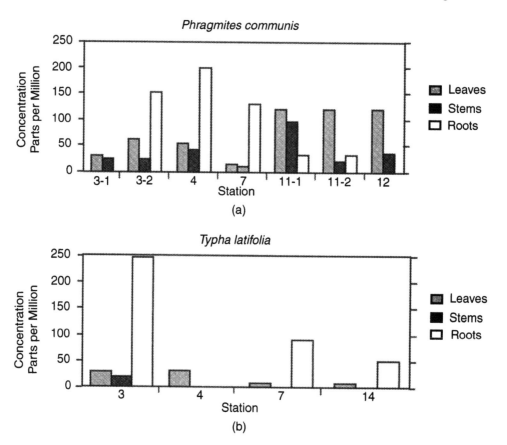

Figure A9.4　Lead concentrations in different parts of wetland plants. (a) *Phragmites* (reeds); (b) *Typha* (cattails). (*continued*)

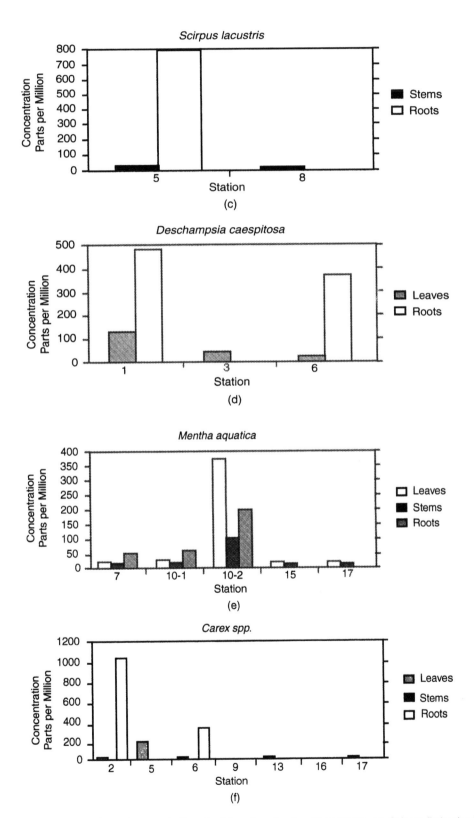

Figure A9.4 (continued) Lead concentrations in different parts of wetland plants. (c) *Scirpus* (bulrushes); (d) *Deschampsia*; (e) *Mentha* (water mint); (f) *Carex* (sedge).

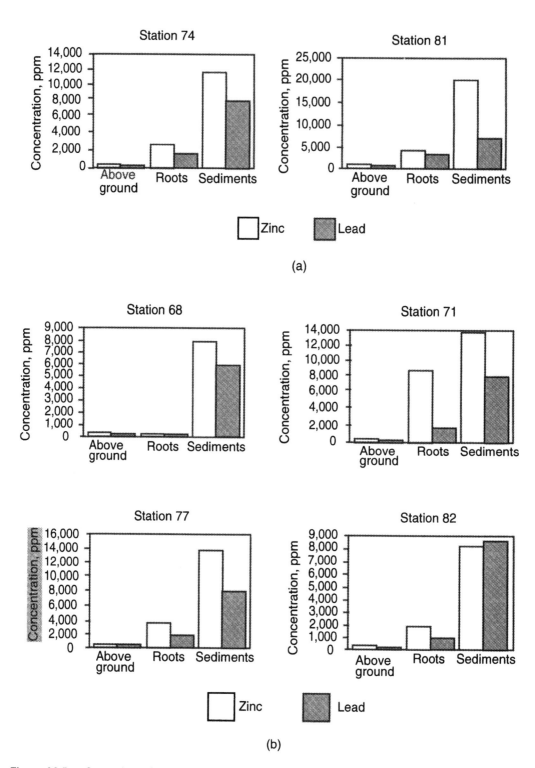

Figure A9.5   Comparison of concentrations of lead (Pb) and zinc (Zn) in plants and their sediments at ten stations in the Biala River wetland (stations mapped in Chapter 9, Figure 9.4). (a) Horsetails (*Equisetum limosum*); (b) sedge (*Carex* sp.); (c) water mint (*Mentha aquatica*). (*continued*)

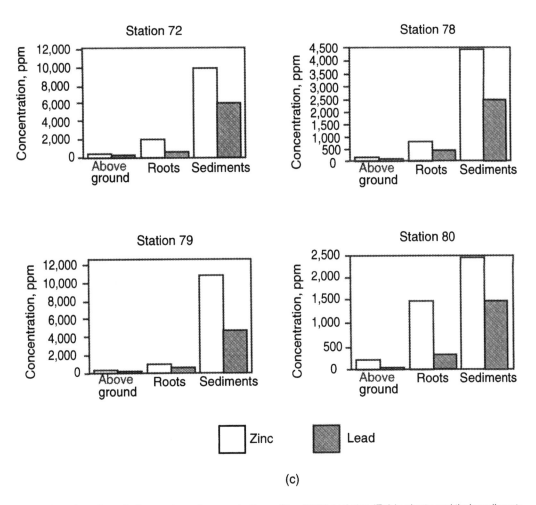

Figure A9.5 (continued) Comparison of concentrations of lead (Pb) and zinc (Zn) in plants and their sediments at ten stations in the Biala River wetland (stations mapped in Chapter 9, Figure 9.4). (a) Horsetails (*Equisetum limosum*); (b) sedge (*Carex* sp.); (c) water mint (*Mentha aquatica*).

Table A9.16   Annual Regrowth of Aboveground Biomass on Cleared Plots

Plot No.	Biomass Production (g dry matter/m²)
1	522
2	1798
3	553
4	834
5	694
6	664
7	1201
8	764
9	586
Average	846

Table A9.17   Changes of Pb and Zn Concentrations along the Biala River Wetland

Date	Concentration Zn/Pb (g/m³) on the Sampling Stations									
	1	2	3	4	5	6	7	8	9	10
4/26	1.21	1.21	—	1.45	—	1.29	—	0.82	1.2	1.18
1989	0.33	0.37	—	0.53	—	0.5	—	0.4	0.5	0.5
5/31	1.67	—	—	1.4	—	2.09	—	—	1.33	2.07
1989	0.64	—	—	0.64	—	0.9	—	—	0.62	0.83
6/28	1.24	—	—	—	—	0.94	—	—	0.49	0.83
1989	0.27	—	—	—	—	0.37	—	—	0.22	0.25
7/4	4.32	0.34	—	0.31	—	0.26	—	—	1.28	0.6
1989	1.4	0.12	—	0.14	—	0.18	—	—	0.52	0.16
9/29	1.24	0.73	1.3	1.59	—	—	—	—	1.38	—
1989	0.22	0.26	0.33	0.43	—	—	—	—	0.3	—
1/6	1.68	1.71	1.72	1.58	1.55	1.42	1.44	1.01	0.96	0.91
1990	0.34	0.35	0.33	0.34	0.31	0.28	0.29	0.25	0.26	0.21
3/10	1.48	1.41	1.28	1.35	1.25	1.18	1.01	1.01	1.02	1.05
1990	0.34	0.44	0.38	0.33	0.33	0.28	0.25	0.19	0.21	0.19
7/22	5.68	5.56	5.52	4.05	3.19	3.32	1.54	0.84	0.82	0.88
1990	1.61	1.51	1.35	1.01	0.95	0.91	0.96	0.79	0.78	0.78
9/2	1.57	1.51	0.99	1.04	1.04	0.85	0.95	0.71	0.55	0.62
1990	1.52	1.62	1.41	1.02	1.02	1.01	0.71	0.92	0.8	0.73
Data for 1978	7.5	9.96	1.95	5.75	3.37	4.46	2.4	1.51	0.96	—
	2.63	3.8	1.68	4.55	2.12	4.31	1.97	0.72	0.27	—

Table A9.18   Reduction of Zn and Pb in Waters Flowing across Experimental Plot No. 1

Time of flow (t[min])	Reduction of Zn (%)	Reduction of Pb (%)	Time of flow (t[min])	Reduction of Zn (%)	Reduction of Pb (%)
25	8	0	40	26	12
25	24	0	40	57	64
25	26	0	41	40	46
25	37	19	43	49	58
25	48	38	45	41	57
25	50	48	55	86	84
25	54	60	55	86	85
25	60	60	55	1	78
25	11	11	60	71	71
30	57	51	65	79	5
30	11	37	65	94	77
30	18	16	65	93	93
30	56	49	65	94	91
30	28	29	65	91	84
30	18	11	75	76	71
30	19	50	75	91	89
35	19	54	75	89	86
38	42	7	75	91	89

Table A9.19 Reduction of Contaminants in Waters Flowing across Experimental Plot No. 2 in 1989

Parameter		Date	Influent	Effluent	Reduction (%)
Zn	mg/l	June 28	1.37	0.57	58
		July 4	1.28	0.72	44
		Sept 29	1.38	0.86	39
		Nov 29	1.24	0.72	42
Pb	mg/l	June 28	0.50	0.21	58
		July 4	0.52	0.32	59
		Sept 29	0.30	0.17	44
		Nov 29	0.53	0.47	12
Cd	mg/l	June 28	0.01	0.004	60
		Sept 29	0.007	0.004	43
		Nov 29	0.014	0.008	43
Fe	mg/l	June 28	6.0	3.9	45
		July 4	4.4	3.5	21
		Sept 29	3.8	3.5	8
		Nov 29	4.14	2.76	33
Mn	mg/l	June 28	0.26	0.11	68
		July 4	0.2	0.07	65
		Sept 29	0.26	0.18	21
		Nov 29	0.29	0.16	45
Cu	mg/l	June 28	0.013	0.01	23
		July 4	0.011	0.007	46
		Sept 29	0.006	0.006	0
		Nov 29	0.008	0.008	0

Measurements of water flux were made on experimental plot 1 (Chapter 9, Figure 9.5) where average plant density was 91 plants per square meter. The profiles of ten cross sections are given in Table A9.20. Water flow was measured with a wire installed at the cross section 10. Changes of water elevations were monitored on gauges between the cross sections 1 and 10. The coefficient "n" studied with this method was 0.392. The highest coefficient "n" found in literature was 0.112 "for slow flow in heavily vegetated creeks."

Table A9.20 Measurements of Cross Sections in the Experimental Plot, June 6, 1990

Cross Section No	Distance from Bank (m) Depths (m)																	Width Bzw (m)	Area F (m²)	Av depth Hsr (m)
	0	0.5	1.0	1.5	2.0	2.5	3.0	3.5	4.0	4.5	5.0	5.5	6.0	6.5	7.0	7.5	8.0			
1	0	0.10	0.12	0.13	0.10	0.10	0.10	0.11	0.15	0.20	0.12	0.19	0.08					6.25	0.738	0.119
2	0	0.07	0.06	0.11	0.08	0.09	0.10	0.12	0.13	0.12	0.07	0.24	0.21	0.09				6.87	0.736	0.108
3	0	0.11	0.14	0.11	0.12	0.18	0.10	0.12	0.15	0.21	0.16	0.10	0.06	0.03				7.04	0.795	0.114
4	0	0.11	0.07	0.08	0.11	0.12	0.08	0.03	0.05	0.15	0.13	0.08	0.04	0.03	0.01			8.34	0.575	0.069
5	0	0.02	0.08	0.09	0.07	0.07	0.09	0.10	0.28	0.16	0.16	0.12	0.12	0.16	0.05	0.02	0.05	7.18	0.775	0.109
6	0	0.10	0.07	0.10	0.14	0.18	0.20	0.24	0.20	0.22	0.20	0.08						5.66	0.849	0.152
7	0	0.05	0.49	0.30	0.18	0.20	0.08	0.11	0.10	0.37	0.35	0.26						6.14	1.206	0.212
8	0	0.35	0.47	0.40	0.28	0.26	0.31	0.21										4.02	1.119	0.294
9	0	0.20	0.20	0.33	0.35	0.38	0.26	0.23	0.12									4.41	1.023	0.238
10	0	0.18	0.13	0.23	0.21	0.30	0.28	0.24	0.33	0.22								4.89	1.027	0.219

Details on Economic Valuation Methods

Lowell Pritchard, Jr.

Two economic methods used for valuating environment in Chapter 11 are explained in more detail here.

PRODUCER SURPLUS MEASURED BY REPLACEMENT COST METHOD

Derived demand is similar to final demand in that it may usually be represented by a downward-sloping curve, because each additional unit of the intermediate good has a lower marginal value to the firm (or, at low prices, the firm will demand more of the good). The total value to the firm is the sum of their marginal values for each additional unit of the good — the area beneath the demand curve. The level of an intermediate good demanded is found at the intersection of the demand curve and the supply curve for that good, where the supply curve is the marginal cost of production for the good. The area under the supply curve is the cost of production of the intermediate good. The area between the curves up to the point of intersection is the net benefit to the final-good producer for using the intermediate good (area A in Figure A11A.la) and the net benefit to the producer of the intermediate good (area B in Figure A11A.la). If the company in question produces both the final good and the intermediate good (for example, a lead refinery which also produces its own pollution treatment), then the area A + B goes to the same firm.

Figure A11A.1b shows a similar situation but with the marginal cost of pollution treatment constant over the range of quantities in question. This may be the case where operating costs are in constant proportion to the amount of lead to be treated. The area A remains as producer surplus (the difference between what the producer would have been "willing to pay" and what he actually did pay for treatment). Area C is the total cost of chemical treatment.

In Figure A11A.1c the supply curve of a less costly alternative treatment system is added, in this case wetland treatment. If the use of the service (for instance, heavy metal assimilation) had no effect on the wetland, the marginal cost would be zero and the supply curve would be a horizontal line along the x-axis. In the case shown in Figure A11A.1c there is a cost to using a wetland (loss of wetland timber, wildlife, recreational opportunities). Taking these costs into account, a firm would choose to consume q_w units of wetland heavy metal retention, or if pollution is on the axis, it would generate q_w units of heavy metal pollution. The cost of using the wetland is the area E under the supply curve for wetland treatment. If the producer had to pay the cost of using the wetland, the benefit he gets is the area A + D, representing a change in D — the cost savings from not having to use the expensive chemical treatment, plus the benefit of being able to produce at q_w instead of q_c. This area D is the total value of the wetland to the producer — the gain in producer surplus from switching to wetland treatment.

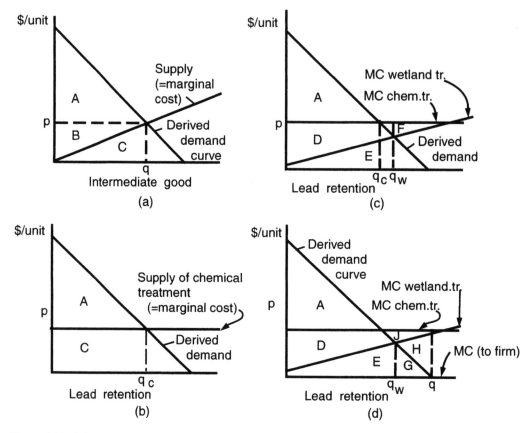

Figure A11A.1 Idealized cost curves showing producer surplus from chemical and wetland treatment of heavy metal pollution.

If the producer did not pay the cost of using the wetland (if, for instance, he used a natural wetland for "free"), then the area E is the "social" cost of using wetlands as a treatment system, since society suffers from the loss while the producer gets the benefit. In this case the producer's increase in benefit from using wetland treatment is area D + E, and society's cost is area E. The net benefit to all of society is still only area D.

However, maximum profit comes from equating marginal value (shown by the derived demand curve) and the marginal cost of treatment. If the use of the wetland is free to the producer, he will produce more pollution and demand more treatment up to the point q in Figure A11A.1d. His additional gain in moving from point q_w to point q is area G, but this area is more than offset from the additional cost to society of area G + H. Although it looks appealing to the producer, from society's point of view, the move would be more costly than it would be beneficial to the producer.

In practice it is difficult to know the derived demand curve of a firm for an intermediate service such as wetland waste assimilation, since it would involve knowing the total cost functions for a number of firms and the demand curve of society for their final products. An upper bound may be placed on the value of wetland if the notion of replacement cost is employed. Since the level of output (of pollution) for a firm may be known, replacement cost can be calculated as the difference in inexpensive wetland treatment and expensive chemical treatment. This assumes a vertical derived demand curve — if a firm using a natural wetland were treating q units of waste (as in Figure A11A.1d), it is assumed for the sake of calculation that they would continue treating q units of waste using chemical treatment, although it is recognized intuitively that they would have instead demanded less treatment at the higher cost. Using the notion of replacement cost, the measured "benefit" to the firm in Figure A11A.1d would be the area D + E + G + H + J, while the cost to

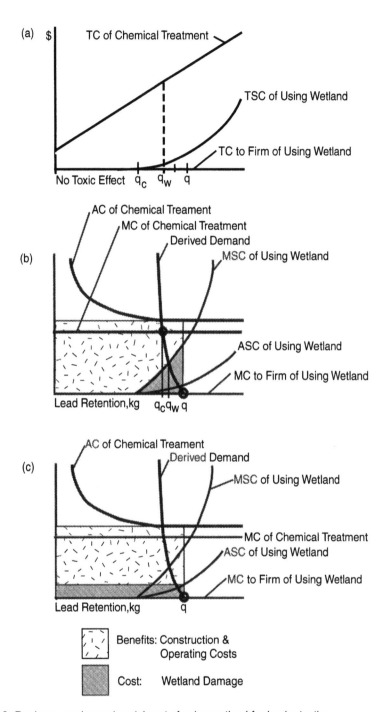

Figure A11A.2 Producer surplus and social cost of using wetland for lead retention.

society would be area E + G + H, so that the net "benefit" measured would be area D + J — overestimating the true benefit from using wetlands by the unknown area J.

Figure A11A.2 shows total (TC), average (AC), and marginal (MC) cost curves for chemical treatment where there is a cost to building a treatment plant, and then a constant unit operating cost to run the plant. The cost curves shown for wetland treatment are for total, average, and marginal "social" cost (TSC, ASC, MSC) since they represent wetland damage (a cost to society rather than the firm) from use for lead treatment. The cost to the firm of using a natural wetland for lead retention is assumed to be zero. Figure A11A.2b is analogous to Figure A11A.1a, showing that the firm would proceed to point q if the use of the wetland were free. In Figure A11A.2b the hatched area between the average and marginal cost of chemical treatment is the construction cost for a treatment plant. The total benefit (to the firm) of wetland treatment is shown as hatched rectangles in Figures A11A.2b and A11A.2c. The social cost is shown as the area under the marginal social cost curve in Figure A11A.1 and as a rectangle (average social cost multiplied by quantity) in Figure A11A.2c. These have the same value, but the area in Figure A11A.2c is easier to calculate.

CALCULATION OF CAPITALIZED VALUE OF WETLAND SERVICE

The economic idea of present value says that the right to an income stream has a price today, which is "no more or less than the sum of money now that could earn the stream of income if invested in a bank at the going rate of interest" (McCloskey, 1985). Where society is concerned, the social rate of discount is used in place of the interest rate.

In our case study, the annual benefit of a wetland for lead treatment was calculated. The capitalized value is the money (financial capital) that the right to the wetland service could be sold for. If lead treatment was the only valuable service the wetland performed, then the present value of the lead treatment would determine the price.

The annual benefit of wetland lead treatment was calculated using the replacement cost based on chemical treatment. By allowing the wetland to treat the lead waste, society earned (i.e., saved) the money that would have gone toward chemical treatment. The present value of the wetland (at the beginning of its use for lead treatment) was the amount of money which, invested elsewhere (read "put in the bank"), would have earned enough return (read "interest") to pay for the annual cost of chemical treatment. Society, or an individual, should be willing to pay no more than the present value for the right to use the wetland for lead treatment (although, if the wetland performs other independently valuable functions, society may be willing to pay for them, also).

The formula for calculating the present value PV of an annual net benefit NB_t extending N years into the future is

$$PV = \sum_{t=0}^{N} \frac{NB_t}{(1+i)^t}$$

where i is the rate of interest (for private investors) or the discount rate (where society is concerned) (Hufschmidt et al., 1983). For convenience of calculation where the annual net benefit is constant the formula can also be written

$$PV = \frac{NB_t}{i}[1] = \frac{1}{(1+i)^N}$$

(McCloskey, 1985).

Where the net benefit extends into perpetuity, the formula simplifies to

$$PV = \frac{NB_t}{i}$$

CALCULATIONS OF COST AND BENEFITS OF THE SWAMP

Economic valuation of benefits and costs to the economy from the swamp work in absorbing lead was summarized in Table 11.1. The year-by-year details of these calculations are in Table A11A.1.

Table A11A.1 Annual Costs and Benefits of Wetland Lead Treatment (1990 Dollars)

Year	Lead Retained (kg)	Benefits: Replacement Costs			Costs: Timber Value				Net Benefit: Total Benefits – Total Costs
		Capital	Operating	Total	Trees Lost	Lost Production	Lead	Total	
1970	23.1698	$324,107	$16,682	$340,789	$0	$0	$20	$20	$340,768
1971	68.598		49,391	49,391	13,200	98	60	13,357	36,033
1972	114.027		82,100	82,100	13,200	195	99	13,494	68,605.3
1973	159.456		114,809	114,809	13,200	293	139	13,631	101,177
1974	204.885		147,518	147,518	13,200	390	178	13,768	133,749
1975	250.315		180,266	180,266	13,200	488	218	13,905	166,321
1976	295.744		212,935	212,935		585	257	842	212,093
1977	341.173		245,644	245,644		682	297	979	244,665
1978	386.602		278,353	278,353		780	336	1,116	277,237
1979	432.031		311,062	311,062		878	376	1,253	309,809
1980						975		975	-975
1981						956		956	-956
1982						936		936	-936
1983						917		917	-917
1984						897		897	-897
1985						878		878	-878
1986						858		858	-858
1987						839		839	-839
1988						819		819	-819
1989						800		800	-800
1990						780		780	-780
1991						761		761	-761
.									
.									
.									
2028						39		39	-39
2029						20		20	-20
2030						0		0	0

Net present value using discount rate:

@4.00%	$324,107	$1,293,268	$1,617,375		$58,764	$1,563	$13,203	$73,530	$1,543,845
@0.00%	$324,107	$1,638,720	$1,962,827		$66,000	$1,980	$29,250	$97,230	$1,865,597

Transformities Used in Calculations

Lowell Pritchard, Jr.

Transformities used were largely from Odum (1992), who summarized a number of previous works. Emergy per mass was estimated for the evaluation of technological treatment of lead. The steps in their estimation are given below.

HYDRATED LIME

Energy and emergy flows for hydrated lime production are given in Figure A11B.3 and Table A11B.2. Terhune (1980) gave energies for (1) mining crushed and broken limestone; (2) manufacturing an average of quicklime, hydrated lime, and burned lime; and (3) transportation of raw materials to manufacturing plant (using 1972 data). It was assumed that these energies were the sum of fuel and electrical energies all converted by Terhune into fuel energies. The resultant value was multiplied in the table by the transformity of oil.

The emergy value of labor in capital and direct labor was calculated from the money paid for hydrated lime multiplied by the U.S. emergy/money ratio. The price of lime came from the Tampa plant data, and the 1988 emergy/money ratio (Odum 1992, p. 159) was used.

The emergy value of the physical capital was assumed to be small relative to direct energies and was not calculated.

CAUSTIC SODA

Figure A11B.4 and Table A11B.3 detail the production of caustic soda. Caustic soda (NaOH) and chlorine gas are produced jointly by electrolysis of brine (Wyandotte Chemicals, 1961). The industry is energy-intensive, using 1% of the country's electric power in 1967 and ranking first in energy use per dollar of shipments among heavy energy-using chemical industries (Wehle, 1974).

Plants are located near solid salt and brine deposits with cheap electricity or cheap natural gas to make it from. Diaphragm cells account for over 90% of caustic production. Effluent is evaporated down to either 50% caustic or solid caustic soda. Mercury cells are being phased out due to metal pollution problems, although some lead pollution results from use of graphite anodes in diaphragm cells (Wehle, 1974).

No data were given for salt used, but the amount was estimated conservatively based on the chemical formula for caustic production,

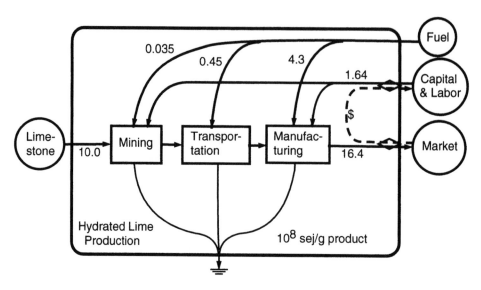

Figure A11B.3 Solar emergy basis of hydrated lime production.

Table A11B.2 Emergy Evaluation of Hydrated Lime Production

Note	Item	Solar Emergy (sej/g)	Raw Units (J, g)	Emergy/Mass (sej/g)
1	Limestone	1.00 E9		
2	Fuel in mining	3.49 E6		
3	Fuel in manufacturing	4.28 E8		
4	Fuel in transportation	4.52 E7		
5	Capital and labor	1.64 E8		
6	Hydrated lime	1.64 E9	1.00 E0 g	1.64 E9

Notes:

1. Limestone. 1.0 g. Emergy/mass 1.0 E9 sej/g (Odum 1992b, p. 156).

    Emergy = (1.0 g)(1.0 E9 sej/g) = 1.0 E9 sej.

2. Fuel in mining. 15.5 kcal/kg hydrated lime (Terhune, 1980, p. 26).

    Fuel transformity (oil) = 54,000 sej/J (Odum, 1992b, p. 88).

    Emergy = (15.5 kcal)(4186 J/kcal)(54,000 sej/J) = 3.49 E6 sej.

3. Fuel in manufacturing. 1890 kcal/kg hydrated lime (Terhune, 1980, p. 26).

    Emergy = (1890 kcal)(4186 J/kcal)(54,000 sej/J) = 4.28 E8 sej.

4. Fuel in transportation. 200 kcal/kg hydrated lime (Terhune, 1980, p. 26).

    Emergy = (200 kcal)(4186 J/kcal)(54,000 sej/J) = 4.52 E7 sej.

5. Labor and services in bought capital. Price of hydated lime $85/ton.

    (1989$) (Neil Oakes, personal communcation) U.S. 1988 emergy/money ratio 1.8 E12 sej/$ (Odum, 1992b, p. 159).

    Emergy = ($85/ton)(1 ton/9.07 E5 g)(1.8 E12 sej/$) = 1.6E8 sej/g hydrated lime.

6. Hydrated lime. Sum of inputs = 1.64 E9 sej. Amount of output = 1 g hydrated lime.

    Emergy/mass = (1.64 E9 sej)/1.0 g) = 1.64 E9 sej/g.

$$2\ NaCl + 2H_2O + electr. \rightarrow 2\ NaOH + H_2{\uparrow} + Cl_2{\uparrow}$$

An excess of salt is used, but residual salt is recycled.

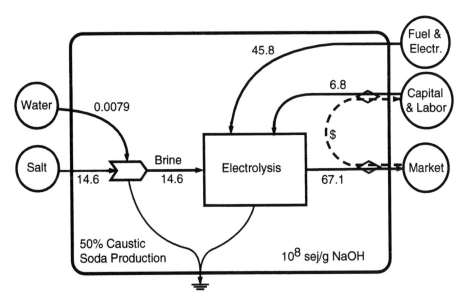

Figure A11ᴮ.4 Solar emergy basis of caustic soda production.

Table A11ᴮ.3 Emergy Evaluation of Caustic Soda Production

Note	Item	Solar Emergy (sej/g)	Raw Units (J, g)	Emergy/Mass (sej/g)
1	Salt	1.46 E9		
2	Water in steam	7.87 E5		
3	Fuel	4.58 E9		
4	Capital and labor	6.75 E8		
5	50% Caustic soda	6.71 E9	1.00 E0 g	6.71 E9
6	Fuel for evaporation	7.3 E8		
7	Solid caustic soda	7.45 E9	1.00 E0 g	7.45 E9

Notes:

1. Salt. Amount necessary from stoichiometric relation = 1.46 g/g NaOH.

   Emergy/gram = 1.0 E9 sej/g (Odum, 1992b, p. 67).

   Emergy = (1.46 g)(1.00 E9 sej/g) = 1.46 E9 sej.

2. Water in steam. 8.85 g/g NaOH produced (Wehle, 1974, p. 197). Gibbs free energy of water = 4.94 J/g. Transformity (rain) = 1.8 E4 sej/J (Odum, 1992b, p. 62).

   Emergy = (8.85 g)(4.94 J/g)(1.8 E4 sej/J) = 7.9 E5 sej.

3. Fuel for electrolysis. 7.29 E7 BTU/ton NaOH (Wehle, 1974, p. 197).

   Transformity of fuel (oil) = 54,000 sej/J. Emergy = (7.29 E7 BTU/ton) (1 ton/9.07 E5 g)(1054.8 J/BTU)(54,000 sej/J) = 4.58 E9 sej.

4. Labor and service in purchased goods. Price of caustic $350/ton.

   U.S. emergy/money ratio (1988) = 1.75 E12 sej/$ (Odum, 1992b, p. 159) Emergy = ($350/ton)(1 ton/9.07 E5 g)(1.75 E12 sej/$) = 6.8 E8 sej/g caustic.

5. 50% caustic soda. Sum of inputs = 6.71 E9 sej.

   Emergy/mass = 6.71 E9 sej/1.0 g NaOH = 6.71 E9 sej/g.

6. Fuel for evaporation to dryness. 1.17 E7 BTU/ton NaOH (Wehle, 1974, p. 197).

   Emergy = (1.17 BTU/ton)(1 ton/9.07 E5 g)(1054.8 J/BTU)(54,000 sej/J) = 7.3 E8 sej.

7. Solid caustic soda. 6.71 E9 sej + 7.3 E8 sej = 7.45 E9 sej.

   Emergy/mass = 7.45 E9 sej/1.0 g NaOH = 7.45 sej/g.

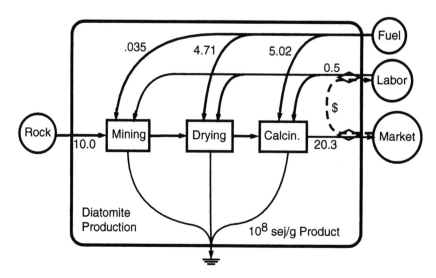

Figure A11B.5 Solar emergy basis of diatomite production.

Table A11B.4 Emergy Evaluation of Diatomite Production

Note	Item	Solar Emergy (sej)	Raw Units (J, g)	Emergy/Mass (sej/g)
1	Diatomite rock	1.00 E9		
2	Fuel in mining	3.49 E6		
3	Fuel in drying	4.71 E8		
4	Fuel in calcination	5.02 E8		
5	Labor	5.00 E7		
6	Diatomaceous earth	2.03 E9	1.00 E0 g	2.03 E9

Notes:

1. Diatomite rock. 1.0 g. Emergy/mass 1.0 E9 sej/g (Odum, 1992b, p. 156).

    Emergy = (1.0 g)(1.0 E9 sej/g) = 1.0 E9 sej.

2. Fuel in mining. Assumed same as for limestone, 15.5 kcal/kg rock (Terhune, 1980, p. 26). Fuel transformity (oil) = 54,000 sej/J (Odum, 1992b, p. 88).

    Emergy = (15.5 kcal)(4186 J/kcal)(54,000 sej/J) = 3.49 E6 sej/g.

3. Fuel in drying. 7.5 E6 BTU/ton rock (Meisinger, 1985, p. 251).

    Emergy = (7.5 E6 BTU/ton)(1 ton/9.07 E5 g)(1054.8 J/BTU)(54,000 sej/J) = 4.7 E8 sej/g.

4. Fuel in calcination. 8 E6 BTU/ton rock (Meisinger, 1985, p. 251).

    Emergy = (8 E6 BTU/ton)(1 ton/9.07 E5 g)(1054.8 J/BTU)(54000 sej/J) = 5.0 E8 sej/g.

5. Labor. Employment = 0.001429 employees/ton (Meisinger, 1985, p. 252).

    U.S. emergy/person = 3.4 E16 sej/employee.

    Emergy = (0.001429 empl/ton)(1 ton/9.07 E5 g)(3.4 E16 sej/empl) = 5.35 E7 sej/g.

6. Diatomaceous earth. Sum of inputs = 2.03 E9 sej.

    Emergy/mass = (2.03 E9 sej)/(1.0 g diatomaceous earth) = 2.03 E9 sej/g.

Fuel is used in two main processes — electrolysis and evaporation. It was assumed that the values given in BTUs were *fuel* values for the electricity used, so they were converted to joules and multiplied by fuel transformities (oil).

Human services in providing physical capital and labor were calculated from the price paid for caustic, which went to pay for human services either embodied in materials and equipment or supplied directly as labor. As with hydrated lime above, the U.S. emergy/money ratio was used to calculate the emergy value. The emergy value of physical capital itself was not calculated.

Some caustic is used at 50% NaOH (by weight) and some is evaporated down to solid to save shipping costs. The treatment plant in question uses 50% NaOH, but both are calculated for the sake of completeness.

DIATOMITE

Figure A11B.5 and Table A11B.4 give the energy and emergy flows associated with diatomite production. In nature diatomite is a soft rocklike material, also known as diatomaceous earth (Meisinger, 1985). It is sedimentary in origin, so the emergy/gram of sedimentary rock (1E9 sej/g; Odum, 1992 p. 67) was used for the natural diatomite.

Energy in mining was assumed similar to limestone mining, and the value from Terhune (1980) was used. Energy used in drying was reported by Meisinger (1985) and was assumed to be fuel energy, as was the energy he reported for calcination (done in large rotary kilns to remove organic matter and other impurities). The transformity of oil was used for the conversion to emergy for both these steps.

Meisinger reported employment for the industry as one production employee per 700 tons of processed diatomite. This value is multiplied by the U.S. emergy/person ratio to get the emergy used in supporting the labor force.

SULFURIC ACID

Figure A11B.6 and Table A11B.5 give the energy and emergy flows involved in production of dilute (20%) sulfuric acid.

LEAD

Figure A11B.7 and Table A11B.6 show the emergy evaluation of lead processing. The emergy per gram of lead metal was calculated from the emergy in lead ore and the emergy used in mining, milling, smelting, and refining.

Economically important U.S. lead deposits occur as cavity fillings or replacements, deposited by the hydrothermal brines generated by intrusive igneous masses (Pedco Environmental, 1980, p. 12) or by compacting sedimentary basins (Skinner, 1979). Ore grade varies from 3 to 8% lead (Pedco Environmental, 1980, p. 19), but the majority of ore used is from Missouri with average ore grades of 6.2% for 1988 (Woodbury, 1988, p. 591). Emergy per gram of lead ore was taken to be the same as that of volcanic intrusive rock (4.5 E9 sej/g; Odum, 1992b).

The emergy/gram of the unrefined lead metal in the ore was calculated from the emergy of the ore. When several materials are processed together as a single product (in this case, lead cycling with sulfur, carbonates, and silicates), each material has the emergy of the product (Odum, 1991). The entire emergy of the carrier was assigned to the lead metal therein. An ore that is 6.5% lead by weight will require (1/.065 =) 15.3 g of ore to make 1 g of lead metal (assuming no losses in production). Thus, the emergy of 15.3 g of ore would be necessary to make 1 g of metal (line 1 of Table A11B.6).

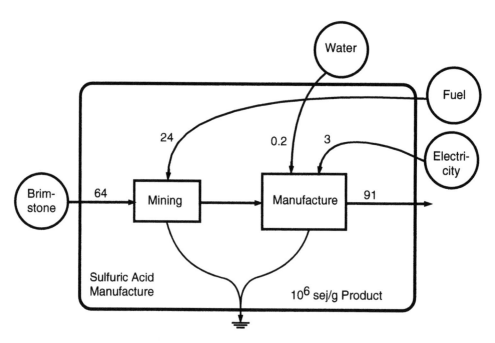

Figure A11B.6 Solar emergy basis of sulfuric acid production.

Table A11B.5 Emergy Evaluation of Sulfuric Acid Production

Note	Item	Solar Emergy (sej)	Raw Units (J, g)	Emergy/Mass (sej/g)
1	Brimstone	6.40 E7		
2	Fuel in mining	2.41 E7		
3	Electricity in production	3.17 E6		
4	Water, dilution	1.92 E5		
5	20% sulfuric acid	9.13 E7	1.00 E0 g	9.13 E7

Notes

1.   Brimstone. 0.32 g S/g H_2SO_4. Emergy/mass = 1.0 E9 sej/g (Odum, 1992b, p. 156). 20% strength acid = 0.2 g H_2SO_4/g acid.

     Emergy = (0.32 g)(.2)(1.0 E9 sej/g) = 6.4 E7 sej.

2.   Fuel in mining. 3000 BTU/lb sulfur (Lee et al., 1978, p. V4). Transformity of fuel (oil) = 54000 sej/J. 20% acid. 0.32 g S/g H_2SO_4.

     Emergy = (3000 BTU/lb)(1054.8 J/BTU)(1 lb/454 g)(0.32 g)(0.20)(54,000 sej/J) = 2.4 E7 sej.

3.   Electricity. 20 kwh/ton acid (Kastens and Hutchison, 1950, p. 13). 20% acid.

     Electricity transformity = 2 E5 sej/J.

     Emergy = (20 kwh/ton)(1 ton/9.07 E5 g)(0.20)(2.0 E5 sej/J) = 3.17 E6 sej.

4.   Water for dilution. 20% strength acid = 0.8 g water/g acid. Gibbs free energy of water = 4.94 J/g (Odum, 1992b). Groundwater transformity = 48,000 sej/J.

     Emergy = (0.8 g)(4.94 J/g)(48,000 sej/J) = 1.92 E5 sej.

5.   20% sulfuric acid. Sum of inputs = 9.13 E7 sej. 1.0 g of product.

     Emergy/mass = 9.13 E7 sej/g.

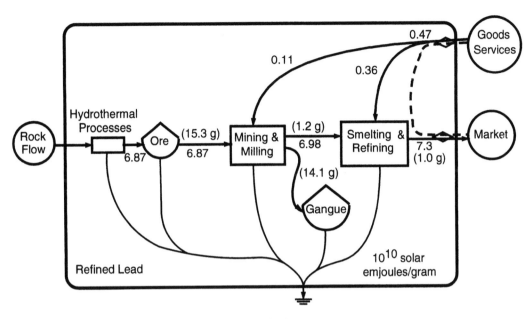

Figure A11B.7 Solar emergy basis of refined lead production.

Table A11B.6 Emergy Evaluation of Lead Processing (1 g)

Note	Item	Solar Emergy (sej)	Raw Units (J, g)	Emergy/Mass (sej/g)
	Mining and Milling			
1	Ore for 1 g metal	6.87 E10		
2	Fuel	2.78 E8		
3	Services	7.92 E8		
4	Emergy/gram Pb concentrate	6.98 E10	1.15 E0	6.07 E10
	Smelting and Refining			
5	Electricity, low estimate	2.85 E9		
6	Services	7.54 E8		
7	Emergy/gram refined Pb	7.34 E10	1.00 E0	7.34 E10

Notes:

1. Lead ore. 6.55% Pb (Kesler, 1978). Emergy/mass ore = 4.5 E9 sej/g (Odum, 1991).

    Emergy = (1.0g Pb)(1 g ore/0.0655 g Pb)(4.5 E9 sej/g ore) = 6.87 E10 sej.

2. Fuel. Assumed oil. Average of 0.3 to 1.0 kwh/lb Pb (Dammert and Chhabra, 1990, p. 52).
    Transformity of oil = 5.4 E4 sej/J.

    Emergy = (1.0 g Pb)(0.65 kwh/lb Pb)(3.6 E6 J/kwh)(1 lb/454 g)(5.4 E4 sej/J) = 2.78 E8 sej.

3. Services: (Dammert and Chhabra, 1990, p. 52); 1981 U.S. emergy/money ratio = 2.70 E12 sej/$.

    Fuel = 8.81 E-5 $/g; concentrator = 16.83 E-5 $/g; mining equipment = 3.70 E-5 $/g.

    Total = 29.34 E-5 $/g.

    Emergy = (1.0 g)(29.34 E-5 $/g)(2.70 E12 sej/$) = 7.92 E8 sej.

4. Concentrate. 87% Pb (Kesler, 1978).

5. Electricity. Average of 1.0 to 2.6 kwh/lb Pb (Dammert and Chhabra, 1990, p. 55).

    Transformity of electricity = 2.0 E5 sej/J (Odum, 1992b).

    Emergy = (1.0 g Pb)(1.8 kwh/lb Pb)(3.6 E6 J/kwh)(1 lb/454 g)(2.0 E5 sej/J) = 2.85 E9 sej.

continued

Table A11ʙ.6 (continued)   Emergy Evaluation of Lead Processing (1 g)

6.   Services: (Dammert and Chhabra, 1990, p. 52); 1981 U.S. emergy/money ratio = 2.70 E12 sej/$.
       Electricity = 8.81 E-5 $/g; plant and equipment = 13.20 E-5 $/g; management = 5.89 E-5 $/g.
       Total = 27.90 E-5 4/g.
       Emergy = (1.0 g)(27.90 E-5 $/g)(2.70 E12 sej/$) = 7.53 E8 sej.

7.   Refined lead = 1.0 g.

The energy in mining is fuel energy (line 2). Line 3 is the value of services paid for in mining and milling. Dollar values were converted to emergy using the 1981 U.S. emergy/money ratio (Odum, 1992b). Concentration is the crushing and flotation process by which the lead content of smelter feedstock is raised (Dammert and Chhabra, 1990).

Energy values in smelting and refining (line 5) reported in kilowatt-hours (Dammert and Chhabra, 1990) are multiplied by electricity transformities assuming electrolytic processes.

Some operating costs were not given explicitly in Dammert and Chhabra (i.e., labor, profit), but it is assumed that the total cost given of $0.26/lb lead includes labor and profits, so these are added in (note 6 in Table A11ʙ.6).

The emergy of refined lead metal was 7.3 E10 sej/g, including the work of nature in making ores and the work of the human economy in mining and refining the ores. The emergy in pure lead from only the human industrial economy inputs was 6.1 E9 sej/g.

The emergy per mass of lead ore may be assumed to be that of mountains. The emergy per mass of mountain flux should be the global emergy 9.44 E24 sej/year driving 2.1 E15 g/year, which works out to 4.5 E9 sej/g for mountain flux.

Emergy Evaluation of Poland

Wlodzimierz Wójcik, Jacek Stasik, and Howard T. Odum

In order to put the emergy evaluations of heavy metals in perspective and obtain values of emergy/money ratio for Poland needed in wetland evaluations, a national emergy evaluation was made of Poland. Methods were the same as those used in evaluations of the U.S. (Odum, 1996).

An overview systems diagram for Poland is drawn (Figure A12.1) identifying main resources. National tables were prepared of annual emergy flows (Table A12.1) and of emergy storages (Table A12.2). Results are summarized in Tables A12.3 and A12.4 and Figure A12.2.

Poland is compared with the U.S. in Table A12.5. In 1995 a dollar in Poland bought twice as much real wealth as the dollar in the U.S. With international monetary transactions between Poland and the U.S., the difference in buying power makes exchange of real wealth unequal.

Table A12.1 Evaluation of Emergy Flows in Poland (Area 345, 211 E6 m² [9]; see Figure A12.1)

Note	Type of Flow, units	Annual Flow (units/year)	Emergy/Unit (sej/unit)	Emergy/year (E20 sej/year)
	Environment			
1	Direct sunlight, J	1.30 E21	1	13.0
2	Rain geopotential, J	7.29 E16	10,48 9	7.7
3	Rain chemical potential, J	1.027 E18	18,189	186.8
4	Rivers geopotential, J	4.86 E17	27,806	135.2
5	Net formation of earth, t	1.62 E6	1.71 E15	27.7
6	Net loss of topsoil, J	1.43 E16	73,750	10.6
7	Earth cycle, J	3.12 E17	25,514	79.6
8	Wood use internally, J	1.81 E17	38,280	69.7
9	Sulfur, t	−2.27 E6	4.95 E6	<0.01
10	Foods	—	—	−16.9
	Fuels			
11.1	Coal, coke: import–export, J	−7.73 E17	40,000	−309.11
	production, J	5.42 E18		2,168.68
11.2	Oil, J	6.14 E17	54,000	331.31
11.3	Gas, J	1.98 E17	48,000	94.86
	Electric Power			
12	Electricity, J	−8.68 E15	2.00 E5	17.36
12.2	Electricity production, J	4.82 E17	2.00 E5	963.81
	Minerals, Import–Export			
13.1	Iron ore, J	2.92 E14	6.00 E7	175.05
13.2	Raw iron and steel, J	1.10 E13	7.83 E7	8.63
13.3	Refined iron in steel, J	−1.63 E11	15.2 E7	−0.25
13.4	Iron, steel end products, J	−3.32 E14	1.39 E8	−461.36
13.5	Aluminum: ore, net import, t	2.64 E4	1.0 E15	0.26
	alumina, t	1.02 E5	1.0 E15	1.02
	ingots, t	2.94 E4	1.6 E16	4.7
	sum	—	—	5.98
13.6	Magnesium ore, t	6.49 E2	—	<0.1
13.7	Chrome ore, t	5.54 E4	—	<1
13.8	Zinc, t	−1.48 E5	6.8 E16	−100.37
13.9	Copper, t	−2.59 E5	6.8 E16	−175.98
13.10	Titanium, t	6.97 E4	—	<1
13.11	Lead, t	−3.62 E4	7.8 E15	−2.80
13.12	Manganese, t	2.02 E5	?	?
13.13	Chrome ore, t	5.54 E4		
13.14	Silver, t	−7.73 E2	3 E20	−2,349.00
13.15	Gold, t	−0.043	4.4 E20	−0.19
13.16	Rubber, t	4.1 E4	4.30 E15	1.76
13.17	Fertilizers, t	—	—	−22.48
13.18	Cement, t	−2.37 E6	9.26 E15	−219.92

Abbreviations: t, metric ton; J, Joule; minus indicates exports in excess of imports.

References for Tables A12.1 and A12.2:

[1]	H.T. Odum, 1988a.
[2]	H.T. Odum, 1983a.
[3]	H.T. Odum, 1994.
[4]	H.T. Odum, 1992a.
[5]	H.T. Odum, 1992b.

Table A12.1 (continued)   Evaluation of Emergy Flows in Poland (Area 345, 211 E6 m² [9]; see Figure A12.1)

[6]	Glowny Urzad Statystyszny Zaklad Wydawnictw Statystycznych, 1994.
[7]	Centrum Posdstawowych Problemow Gospodarki Surowcami i Energia PAN, 1995.
[8]	Glowny Urzad Statystyszny, 1994b.
[9]	Glowny Urzad Statystyszny, 1995.
[10]	E.P. Odum, 1982.
[11]	Dziennik Ustaw Rzeczpospolitej Polskiej Nr 32. Warszawa dnia 18 kwietnia 1991r, Ustawa 131 z dnia, 21 marca 1991r.
[12]	Odum, 1996; Brown and Arding, 1997.
[13]	Szewczynski and Skrodzka, 1997.

Notes:

1. Direct sunlight, 1 sej/J [2]; annual solar energy:

   (90 E3 cal/cm²/year)(4.1868 J/cal)(1 E4 cm²/m²) = 376,812 E4 J/m²/year) [2]; (345,211 E6 m²)(376,812 E4 J/m²/year) = 1.30 E21 J/year.

2. Rain, geopotential, 10,489 sej/J [12]; average elevation, 173 m n.p.m. [9]; runoff 43 E9 m³ [9]; density, 1 E3 kg/m³ [5]; gravity, 9.8 m/s² [2];

   (173 m)(43 E9 m²)(1 E3 kg/m³)(9.8 m/s²) = 7.29022 E16 J/year.

3. Rain, chemical potential, 18,189 sej/J [12]; precipitation, 602.2 mm/year [8]; Gibbs free energy, 4.94 J/g [5]; density, 1 E6 g/m³ [2];

   (345,211 E6 m²)(602.2 E−3 m/year)(4.94 J/g)(1 E6 g/m³) = 1.027 E18 J/year.

4. River geopotential, 27,806 sej/J [12]; volume flow of the major rivers 1951–1990 [9]:

	Volume Flow		Height Change	Height × Volume
	(m³/s)	(km³/year)	(m)	(km⁴/year)
Odra	575.0	18.133	634	11.496
Rega	21.2	0.669	146	0.098
Parseta	28.9	0.911	137	0.125
Wieprza	23.6	0.744	154	0.115
Wisla	1080.0	34.059	1106	37.669
Parsleka	18.8	0.593	156	0.092
Lyne	35.1	0.107	155	0.142
Sum				49.595

   Density, 1 E3 kg/m³ [2]; gravity, 9.8 m/s² [2];
   (49.595 E12 m³/year)(1 E3 kg/m³)(9.8 m/s²) = 4.86 E17 J/year.

5. Net formation of earth, 1.71 E15 sej/t [2]; land area, 311,904 E6 m² [8]; formation rate, 31.2 g/m²/year [2]; erosion rate, 26 g/m²/year [2];

   (26 g/m²/year)(311,904 E6 m²) − (31.2 g/m²/year)(311,904 E6 m²) = −1.62 E12 g/year.

6. Net loss of topsoil, 73,750 sej/J [12]; agricultural land, 187,128 E6 m² [8]; erosion rate, 700 g/m²/year [2]; successional area, 8.72 E10 m² [9]; formation rate, 1260 g/m²/year; organic matter, 3%.

   Erosion: (187,128 E6 m²)(700 g/m²/year) − (8.72 E10 m²)(1260 g/m²/year) = 2.11176 E13 g/year; net erosion (2.11176 E13 g/year)(0.03)(5.4 E6 kcal/g)(4186 kcal/J) = 1.43 E16 J/year.

7. Earth cycle with heat upflow, 25,514 sej/J; land area, 311,904 E6 m² [8]; heat flow of old stable land, 1 E6 J/m²/year [2];

   (311,904 E6 m²)(1 E6 J/m²/year) = 3.12 E17 J/year.

8. Wood use, 38,280 sej/J [12]; total production, 21,631 E3 m³ [6]; density, 0.7 E6 g/m³ [12];

   (21,631 E3 m³)(0.7 E6 g/m³) = 1.5141 E13 g = 15.1417 E6 t.

   Export, 2 E6 t [9]; production — export, 13.1414 E6 t/year; energy, 13.8 J/t [2]; production — export: (13.1414 E6 t)(13.8 E9 J/t) = 1.81 E17 J/year.

9. Sulfur, 4.95 E6 sej/t; import, 28.0 tys.t [7]; export, 2293.4 tys.t [7]; import–export, −2.27 E6 t/year.

Table A12.1 (continued)   Evaluation of Emergy Flows in Poland (Area 345, 211 E6 m² [9]; see Figure A12.1)

10.	Foods: Energy J/year: (import–export mass)(energy/mass)

        Meat (70 thousand t)(15.8 E9 J/t) = 1.106 E15 J/year [2]

        Fish (62.488 thousand t)(3.642516 E9 J/t) = 2.276135398 E14 J/year [13]

        Grains (2386.5 thousand t)(13.9 E9 J/t) = 3.317236 E16 J/year [2]

        Milk (−429 million l)(2.0264112 E6 J/dm³) = −8.693304048 E14 J/year [2]

        Wood (−1895.5 thousand t)(13.8 E9 J/t) = −2.61579 E16 J/year [2]

        Sugar (−112.2 thousand t)(16.7 E9 J/t) = −1.87374 E15 J/year [2]

        Potatoes (−365 thousand t)(2.76 E9 J/t) = −1.0086 E15 J/year [13]

        Butter (−18,517 t)(3.1 E10 J/t) = −5.799 E14 J/year [13]

        Alcohol (128.6 thousand t)(16.7 E9 J/t) = −2.14762 E15 J/year [2]

    Emergy flow = (energy/year)(emergy/energy)

        Meat (1.106 E15 J/year)(4.00 E6 sej/J) = 4.4 E21 sej/year

        Fish (2.276135398 E14 J/year)(2.00 E6 sej/J) = 4.6 E20 sej/year

        Grains (3.317236 E16 J/year)(68,000 E6 sej/J) = 2.3 E21 sej/year

        Wood (−2.61579 E16 J/year)(38,200 E6 sej/J) = −1.0 E21 sej/year

        Sugar (−1.87374 E15 J/year)(85,000 E6 sej/J) = −1.6 E20 sej/year

        Butter (−5.799 E14 J/year)(1.30 E6 sej/J) = −7.5 E21 sej/year

        Alcohol (−2.14762 E15 J/year)(6.0 E4 sej/J [2]) = −1.3 E20 sej/year

        Sum   −16.9 E20 sej/year

Fuels:

11.1	Coal and coke. 40,000 sej/J [12].

(a) Hard coal: import,129 E3 t/year [6]; export 22,968 E3 t/year [6]; energy, 30.65 E9 J/t [2]; import–export, −2.2839 E7 t/year; import–export energy: (−2.28 E7 t/year)(30.65 E9 J/t) = −7.00 E17 J/year.

Hard coal production, 130,479 E3 t [6];

Energy (130,479 E3 t)(30.65 E9 J/t) = 4.00 E18 J/year.

(b) Lignite: import, 1 E3 t/year [6]; export, 909 E3 t/year [6]; energy, 6.26 E9 J/t [2]; import–export, −908 E3 t/year; energy: (−908 E3 t/year)(16.29 E9 J/t) = −1.48 E16 [J/year].

Lignite production, 68,105 E3 t [6];

energy: (68,105 E3 t)(16.26 E9 J/t) = 1.107 E18 J/year.

(c) Coke: import, 3 E3 t/year [6]; export, 1892 thousand t = −1892 E3 t/year [6]; import–export, −1891 E3 t/year; actual energy, 30.65 E9 [J/t] [2]; (−1891 E3 t)(30.65 E9 J/t) = −5.80 E16 J/year

Coke production, 0.10,282 E3 t [6];

energy (10,282 E3 t)(30.65 E9 J/t) = 3.15 E17 J/year.

11.2	Oil, 54,000 sej/J [12]; import, 13,674.3 E3 t/year [7]; export, 40 E3 t/year [7]; import–export, 13,607.3 E3 t/year; energy, 4.5 E10 J/t [2]; (import–export)(energy);

(13,634.3 E3 t/year)(4.5 E10 J/t) = 6.14 E17 J/year.

11.3	Gas, 48,000 sej/J [12]; import, 198,060 E12 J/year [6];export, 548 E12 J/year [6];

(import–export) = 1.98 E17 J/year.

Electric power, 2.0 E5 sej/J [12].

12.1	Electricity exchange.

Import, 5600 E6 kWh [6]; export, 8011 E6 kWh [6]; energy, 3.6 J/kWh [1]; import–export, −2411 E6 kWh;

(−2411 E6 kWh)(3.6 E6 J/kWh) = −8.68 E15 J/year.

Table A12.1 (continued)   Evaluation of Emergy Flows in Poland (Area 345, 211 E6 m² [9]; see Figure A12.1)

12.2	Electric power production.
	(133,863 E6 kWh)(3.6 E6 J/kWh) = 4.82 E17 J/year.

Minerals.

13.1	Iron ore, 1.39 E8 sej/J [12]; import, 8776.2 E3 t/year [7]; export, 119 E3 t/year [7]; import–export, 8657 E3 t/year; energy, 3.37 E7 J/t [1];
	(8657 E3 t/year)(3.37 E7 J/t) = 2.92 E14 J/year.
13.2	Raw iron and steel, 7.83 E7 [12]; import 139.2 E3 t/year [7]; export, 17.3 E3 t/year [7]; energy, 9.04 E7 J/t [1]; import–export, 121.9 E3 t/year;
	(121.9 E3 t/year)(9.04 E7 J/t) = 1.10 E13 J/year.
13.3	Refined iron and steel, 15.2 E7 sej/J; import, 200 t/year [7]; export, 2 E3 t [7]; import–export, –1800 t/year; (–1800 t/year)(9.04 E7 J/t) = –1.63 E11 J/year.
13.4	Iron and steel end products, 1.3 E9 sej/J; import, 565,885 t/year [9]; export, 4,237,500 t/year [9]; energy, 9.04 E7 [J/t] [1]; import–export, –3,671,615 t/year; (–3,671,615 t/year)(9.04 E7 J/t) = –3.32 E14 J/year.
13.5	Aluminum ore, bauxite, 1 E15 sej/t [12]; import, 26.4 E3 t [7]; export, 0 [7]; import–export, 26.4 E3 t/year.
	Alumina, 1 E15 sej/t; import, 10.24 E5 t [7]; export, 1 E2 t [7]; import–export, 102.3 E3 t/year.
	Aluminum ingots, 1.6 E16 sej/t [12]; import, 32.9 E3 t [7]; export, 3.5 E3 t [7]; import–export, 29.4 E3 t/year.
13.6	Magnesium, >1 E9+ sej/t; import, 651 t [7]; export, 2 t [7]; import–export, 649 t/year.
13.7	Chrome ore; import, 59.1 E3 t [7]; export, 3.7 E3 t [7]; import, 11 t [7]; import–export, 55,411 t/year.
13.8	Zinc, 6.8 E16 sej/t [12]; import, 1.5 E3 t [7]; export, 69.5 E3 t [7]; import–export, –147.6 E3 t.
13.9	Copper, 6.8 E16 sej/t [12]; import, 2.2 E3 t [7]; export, 261 E3 t [7]; import–export, –258.8 E3 t.
13.10	Titanium; ore import, 69 E3 t [7]; Ti oxide import, 7 E4 t [7]; sum 69.7 E3 t/year.
13.11	Lead ore and metal, 7.8 E15 sej/t; import, 7.3 E3 t/year; export, –43.5 E3 t/year; import–export, –36.2 E3 t/year.
13.12	Manganese ore and metal; import, 201.9 E3 t [7] and 198 t [7]; import–export, 202,098 t/year.
13.13	Silver, 3 E20 sej/t [12]; import, 11 t [7]; export, 784 t [7]; import–export, –773 t/year.
13.14	Gold 4.4 E20 sej/t [12]; import, 16 kg [7]; export, 59 kg [7]; import–export, –43 kg/year.
13.15	Rubber, 4.3 E15 sej/t [12]; import, 90,924 t/year [9]; export, 49,961 t/year [9]; energy, 1.47 E10 J/t [2]; import–export, 40,963 t/year. (40,963 t/year)(1.47 E10 J/t) = 6.02 E14 J/year.
13.16	Fertilizers
	(Mass import–export)(emergy/mass) [12]

Nitrogen: (–424.5 E3 t/year)(3.80 E15 sej/t) = –16.2 E20 sej/year

Phosphatic: (–136.7 E3 t/year)(3.50 E15 sej/t) = –4.78 E20 sej/year

Potassic: (–158.29 E3 t/year)(9.5 E14 sej/t) = –1.50 E20 sej/year

Sum   –22.48 E20 sej

13.17	Cement, 9.26 E15 sej/t [12]; import, 26 E3 t/year [6]; export, 2401 E3 t/year [6]; import–export, –2375 E3 t/year.

Table A12.2 Solar Emergy Storages in Poland

Note	Item	Energy Stored (J)	Emergy/Unit (sej/J)	Emergy (E23 sej)
1	Fuel resources			
	Gas	7.064 E18	4.8 E4	3.39
	Oil	2.201 E17	5.4 E4	0.118
	Lignite	2.342 E20	3.7 E4	86.6
	Hard coal	1.989 E21	4.0 E4	795
	Sum	—	—	885.1
2	Soils			
	Farmed land	1.18 E19	6.25 E4	7.4
	Forested land	1.60 E19	6.25 E4	10.0
	Sum	2.78 E19	6.25 E4	17.4
3	Wood	2.62 E19	3.82 E4	10.0

Abbreviations: sej = solar emjoules; t = metric ton; J = Joule; E3 = $\times 10^3$ = $\times 1000$; E–3 = $\times 10^{-3}$ = $\times 0.001$.

References
[1]   H.T. Odum, 1983a.
[2]   Glowny Urzad Statystyczny Zaklad Wydawnictw Statystycznych, 1994.
[3]   Centrum Podstawowych Problemow Gospodarki Surowcami i Energia PAN, 1995.

Notes:

1.   Fuel resources and reserves.

Gas [3]:

$(200,859.5 \text{ mln.m}^3)(35,169 \text{ E3 J/m}^3) = 7.064 \text{ E18 J}$

Oil [3]:

$[4,893 \text{ E3 t}](4.5 \text{ E10 J/t}) = 2.201 \text{ E17 J}$

Lignite [3]:

$(14,401,897 \text{ E3 t})(16.26 \text{ E9 J/t}) = 2.342 \text{ E20 J}$

Hard coal [3]:

$(64,889,459 \text{ E3 t})(30.65 \text{ E9 J/t}) = 1.989 \text{ E21 J}$

2.   Soils [2].

Farmed soils: 18,713 E3 ha; 27.9 t organics/ha; 3.0% organic matter; 5.4 kcal/g = 2.26 E10 J/t [1].

Energy stored: (18,713 E3 ha)(27.9 t organics/ha)(2.26 E10 J/t) = 1.18 E19 J.

Forested soils: 8917 E3 ha [2]; 0.18 m thick; organic matter, 3.0%; 5.4 kcal/g organic = 2.26 E10 J/t [1]; density 2.47 T/m³.

Energy stored: (8917 E7 m³)(0.18 m)(0.03)(1.47 t/m³)(2.26 E10 J/t organics) = 1.60 E19 J.

3.   Wood

Total forested area, 8917 E3 ha [2]; biomass, 20,000 g/m² = 200 t/ha [1]. Energy content: (3.5 kcal/g)(4186 J/kcal)(1 E6 g/t) = 1.47 E10 J/t [1]. Energy storage: (8917 E3 ha)(200 t/ha)(1.47 E10 J/t) = 2.62 E19 J.

Table A12.3 Summary of Energy Flows for Poland (see Figure A12.2)

Letter		Sej/Year
R	Emergy use from renewable resources	186.79 E20
N_0	Dispersed rural	10.56 E20
N_1	Concentrated use	1,859.57 E20
N_2	Export without use	309.11 E20
N	Sum	2,179.24 E20
F	Import minerals and fuels	603.48 E20
G	Import goods	38.57 E20
P_2I_3	Import services	7.58 E21
I	Money paid for imports	18,834.4 million $
I_1	Money paid for minerals and fuels	2,162 million $
I_2	Money paid for food and goods	13,757 million $
I_3	Money paid for services	2,916 million $
E	Money paid for exports	14,143.1 million $
P_1E_3	Exported services	1.10 E22
B	Exported products	960.68 E20
X	Gross National Product	85.853 E9
	Money per person	$2,232 $/ind
P_2	Ratio emergy/$	2.60 E12 sej/$
P_1	Ratio emergy/$ and for its exports	3.41 E12 sej/$

Notes:

R   Rain chem potential — Table A12.1, footnote 3

N_0 Net loss of topsoil — Table A12.1, footnote 6

N_1 Coal internal use

        Production: 2168.68 E20 sej/year

        Import–export: –309.11 E20 sej/year

        Sum: 1859.57 E20 sej/year

N_2 Coal (export–import): 309.11 E20 sej/year

F   Imported minerals and fuels

        Oil: 331.31 E20 sej/year

        Gas: 94.86 E20 sej/year

        Iron ore: 175.05 E20 sej/year

        Bauxite: 2.64 E19 sej/year

        Sum: 603.48 E20 sej/year

G   Imported goods

        Raw Fe and steel: 8.63 E20 sej/year

        Aluminum: 5.72 E20 sej/year

        Rubber: 1.76 E20 sej/year

        Fertilizer: 22.49 E20 sej/year

        Sum: 38.57 E20 sej/year

P_2I_3 Imported services: 7.58 E21

I   Import: 18,834.4 million $

I_1 Money paid for minerals and fuels

        Oil and gas: 3487.9 million ZL

        Petroleum: 44.0 million ZL

        Iron and nonferrous ore: 381.7 million ZL

        Sum: 4613.6 million ZL

        4613.6 million ZL divided by 2.134 ZL/$ = 2161.949 million $

Table A12.3 (continued)   Summary of Energy Flows for Poland (see Figure A12.2)

I_2	Money paid for food and goods: 29,357 million ZL = 13,756.79 million $
I_3	Money paid for services: $I_3 = I - (I_1 + I_2)$
	18,834.4 million $ − (2161.949 million $ + 13,756.79 million $) = 2,915.661 million $
E	$ paid for export: 14,143.1 million $
$E_1 + E_2$	$ paid for products: 23,303.1 million ZL = 10,919.9 million $
E_3	$ paid for services: $E3 = E - (E1 + E2)$
	14,143.1 million $ − 10,919.9 million $ = 3223.2 million $
P_1E_3	Exported services: 1.10 E22
B	Exports
	Refined Fe and steel: 0.25 E20 sej/year
	Fe and steel products: 461.36 E20 sej/year
	Zinc: 100.37 E20 sej/year
	Copper: 175.98 E20 sej/year
	Lead: 2.8 E20 sej/year
	Cement: 219.92 E20 sej/year
	Sum: 960.68 E20 sej/year
P_1	Embodied energy to $ of country for its exports
	P1 = (R + N0 + N1 + F + G + P2I3)/X = 3.41 E12 [sej/$]
P_2	Average emergy/money ratio of other nations
B'	Emergy which goes out in exported machines, steel, etc. calculated as total coal used minus that used for power plants and home heating
	(2168.68 E20 sej/year − 309.11 E20 sej/year − 963.81 E20) = 895.76 E20

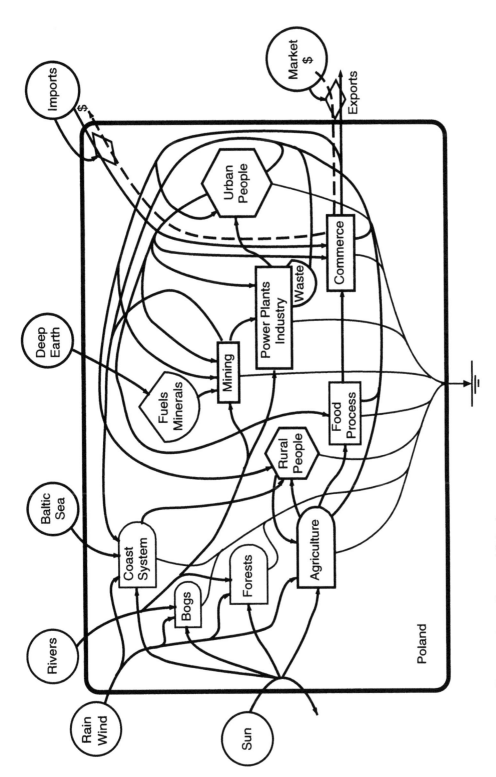

Figure A12.1 Energy systems diagram of Poland.

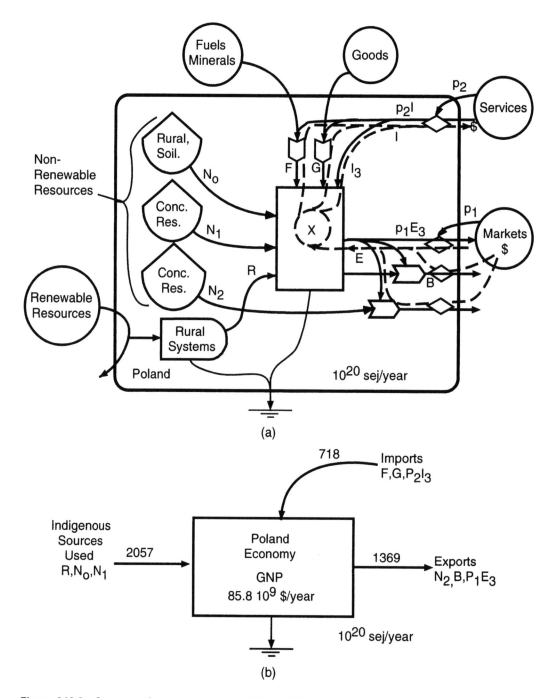

Figure A12.2 Summary of emergy evaluation of Poland (Table A12.1).

Table A12.4 Indices for Emergy Overview of Poland (See Figure A12.2)

Note	Item	Letter	Value
1	Renewable embodied emergy flow	R	186.79 E20 sej/year
2	Flow: indigenous nonrenewable reserves	N	2179.24 E20 sej/year
3	Flow of imported embodied emergy	F + G + P2l3	717.85 E20 sej/year
4	Total embodied emergy inflows	R + N + F + G + P2l3	3083.88 E20 sej/year
5	Total embodied emergy used	U = N0 + N1 + R + F + G + P2l3	2774.77 E20 sej/year
6	Total exported embodied emergy	B + P1E3	1070.68 E20 sej/year
7	Fraction of emergy from home sources	(N0 + N1 + R)/U	0.74
8	Export–import	(N2 + B + P1E3) – (F + G +P2l3)	661.94 E20
9	Ratio export/import	(N2 + B + P1E3)/F + G + P2l3)	1.92
10	Fraction used locally renewable	R/U	0.07
11	Fraction of use purchased	(F + G + P2l3)/U	0.727
12	Fraction used that is imported service	(P2l3)/U	0.26
13	Fraction of use that is free	(R + N0)/U	0.07
14	Ratio of concentrated to rural	(F + G + P2l3 + N1)/(R + N0)	13.06
15	Use per unit area	U/area	2.18 E12 sej/m^2
16	Use per capita[a]	U/population	7.21 E15
17	Renewable capacity at present standard	(R/U)(population)	2.53 E6
18	Developed capacity at same standard	(8)(R/U)(population)	20.71 E6
19	Ratio of use to GNP	P1 = U/PKB	3.23 E12 sej/$
20	Electricity/per capita	963.81 E20/38,459 thousand	2.51 E15 sej/person

[a] Population of Poland: 38,459 million.

Table A12.5 Comparison of Emergy Flows and Indices for the U.S. and Poland

No.	Symbols	Units	U.S. 1983	Poland 1983	Poland 1993
1	R	sej/year	82.4 E22	15.9 E21	18.679 E21
2	N	sej/year	534.6 E22	254.25 E21	217.924 E21
3	F + G + P_2l_3	sej/year	193.6 E22	112.91 E21	71.785 E21
4	R + N + F + G + P_2l_3	sej/year	810.6 E22	383.06 E21	308.388 E21
5	U = No + N1 + R + F + G + P_2l_3	sej/year	785.1 E22	330.46 E21	277.477 E21
6	B + P + 1E$_3$	sej/year	87.0 E22	39.55 E21	107.068 E21
7	(No + N1 + R)/U		0.76	0.66	0.74
8	(N2 + B + P_1E_3) – (F + G + P_2l_3)	sej/year	4.1 E22	61.84 E21	66.194 E21
9	(N2 + B + P_1E_3)/(F + G + P_2 + l_3)		0.57	1.55	1.92
10	R/U		0.10	0.05	0.07
11	(F + G + P_2l_3)/U		0.25	0.34	0.727
12	(P_2l_3)/U		0.18	0.07	0.26
13	(RNO)/U		0.22	0.05	0.07
14	(F + G + P_2l_3 + N1)/(R + No)		3.4	18.5	13.06
15	U/area	sej/m^2	8.4 E11	1.06 E12	2.18 E12
16	U/population	sej/person	3.4 E16	9.58 E15	7.21 E15
17	(R/U)(population)		23.4 E16	1.66 E6	2.59 E6
18	(8)(R/U)(population)		187.2 E6	13.3 E6	20.71 E6
19	P1 = U/PKB	sej/$	2.4 E12	6 E12	3.23 E12
20	Electricity per capita	sej/person	1.5 E16	1.8 E15	2.51 E15

References to Laws Cited in Chapter 13

Jay D. Patel

Note 1
Title 30, United States Code Annotated, Section 22, 1986, West Publishing, St. Paul, MN.
Note 2
Title 30, United States Code Annotated, Sections 21–54, 1986, West Publishing, St. Paul, MN.
Note 3
Title 43, United States Code Annotated, Sections 1701–1782, 1986, West Publishing, St. Paul, MN.
Note 4
Title 43, United States Code Annotated, Section 1702(c), 1986, West Publishing, St. Paul, MN.
Note 5
Title 43, Code of Federal Regulations, Section 3809.0-5(k), U.S. Government Printing Office, Washington, D.C.
Note 6
Title 43, Code of Federal Regulations, Section 3809.1-4, U.S. Government Printing Office, Washington, D.C.
Note 7
Title 43, Code of Federal Regulations, Section 3809.2-1(a), U.S. Government Printing Office, Washington, D.C.
Note 8
Title 43, Code of Federal Regulations, Section 3809.1-6(a), U.S. Government Printing Office, Washington, D.C.
Note 9
Title 36, Code of Federal Regulations, Section 228.4(a), U.S. Government Printing Office, Washington, D.C.
Title 36, Code of Federal Regulations, Section 3809.1-3, U.S. Government Printing Office, Washington, D.C.
Note 10
Title 36, Code of Federal Regulations, Section 228.5(a), U.S. Government Printing Office, Washington, D.C.
Note 11
Title 42, United States Code Annotated, Section 4331(a), 1994, West Publishing, St. Paul, MN.
Note 12
Title 42, United States Code Annotated, Section 331(b)(2), 1994, West Publishing, St. Paul, MN.
Note 13
Title 42, United States Code Annotated, Sections 4332–4333, 1994, West Publishing, St. Paul, MN.
Note 14
Title 42, United States Code Annotated, Section 4333(C), 1994, West Publishing, St. Paul, MN.

Note 15

Title 40, Code of Federal Regulations, Section 1508.18, U.S. Government Printing Office, Washington, D.C.

Note 16

Stryker's Bay v. Karlen, 444 U.S. 223 (1980).

Note 17

Kleppe v. Sierra Club, 427 U.S. 390 (1976).

Note 18

Title 42, United States Code Annotated, Section 4321, 1994, West Publishing, St. Paul, MN.

Note 19

Title 40, Code of Federal Regulations, Section 268.40, U.S. Government Printing Office, Washington, D.C.

Note 20

Title 40, Code of Federal Regulations, Section 266.106, U.S. Government Printing Office, Washington, D.C.

Note 21

Title 40, Code of Federal Regulations, Section 268.35, U.S. Government Printing Office, Washington, D.C.

Note 22

Title 42, United States Code Annotated, Sections 7401–7671(q), 1994, West Publishing, St. Paul, MN.

Note 23

Title 42, United States Code Annotated, Section 7401(b)(1), 1994, West Publishing, St. Paul, MN.

Note 24

Title 33, United States Code annotated, Sections 1251–1387, 1986, West Publishing, St. Paul, MN.

Note 25

Title 33, United States Code Annotated, Section 1251(a), 1994, West Publishing, St. Paul, MN.

Note 26

Title 33, United States Code Annotated, Sections 1311(a) and 1342, 1986, West Publishing, St. Paul, MN.

Note 27

Title 40, Code of Federal Regulations, Section 141.11(b), U.S. Government Printing Office, Washington, D.C.

Note 28

Title 33, United States Code Annotated, Section 1329(a)(1), 1987, West Publishing, St. Paul, MN.

Note 29

Title 16, United States Code Annotated, Sections 703–711, 1986, West Publishing, St. Paul, MN.

Note 30

Title 16, United States Code Annotated, Sections 1531–1543, 1986, West Publishing, St. Paul, MN.

References

Adriano, D.C., 1986. *Trace Elements in the Terrestrial Environment*. Springer-Verlag, New York, 219–262.

Adriano, D.C., 1992. *Biogeochemistry of Trace Metals*. Lewis Publishers, Boca Raton, FL, 513 pp.

Adriano, D.C. and Brisben, I.L. (Editors), 1978. Environmental Chemistry and Cycling Process. Division of Technical Information, Department of Energy, Oak Ridge, TN, 911 pp.

Ahrens, L.H., 1954. The lognormal distribution of the elements I. *Geochimica et Cosmochimica Acta*, 5: 49–73; (Part 2) 6: 121–131.

Aiken, G.R., McKnight, D.M., Wershaw, R.L., and MacCarthy, P. (Editors), 1985. *Humic Substances in Soil, Sediment and Water*. Wiley-Interscience, New York.

Alberts, J.J. and Giesy, J.P., 1983. Conditional stability constants of trace metals and naturally occurring humic materials: application of equilibrium models and verification with field data. In: R.F. Christman and E.T. Gjessing (Editors), *Aquatic and Terrestrial Humic Materials*. Ann Arbor Science, Ann Arbor, MI, 333–348.

Alcamo, J., 1991. Heavy Metals in the Atmosphere. In: *Options*, September, International Institute of Applied Systems Analysis, Laxenburg, Austria, 11–12.

Allen, H.E. (Editor), 1995. *Metal Contaminated Aquatic Sediments*. Ann Arbor Press, Ann Arbor, MI, 290 pp.

Allen, H.E., Perdue, E.M., and Brown, D., 1991. *Metals in Groundwater*. Lewis Publishers, Boca Raton, FL, 300 pp.

Alloway, B.J. (Editor), 1990. *Heavy Metals in Soils*. John Wiley & Sons, New York.

Althaus, H., 1966. Biologische mit Flechtbinsen. *Das Gas und Wasserfach* 107, Heft 18, 486–488.

American Public Health Association, 1985. *Standard Methods for the Examination of Water and Wastewater*, 16th ed. American Public Health Association, New York.

Anderson, R. and Rockel, M., 1991. Economic Valuation of Wetlands. Discussion Paper No. 065. American Petroleum Institute, Washington, D.C.

Ankley, G.T., Phipps, G.I., Leonard, E.N., Benoit, D.A., Mattson, V.R., Kosian, P.S., Cotter, A.M., Dierkes, J.R., Hansen, D.J., and Mahoney, J.D., 1991. Acid volatile sulfide as a factor mediating cadmium and nickel bioavailability in contaminated sediments. *Environmental Toxicology Chemistry*, 10: 1299–1307.

Asami, T., Masatsugu, K., and Orikasa, K., 1995. Distribution of different fractions of cadmium, zinc, lead and copper in unpolluted and polluted soils. *Water, Air, and Soil Pollution*, 83: 187–194.

Assche, F.V. and Clijsters, H., 1990. A biological test system for the evaluation of the phytotoxicity of metal-contaminated soils. *Environmental Pollution*, 66: 157–172.

Baker, K.A., Fennessy, M.S., and Mitsch, W.J., 1991. Designing wetlands for controlling coal mine drainage: an ecologic-economic modelling approach. *Ecological Economics*, 3: 1–24.

Bansal, M.K., 1973. Atmospheric reaeration in natural streams. *Water Research*, 7: 769–782.

Banus, M.D., Valiela, I., and Teal, J.M., 1975. Lead, zinc and cadmium budgets in experimentally enriched salt marsh ecosystems. *Estuarine Coastal Marine Science*, 3: 421–430.

Baudo, R., 1987. Heavy metal pollution and ecosystem recovery. In: O. Ravera (Editor), *Ecological Assessment of Environmental Degradation, Pollution and Recovery*. Elsevier, Amsterdam, 325–352.

Baumol, W.J. and Blinder, A.S., 1985. *Economics: Principles and Policy*, 3rd ed. Harcourt Brace Jovanovich Publishers, New York.

Beaty, R.D., 1988. *Concepts, Instrumentation and Techniques in Atomic Adsorption Spectrophotometry*. Perkin-Elmer Corporation, Norwalk, CT.

Bechtel Environmental, Inc., 1991. Final design report for the remedial design at the Sapp Battery Site, Vol. I–III. Bechtel Environmental, Inc., Oak Ridge, TN.

Becker, J., 1981. Photographic Vegetation Damage Assessment, Sapp Battery Salvage Site. Report of Remote Sensing Center, Florida Department of Transportation, Florida Department of Environmental Regulation, Tallahassee, 11 pp.

Behmanesh, N., Allen, D.T., and Warren, J.L., 1992. Flow rates and compositions of incinerated waste streams in the United States. *Journal of Air Waste Management Association*, 42(4): 437–442.

Bell, F.W., 1989. Application of Wetland Valuation Theory to Commercial and Recreational Fisheries in Florida. Report No. 95. Florida Sea Grant College, Tallahassee.

Bender, M.E., 1973. Water quality models and aquatic ecosystems, status, problems, and perspectives. In: R.A. Deninger (Editor), *Models for Environmental Pollution Control*. Ann Arbor Science, Ann Arbor, MI, 137–153.

Benjamin, M.M. and Honeyman, B.D., 1992. Trace metals. In: S.S. Butcher, R.J. Charlson, G.H. Orians, and G.V. Wolfe (Editors), *Global Biogeochemical Cycles*. Academic Press, London, 317–352.

Berriman, L.P., 1995. Electric car may break taxpayer's bank. Feature article, *New Haven Register* (Connecticut), June 14, 1995, p. A17.

Berti, W.R. and Jacobs, L.W., 1998. Distribution of trace elements in soil from repeated sewage sludge applications. *Journal of Environmental Quality*, 27: 1280–1286.

Bertine, K.K. and Goldberg, E.D., 1971. Fossil fuel combustion and the major sedimentary cycle. *Science*, 173: 233–235.

Best, G.R., 1987. Natural wetland — southern environment: wastewater to wetland, where do we go from here? In: K.R. Reddy and W.H. Smith (Editors), *Aquatic Plants for Water Treatment and Resource Recovery*. Magnolia Publishing, Orlando, FL, 99–120.

Best, G.R., Tuschall, J.R., Brezonik, P.L., Butner, J.R., DeBusk, W.F., Ewel, K.C., Hernandez, A., and Odum, H.T., 1982. Fate of Selected Heavy Metals in a Forested Wetland Ecosystem. Report by Center for Wetlands, University of Florida, to Corvallis Environmental Research Laboratory, Corvallis, OR, 97330.

Beyer, W.M., Audet, D.J., Morton, A., Campbell, J.K., and LeCaptain, L., 1998. Lead exposure of waterfowl ingestion in Coeur D'Aleve River basin sediments. *Environmental Quality*, 27: 1533–1538.

Beyer, W.N., Chaney, R.L., and Mulhern, B.M., 1982. Heavy metal concentration in earthworms from soil amended with sewage sludge. *Journal of Environmental Quality*, 11: 381–385.

Beyers, R. and Odum, H.T., 1993. *Ecological Microcosms*. Springer-Verlag, New York.

Bindler, R., Brannvall, M., Renberg, I., Emteryd, O., and Grip, H., 1999. Natural lead concentrations in pristine boreal forest soils and past pollution trends: a reference for critical load models. *Environmental Science and Technology*, 33: 3362–3367.

Blancher, P. and McNicol, D., 1987. Peatland water chemistry in central Ontario in relation to acidic deposition. *Water, Air, and Soil Pollution*, 35: 217–232.

Bohn, H.L., McNeal, B.L., and O'Connor, G.A., 1985. *Soil Chemistry*, 2nd ed. Wiley-Interscience, New York.

Bolewski, A., Ney, R., and Smakowski, T. (Editors), 1995. Bilans surowcow mineralnych. CPPGSM PAN. Wydawnictwo Geologiczne, Warsaw.

Bond, A.M., Reust, V., Hudson, H.A., Arnup, R., Hanna, P.J., and Strother, S., 1988. The effects of temperature, salinity and seagrass species on the uptake of lead from seawater by excised leaves. *Marine Chemistry*, 24: 253–260.

Bonnevie, N.L., Gunster, D.G., and Wenning, R.J., 1992. Lead contamination in surficial sediments from Newark Bay, New Jersey. *Environment International*, 18: 497–508.

Borg, H., 1995. Trace elements in lakes. In: R. Salbu and E. Steinnes (Editors), Trace Elements in Natural Waters. CRC Press, Boca Raton, FL, 177–201.

Borg, H. and Johansson, K., 1989. Metal fluxes to Swedish forest lakes. *Water, Air, and Soil Pollution*, 47: 427–440.

Borovic, A.S., 1990. Characterizations of metal ions in biological systems. In: A.J. Shaw (Editor), *Heavy Metal Tolerance in Plants: Evolutionary Aspects*. CRC Press, Boca Raton, FL, 3–5.

Boudou, A. and Ribeyre, F., 1989. Fundamental concepts in aquatic ecotoxicology. In: A. Boudou and F. Ribeyre (Editors), *Aquatic Ecotoxicology: Fundamental Concepts and Methodologies*, Vol. 1. CRC Press, Boca Raton, FL, 35–75.

Bourgoin, B.P., Risk, M.J., and Aitken, A.E., 1989. Possible effect of sedimentary phosphorus on the accumulation of lead in *Mytilus edulis*. *Bulletin of Environmental Contamination Toxicology*, 43: 635–640.

Bowen, H.J.M., 1966. *Trace Elements in Biochemistry*. Academic Press, New York.

Bowen, J.J.M., 1979. *Environmental Chemistry of the Elements*. Academic Press, London.

Boyt, F.L., Bayley, S.E., and Zoltek, J., 1977. Removal of nutrients from treated municipal wastewater by wetland vegetation. *Journal of the Water Pollution Control Federation*, 48: 798–799.

Bradbury, I.K. and Grace, J., 1983. Primary production in wetlands. In: A.J.P. Gore (Editor), *Mires: Swamp, Bog, Fen and Moor*. Elsevier Science, New York, 285–310.

Bradshaw, A.D., 1987. The reclamation of derelict land and the ecology of ecosystrems. In: W.R. Jordan, III, M.E. Gilpin, and J.D. Aber (Editors), *Restoration Ecology — A Synthetic Approach to Ecological Research*. Cambridge University Press, Cambridge, U.K., 52–74.

Breen, J.J. and Stroup, C.R. (Editors), 1995. *Lead Poisoning — Exposure, Abatement, Regulation*. Lewis Publishers, Boca Raton, FL, 304 pp.

Brightman, R.S., 1984. Benthic macroinvertebrate response to secondarily treated wastewater in north-central Florida cypress domes. In: K.C. Ewel and H.T. Odum (Editors), *Cypress Swamps*. University of Florida Press, Gainesville, 186–196.

Bromley, D.W., 1986. Markets and externalities. In: D.W. Bromley (Editor), *Natural Resource Economics, Policy Problems and Contemporary Analysis*. Kluwer-Nijhoff, Boston, 37–68.

Brookins, D.G., 1988. *Eh-pH Diagrams for Geochemistry*. Springer-Verlag, New York.

Brooks, R.R. (Editor), 1998. Plants that Hyperaccumulate Heavy Metals, Their Role in Phytoremediation, Microbiology, Archaeology, Mineral Exploration and Phytomining. CAB International, University Press, Cambridge, U.K., 380 pp.

Brown, D.S. and Allison, D.J., 1987. MINTEQA1: An Equilibrium Speciation Model, User's Manual. EPA-600/3-87-012.

Brown, K.S., 1995. The green clean. *Bioscience*, 45: 579–582.

Brown, M.T. and Arding, J., 1991. Transformities — prepared from many sources. Center for Environmental Policy, Department of Environmental Engineering Sciences, University of Florida, Gainesville. Unpublished paper.

Brown, M.T. and Arding, J., 1997. Compilation of Transformitites. Unpublished manuscript.

Brown, M.T. and Ulgiati, S., 1999. Emergy evaluation of the biosphere and natural capital. *Ambio*, 28(6): 468–493.

Brown, S.L., Flohrschutz, E.W., and Odum, H.T., 1984. Structure, productivity, and phosphorus cycling of the scrub cypress ecosystem. In: K.C. Ewel and H.T. Odum (Editors), *Cypress Swamps*. University of Florida Press, Gainesville, 304–317.

Burns, L.A., 1985. Validation and predictability of laboratory methods for assessing the fate and effects of contaminants in aquatic ecosystems. In: T.P. Boyle (Editor), ASTM, STP 865, American Society for Testing and Materials, Philadelphia, PA, 176–182.

Cairns, J., 1977. Quantification of biological integrity. In: R.K. Ballentine and L.J. Guarraia (Editors), The Integrity of Water. U.S. Environmental Protection Agency, Office of Water and Hazardous Materials, Washington, D.C., pp 171–187.

Cairns, J., Jr. (Editor), 1988. Restoration ecology: the new frontier. In: J. Cairns, Jr. (Editor), *Rehabilitating Damaged Ecosystems*, Vol. 1. CRC Press, Boca Raton, FL, 1–11.

Callander, E. and Van Metre, P.C., 1997. Reservoir sediment cores show U.S. lead declines. *Environmental Science and Technology News*, American Chemical Society, 31(9): 424A–428A.

Calmano, W., Förstner, U., and Kersten, M., 1986. Metal associations in anoxic sediments and changes following upland disposal. *Toxicology and Environment Chemistry*, 12: 313–321.

Campbell, C.J., 1997. Depletion patterns show change due for production of conventional oil. *Oil and Gas Journal*, December: 33–39.

Campbell, G.C. and Tessier, A., 1996. Ecotoxicology of metals in the aquatic environment: geochemical aspects. In: M.C. Newman and C.H. Jagoe (Editors), *Ecotoxicology*. Lewis Publishers, Boca Raton, FL, 11–58.

Carriker, N.E., 1977. Heavy Metal Interactions with Natural Organics in Aquatic Environments. Ph.D. dissertation, Department of Environmental Engineering Sciences, University of Florida, Gainesville, 154 pp.

Carter, V., 1986. An overview of the hydrological concerns related to wetlands in the United States. *Canada Journal of Botany*, 64: 364–374.

Cassagrande, G.L. and Erchull, L.D., 1976 and 1977. Metals in plants and waters in the Okefenokee Swamp and their relationship to constituents in coal. *Geochimica et Cosmochimica Acta*, 40: 387–393; 41: 1391–1394.

Centrum Podstawowych Problemow Gospodarki Surowcami i Energia PAN, 1996. Bilans Gospodarki Surow-
    cami Mineralnymi Polski na Tle Gospodarki Swiatowej. Warszawa.

Cervinka, V., 1980. Fuel and energy efficiency. In: D. Pimental (Editor), *Handbook of Energy Utilization in
    Agriculture*. CRC Press, Boca Raton, FL, 15–21.

CH2M Hill, Inc., 1991. *Work Plan for Ecological RI/FS Steel City Bay Operable Unit Sapp Battery Project,
    Jackson County, Florida*. CH2M Hill, Inc., Gainesville, FL.

Chadwick, M.J., 1973. The cycling of materials in disturbed environments. In: *The Ecology of Resource
    Degradation and Renewal*. John Wiley, New York, 3–16.

Chan, E., Bursztynsky, T.A., Hantzsche, N., and Litwin, Y.J., 1982. The Use of Wetlands for Water Pollution
    Control. EPA-600/2–82–086, Association of Bay Area Governments, Berkeley, CA.

Chang, A.C., Page, A.L., Warneke, J.E., and Grgurevic, E., 1984. Sequential extraction of soil heavy metals
    following a sludge application. *Journal of Environmental Quality*, 13: 33–38.

Chang, A.C., Warneke, J.E., Page, A.L., and Lund, L.J., 1984. Accumulation of heavy metals in sewage sludge-
    treated soil. *Journal of Environmental Quality*, 13: 87–91.

Chapman, J. and Hall, C.A.S., 1986. Forest resources and energy use in the forest products industry. In: C.A.S.
    Hall, C.J. Cleveland, and R. Kaufmann (Editors), *Energy and Resource Quality: The Ecology of the
    Economic Process*. John Wiley, New York, 461–479.

Chan, Y.K. and Wong, P.T.S., 1984. In: P. Grandjean (Editor), *Biological Effects of Organolead Compounds*.
    CRC Press, Boca Raton, FL.

Chen, Z., 1992. Metal contamination of flooded soils, rice plants, and surface waters. In: D.C. Adriano (Editor),
    Biogeochemistry of Trace Metals. Lewis Publishers, Boca Raton, FL, 85–108.

Chereminisoff, P.N., 1993. *Lead: A Guidebook to Hazard Detection, Remediation and Control*. Prentice-Hall,
    Englewood Cliffs, NJ, 287 pp.

Chester, R., 1990. *Marine Geochemistry*. Unwin Hyman, London, U.K.

Chow, T.J., 1978. Lead in natural waters. In: J.O. Nriagu (Editor), *The Biogeochemistry of Lead in the
    Environment*. Elsevier, Amsterdam, 185–218.

Chow, T.J. and Patterson, C.C., 1962. The occurrence and significance of lead isotopes in pelagic sediments.
    Geochimica et Cosmochimica Acta, 26: 263–308.

Chumbley, C.G. and Unwin, R.J., 1982. Cadmium and lead content of vegetable crops grown on land with a
    history of sewage sludge application. *Environmental Pollution (Series B)*, 4: 231–237.

Churchill, M.A., Elmore, H.L., and Buckingham, R.A., 1962. The prediction of stream reaeration rates. *Journal
    of the Sanitary Engineering Division*, ASCE, 88: 1–46.

Clesceri, L.S., Greenberg, A.E., Trussell, R.R., and Franson, M.A., 1989. Aggregate organic constituents. In:
    Standard Methods for the Examination of Water and Wastewater, 17th ed. APHA, Washington, D.C.,
    5.37–5.4l.

Clevenger, T.E. and Rao, D., 1996. Mobility of lead in minetailings due to landfill leachate. *Water, Air, and
    Soil Pollution*, 91: 197–207.

Cohen, R.R.H., n.d. Passive Mine Drainage Treatment Systems: A Simple Solution to the Complex Problem
    of Acid Mine Drainage. Colorado School of Mines, Golden, CO.

Cole, G.A., 1975. *Textbook of Limnology*. C.V. Mosby, St. Louis.

Colinveaux, P., 1986. *Ecology*. John Wiley & Sons, New York.

Cornel, P.K., Summer, R.S., and Roberts, P.V., 1986. Diffusion of humic acid in dilute aqueous solution.
    Journal of Colloid Interface Science, 110: 149–164.

Costanza, R. and Farber, S.C., 1984. Theories and methods of valuation of natural systems: a comparison of
    willingness-to-pay and energy analysis based approaches. *Man, Environment, Space and Time*, 4: 1–38.

Cowardin, L.M., Carter, V., Golet, F.C., and LaRoe, E.T., 1979. Classification of Wetlands and Deepwater
    Habitats of the United States. FWS/OBS-79/31. U.S. Fish and Wildlife Service, Washington, D.C.

Crist, T.O., Williams, N.R., Amthor, J.S., and Siccame, T.G., 1985. The lack of an effect of lead and acidity
    on leaf decompostion in laboratory microcosms. *Environmental Pollution (U.K.)*, 38(4): 295–303.

Damman, W.H., 1979. Mobilization and accumulation of heavy metals in freshwater wetlands. Research
    project technical report, W8001379 OWRTA-073–CONN(1), Institute of Water Resources, University of
    Connecticut, Storrs.

Dammert, A.J. and Chhabra, J.G.S., 1990. The Lead and Zinc Industries: Long-Term Prospects. World Bank
    Staff Commodity Working Paper No. 22. The World Bank, Washington, D.C.

Darby, D.A., Adams, D., and Niven, W.T., 1986. Artificially created estuarine marsh. In: P.G. Sly (Editor), Sediment and Water Interactions. Springer, New York, 343–351.

Davies, B.E. and Wixson, B.G., 1988. *Lead in Soil: Issues and Guidelines*. Science Reviews, Northwood, U.K., 315 pp.

Davies, B.E., 1990. Lead. In: B.J. Alloway (Editor), Heavy Metals in Soils. John Wiley & Sons, New York, 177–196.

Davies, P.H., Goettl, J.P., Jr., Sinley, J.R., and Smith, N.F., 1976. Acute and chronic toxicity of lead to rainbow trout *Salmo gairdneri*, in hard and soft water. *Water Research*, 10: 199–206.

Davis, J.A. and Leckie, J.O., 1978. Effect of adsorbed complexing ligands on trace metal uptake by hydrous oxides. *Environmental Science Technology*, 12: 1309–1315.

Davis, W.M., 1993. Influence of Humic Substance Structure and Composition on Interactions with Hydrophobic Organic Compounds. Ph.D. dissertation, University of Florida, Gainesville.

Day, F.P., Jr., 1984. Biomass and litter accumulation in the Great Dismal Swamp. In: K.C. Ewel and H.T. Odum (Editors), *Cypress Swamps*. University of Florida Press, Gainesville, 386–392.

Deghi, G.S., 1984. Seedling survival and growth rates in experimental cypress domes. In: K.C. Ewel and H.T. Odum (Editors), *Cypress Swamps*. University of Florida Press, Gainesville, 141–144.

De Gregori, I., Pinochet, H., Gras, N., and Munoz, L., 1996. Variability of cadmium, copper and zinc levels in mollusks and associated sediments from Chile. *Environmental Pollution*, 92: 259–308.

Delfino, J.J. and Enderson, R.E., 1978. Comparative study outlines methods of analysis of total metal in sludge. *Water & Sewage Works*, 125(RN): R32–R34 and R47–R48.

Diaz, G., Azcom-Aquilar, C., and Honrubia, M., 1996. Influence of orbuscular mycorrhizae on heavy metals (Zn & PB) uptake and growth of *Lygeum spartum* and *Anthyllis cytisoides*. *Plant and Soil*, 180: 41–249.

Dierberg, F.E. and Brezonik, P.L., 1984. The effect of wastewater on the surface water and groundwater quality of cypress domes. In: K.C. Ewel and H.T. Odum (Editors), *Cypress Swamps*. University of Florida Press, Gainesville, 83–101.

Dobrovolsky, V.V., 1994. *Biogeochemistry of the World's Land*, translated by B.V. Rassadin and H.T. Shacklette. CRC Press, Boca Raton, FL, 362 pp.

Dong, Y., 1996. Roles of Iron and Dissolved Organic Carbon on Lead Leachability in Contaminated Soils. Ph.D. work in progress, Department of Soil and Water Science, University of Florida, Gainesville.

Dong, Y., 2000. Heavy Metals and Colloid Immobility in Soils. Ph.D. dissertation, Department of Soils and Waters, University of Florida, Gainesville.

Douglas, A.J., 1989. Annotated Bibliography of Economic Literature on Wetlands. Biological Report 89(19). U.S. Fish and Wildlife Service, Washington, D.C.

Drayton, E.R., III and Hook, D.D., 1989. Water management of a baldcypress-tupelo wetland for timber and wildlife. In: D.D. Hook and R. Lea (Editors), Proceedings of the Symposium: The Forested Wetlands of the Southern United States; July 12–14, 1988, Orlando, FL. U.S. Department of Agriculture, Forest Service, Southeastern Forest Experiment Station, Asheville, NC, 54–58.

Drever, J.I., 1988. *The Geochemistry of Natural Waters*, 2nd ed., Prentice-Hall, Englewood Cliffs, NJ, 437 pp.

Drever, J.I., Li, Y.H., and Maynard, J.B., 1988. Geochemical cycles: the continental crust and the oceans. In: C.B. Gregor, R.M. Garrels, F.T. MacKenzie, and J.B. Maynard (Editors), *Chemical Cycles in the Evolution of the Earth*. Wiley, New York, 17–53.

Dunbabin, J.S., Pokorny, J., and Bowmer, K.H., 1988. Rhizosphere oxygenation by *Typha domingensis* Pers. in miniature wetland filters used for metal removal from wastewaters. *Aquatic Botany*, 29.

Dushenkov, V., Kumar, P.B.A.N., Motto, H., and Raskin, I., 1995. Rhizofiltration: the use of plants to remove heavy metals from aqueous streams. *Environmental Science and Technology*, 29(5): 1239–1245.

Dziennik Ustaw Rzeczpospolitej Polskiej Nr 32. Warszawa dnia 18 kwietnia 1991r. Ustawa 131 z dnia 21 marca 1991r. O obszarach morskich Rzeczpospolitej Polskiej i administracji morskiej.

Ecology and Environment, Inc., 1986. Summary Report on the Field Investigations of the Sapp Battery Site, Jackson County, Florida, Vol. 1. Florida Department of Environmental Regulation, Tallahassee.

Ecology and Environment, Inc., 1986. Summary Report on the Field Investigations of the Sapp Battery Site, Jackson County, Florida, Vols. 1 and 2. Ecology and Environment, Inc., Buffalo, NY.

Ecology and Environment, Inc., 1989. Remedial Design Field Investigation Report, Sapp Battery Site, Jackson County, Florida. Ecology and Environment, Inc., Buffalo, NY.

Eisenreich, S.J., Metzer, N.A., and Urban, N.R., 1986. Response of atmospheric lead to decreased use of lead in gasoline. *Environmental Science Technology*, 20: 171–174.

Eisler, R., 1988. Lead Hazards to Fish, Wildlife, and Invertebrates: A Synoptic Review. Biological Report 85(1.14). U.S. Fish and Wildlife Service, Laurel, MD.

Eklund, M., 1995. Cadmium and lead deposition around a Swedish battery plant as recorded in oak tree rings. *Journal of Environmental Quality*, 24: 126–131.

Elliott, H.A. and Shields, G.A., 1988. Comparative evaluation of residual and metal analysis in polluted soils. *Communications in Soil Science*, 19: 1907–1915.

ENR, 1990. Market trends. *Engineering News-Record*, 225: 62.

EPA (U.S. Environmental Protection Agency), 1979. Methods for chemical analysis of water and wastes. EPA-600/4–79–020, Environmental Monitoring and Support Laboratory, Cincinnati, OH.

Ernst, W.H.O., 1990. Mine vegetation in Europe. In: A.J. Shaw (Editor), *Heavy Metal Tolerance in Plants: Evolutionary Aspects*. CRC Press, Boca Raton, FL, 21–37.

Ewel, K.C. and Odum, H.T. (Editors), 1984. *Cypress Swamps*. University of Florida Press, Gainesville.

Ewers, U. and Schliptkoter, H., 1991. Lead. In: E. Merian (Editor), *Metals and Their Compounds in the Environment*. VCH, Weinheim, 67–103.

Farmer, P., 1987. *Lead Pollution from Motor Vehicles 1974–86: A Select Bibliography*. Elsevier, London.

Feijtel, T.C., DeLaune, R.D., and Patrick, W.H., Jr., 1988. Biogeochemical control on metal distribution and accumulation in Louisiana sediments. *Journal of Environmental Quality*, 17: 88–94.

Fergusson, J.E., 1990. *The Heavy Elements: Chemistry, Environmental Impact and Health Effects*. Pergamon Press, New York, 613 pp.

Fernald, E.A. (Editor), 1981. *Atlas of Florida*. Florida State University Foundation, Tallahassee.

Fetter, C.W., 1988. *Applied Hydrogeology*, 2nd ed. Merrill Publishing, Columbus, OH.

Finn, J.T., 1978. Cycing index: a general defininer for cycling in compartment models. In: D.C. Adriano and I.L. Brisben (Editors), Environmental Chemistry and Cycling Process. Division of Technical Information, Department of Energy, Oak Ridge, TN, 911 pp.

Flaig, W., 1973. An introductory review on humic substances: aspects of research on their genesis, their physical and chemical properties, and their effect on organisms. In: D. Povoledo and H.L Golterman (Editors), *Humic Substances: Their Structure and Function in the Biosphere. Proceedings of the International Meeting on Humic Substances, May 29–31, 1972*, Nieuwersluis, Netherlands, 19–42.

Fletcher, J.S., 1990. Use of algae vs. vascular plants to test for chemical toxicity. In: W. Wang, J.W. Gorsuch, and W.R. Lower (Editors), *Plants for Toxicity Assessment. American Society for Testing and Materials*, Philadelphia, PA, 33–39.

Florence, T.M., 1977. Trace metal species in fresh waters. *Water Research*, 11: 681-687.

Florence, T.M. and Batley, G.E., 1980. Chemical speciation in natural waters. *CRC Critical Review in Analytical Chemistry*, 9: 219–296.

Folke, C., 1991. The societal value of wetland life-support. In: C. Folke and T. Kaberger (Editors), *Linking the Natural Environment and the Economy: Essays from the Eco-Eco Group*. Kluwer Academic, Dordrecht, 141–171.

Forbes, R.M. and Sanderson, G.C., 1978. Lead toxicity in domestic animals and wildlife. In: *The Biogeochemistry of Lead in the Environment, Part B, Biological Effects*. Elsevier/North-Holland Biomedical Press, Amsterdam, 225–227.

Förstner, U. and Wittmann, G.T.W., 1979. *Metal Pollution in the Aquatic Environment*. Springer-Verlag, New York.

Förstner, U. and Wittmann, G.T.W., 1983. *Metal Pollution in the Aquatic Environment*, 2nd ed. Springer-Verlag, New York, 486 pp.

Francois, R., 1990. Marine sedimentary humic substances: structure, genesis and properties. *Reviews in Aquatic Sciences*, 3: 41–80.

Friedland, A.J., 1990. The movement of metals through soils and ecosystems. In: A.J. Shaw (Editor), Heavy *Metal Tolerance in Plants: Evolutionary Aspects*. CRC Press, Boca Raton, FL, 7–19.

Friedlander, S.K., 1973. Chemical element balances and identification of air pollution sources. *Environmental Science and Technology*. 7: 235–240.

Fuhr, F., 1987. Non-extractable pesticide residues in soil. In: R. Greenhalgh and T.R. Robertson (Editors), *Pesticide Science Biotechnology*. Blackwell Scientific, Boston, 381–389.

Furness, R.W. and Rainbow, P.S., 1990. *Heavy Metals in the Marine Environment*. CRC Press, Boca Raton, FL, 224 pp.

Gambrell, R., Khalid, R., and Patrick, W., 1980. Chemical availability of mercury, lead, and zinc in Mobile Bay sediment suspensions as affected by pH and oxidation-reduction. *Environmental Science and Technology*, 14: 431–436.

Gambrell, R.P., 1994. Trace and toxic metals in wetlands — a review. *Journal of Environmental Quality*, 23: 883–891.

Garcia, M.A., Alonso, J., Fernandez, N.I., and Melgar, M.J., 1998. Lead content in edible wild mushrooms in northwest Spain as indicator of environmental contamination. *Archives of Environmental Contamination and Toxicology*, 34: 330–335.

Gardner, L.R., Chen, H., and Seillemyr, J.L., 1978. Comparison of trace metals in South Carolina floodplain and marsh sediments. In: D.C. Adriano and I.L. Brisben (Editors), *Environmental Chemistry and Cycling Process. Division of Technical Information*, Oak Ridge, TN, 446–461.

Garlaschi, G., Schlascha, E.B., Vergara, I., and Chang, A.C., 1985. Trace metals in sediments of the Mapocho River, Chile. *Environmental Technology Letters*, 6: 405–414.

Garrels, R.M., Mackenzie, F.T., and Hunt, C., 1975. *Chemical Cycles and the Global Environment: Assessing Human Influences*. William Kaufmann, Los Altos, CA.

Genoni, G.P., 1997a. Influence of the energy relationships of organic compounds on toxicity to the Cladoceran *Daphnia magna* and the fish *Pimephales promelas*. *Ecotechnology and Environmental Safety*, 36: 27–37.

Genoni, G.P., 1997b. Influence of the energy relationships of organic compounds on their specificity toward aquatic organisms. *Ecotechnology and Environmental Safety*, 36: 99–108.

Genoni, G.P., 1997c. Towards a conceptual synthesis in ecotoxicology. *Oikos*, 80: 96–106.

Genoni, G., 1998. The energy dose makes the poison. *EWAG News*, 3 pp.

Genoni, G.P. and Montague, C.L., 1995. Influence of the energy relationships of trophic levels and of elements on bioaccumulation. *Ecotoxicology and Environmental Safety*, 30(2): 203–218.

Georgescu-Roegen, N., 1977. The steady state and economic salvation. A thermodynamic analysis. *Bioscience*, 27: 266–270.

Gersberg, R.M., Lyon, S.T., Elkins, B.V., and Goldman, C.R., 1984. The Removal of Heavy Metals by Artificial Wetlands. EPA-600/D-84-258, Santee, CA.

Ghosh, K. and Schnitzer, M., 1980. Macromolecular structures of humic substances. *Soil Science*, 129: 266–276.

Giblin, A.E., 1985. Comparisons of the processing of elements by ecosystems. II. Metals. In: P. Godfrey, E.R. Kaynor, S. Pelczarski, and J. Benforado (Editors), *Ecological Considerations in Wetlands Treatment of Municipal Wastewaters*. Van Nostrand Reinhold, New York, 158–179.

Glooschenko, W.A., 1986. Monitoring the atmospheric deposition of metals by use of bog vegetation and peat profiles. In: J.O. Nriagu (Editor), *Toxic Metals in the Atmosphere*. Wiley, New York, 507–533.

Glowny Urzad Statystyszny Zaklad Wydawnictw Statystycznych, 1994. Rocznik Statystyczny, 1994.

Glowny Urzad Statystyszny, 1994. Ochrona Srodowiska, 1994. Warszawa.

Glowny Urzad Statystyszny, 1995. Rocznik Statystyczny, 1995. Warszawa.

Goldberg, E.D. and Arrhenius, G.O.S., 1958. Chemistry of Pacific pelagic sediments. *Geochimica et Cosmochimica Acta*, 13: 153–212.

Goldberg, E.D., 1971. Atmospheric dust, the sedimentary cycle and man. *Comments on Earth Sciences: Geophysics*, 1: 117–132.

Gosselink, J.G., Odum, E.P., and Pope, R.M., 1974. The Value of the Tidal Marsh. Publication No. LSU-SG-74-03. Center for Wetland Resources, Louisiana State University, Baton Rouge.

Gottfried, R.R., 1992. The value of a watershed as a series of linked multiproduct assets. *Ecological Economics*, 5: 145–161.

Graedel, T.E. and Allenby, B.R., 1998. *Industrial Ecology and the Automobile*. Prentice-Hall, Upper Saddle River, NJ, 243 pp.

Grandjean, P. (Editor), 1984. *Biological Effects of Organolead Compounds*. CRC Press, Boca Raton, FL.

Grodzinska, K., Godzik, B., Darowska, E., and Pawlowska, B., 1987. Concentration of heavy metals in trophic chains of Niepolomice Forest, *S. Poland. Ekologia Polska*, 35: 327–344.

Gruber, W., 1991. Lead-acid battery recycling. *EI Digest*, 20: 18–27.

Grunwald, C., Iverson, L.R., and Szafoni, D.B., 1988. Abandoned mines in Illinois and North Dakota: toward understanding of revegetation problems. In: J. Cairns (Editor), *Rehabilitating Damaged Ecosystems*, Vol. 1. CRC Press, Bocan Raton, FL, 39–59.

Gumerman, R.C., Culp, R.L., and Hansen, S.P., 1979a. Estimating Water Treatment Costs, Vol. 2. Cost Curves Applicable to 1 to 200 mgd Treatment Plants. EPA 600/2–79–162b. U.S. Environmental Protection Agency, Cincinnati, OH.

Gumerman, R.C., Culp, R.L., and Hansen, S.P., 1979b. Estimating Water Treatment Costs, Vol. 1. Summary. (EPA 600/2–79–162a). U.S. Environmental Protection Agency, Cincinnati, OH.

Gupta, A., 1995. Heavy metal accumulation by three species of mosses in Shillong, Northeastern India. *Water, Air, and Soil Pollution*, 82: 751–756.

Guyette, R.P., Cutter, B.E, and Henderson, G.S., 1991. Long-term correlations between mining activity and levels of lead and cadmium in tree-rings of eastern red cedar. *Journal of Environmental Quality*, 20: 146–150.

Haggin, J., 1986. More awareness sought concerning role of metals in pollution. *C and E News*, September 8, 37–42.

Hall, C.A.S., Cleveland, C.J., and Kaufmann, R., 1986. *Energy and Resource Quality*. Wiley-Interscience, New York, 575 pp.

Hansen, S.P., Gumerman, R.C., and Culp, R.L., 1979. Estimating Water Treatment Costs, Vol. 3. Cost Curves Applicable to 2,500 gpd to 1 mgd Treatment Plants. EPA 600/2-79-162c. U.S. Environmental Protection Agency, Cincinnati, OH.

Harper, H.H., 1985. Fate of Heavy Metals from Highway Runoff in a Stormwater Management System. Ph.D. dissertation, University of Central Florida, Orlando, FL.

Harrison, R.M., 1989. Cycles, fluxes and speciation of trace metals in the environment. In: J.W. Patterson and R. Passino (Editors), *Metals Speciation, Separation and Recovery*. Lewis Publishers, Boca Raton, FL.

Harrison, R.M. and Laxen, D.P.H., 1981. *Lead Pollution Causes and Control*. Chapman & Hall, New York, 168 pp.

Harvey, R.W. and Leckie, J.O., 1985. Sorption of lead onto two gram-negative marine bacteria in seawater. *Marine Chemistry*, 15: 333–344.

Haworth, D.T., Pitluck, M.R., and Pollard, B.D., 1987. Conditional stability constant determination of metal aquatic fulvic acid complexes. *Journal of Liquid Chromatography*, 10: 2877–2889.

Hayes, M.H.B. and Wilson, W.S. (Editors), 1997. *Humic Substances in Soils, Peats, and Sludges: Health and Environmental Aspects*. Royal Society of Chemistry, Cambridge, U.K., 496 pp.

Hellawell, J.M., 1988. Toxic substances in rivers and streams. *Environmental Pollution*, 50: 61–85.

Herrick, G.T. and Friedland, A.J., 1990. Pattern of trace metal concentration and acidity in montane forest soils of northeastern United States. *Water, Air, and Soil Pollution*, 53: 151–157.

Hessen, D.O. and Tranvik, L.J. (Editors), 1998. *Aquatic Humic Substances*. Springer, New York, 346 pp.

Hester, R.E. and Harrison, R.M. (Editors), 1994. *Mining and Its Environmental Impact. Issues in Environmental Science and Technology*, Royal Society of Chemistry, Letchworth, U.K.

Hickey, M.C. and Kittrick, J.A., 1984. Chemical partitioning of cadmium, copper, nickel and zinc in soils and sediments containing high levels of heavy metals. *Journal of Environmental Quality*, 13: 372–376.

Holm, P.R., Christensen, T.H., Tjell, J.C., and McGrath, S.P., 1995. Speciation of cadmium and zinc with application to soil solutions. *Journal of Environmental Quality*, 24: 183–190.

Hoover, T.B., 1978. Inorganic Species in Water: Ecological Significance and Analytical Needs (a literature review). EPA-600/3–78–064, Athens, GA.

Huang, C.P., Elliott, H.A., and Ashmead, R.M., 1977. Interfacial reactions and the fate of heavy metals in soil-water systems. *Journal of the Water Pollution Control Federation*, 49 (May): 745–755.

Huang, S.L. and Odum, H.T., 1991. Ecology and economy: emergy synthesis and public policy in Taiwan. *Journal of Environmental Management*, 32: 313–333.

Hufschmidt, M.M., James, D.E., Meister, A.D., Bower, B.T., and Dixon, J.A., 1983. *Environment, Natural Systems, and Development: An Economic Valuation Guide*. Johns Hopkins University Press, Baltimore, MD.

Isphording, W.C., 1991. Organic and heavy metal chemistry of Mobile Bay sediments: Ala. Geol. Sur. Proj., U.S.G.S. Grant #14–08–001–A0775, Sediment Distribution and Geological Framework of Coastal Alabama, Final Report, 1–43.

Isphording, W.C., Brown, B.I., and Beyers, R.J., 1992. Algal Nutrient Depletion by Mineral Scavenging in Upper Bear Creek Reservoir: Final Report to the Tennessee Valley Authority, AUTRC Project 9018USA, 1–92.

Jacobs, D.E., 1996a. The economics of lead-based paint hazards in housing. *Lead Perspectives*, 6 pp.

Jacobs, D.E., 1996b. The health effects of lead on the human body. *Lead Perspectives*, 10–2, 32.

Jain, S.K., Vasudevan, P., and Jha, N.K., 1990. *Azolla pinnata R.Br.* and *Lemna minor L.* for removal of lead and zinc from polluted water. *Water Research*, 24: 177–183.

Jakeman, A.J., Beck, M.B., and McAleer, M.J. (Editors), 1993. *Modeling Change in Environmental Systems*, John Wiley, New York, 584 pp.

James M. Montgomery Consulting Engineers Inc., 1985. *Water Treatment Principles and Design*. John Wiley, New York.

Jaworski, J.F., Nriagu, J., Denny, P., Hart, B.T., Lasheen, M.R., Subramanian, V., and Wong, M.H., 1987. Group report: lead. In: T.C. Hutchinson and K.M. Meema (Editors), *Lead, Mercury, Cadmium and Arsenic in the Environment*. John Wiley, New York, 3–16.

Jenne, E.A., 1995. Metal adsorption and desorption from sediments. In: H.E. Allen (Editor), *Metal Contaminated Aquatic Sediments*. Ann Arbor Press, Ann Arbor, MI, 80–110.

Jenner, H.A. and Bowmer, T., 1990. The accumulation of metals and their toxicity in the marine intertidal invertebrates *Cerastoderma edule*, *Macoma balthica* and *Arenicola marina* exposed to pulverised fuel ash in mesocosms. *Environmental Pollution*, 66: 139–156.

Jennett, J.C. and Linnemann, S.M., 1977. Disposal of lead and zinc-containing wastes on soils. *Journal of the Water Pollution Control Federation*, 49: 1842–1855.

Jerger, B.E. and Exner, J.H., 1994. *Bioremediation*. CRC Press, Boca Raton, FL, 548 pp.

Jernigan, E.L. (Editor), 1973. *Lead Poisoning in Man and the Environment*. MSS Information Corp., New York, 225 pp.

J.M. Montgomery Consulting Engineers Inc., 1985. *Water Treatment Principles and Design*. John Wiley, New York.

Johnson, M.S., Coate, J.A., and Stevenson, J.K.W., 1994. Revegetation of metalliferous waters and land after metal mining. In: R.E. Hester and R.M. Harrison (Editors), *Mining and Its Environmental Impact. Issues in Environmental Science and Technology*. Royal Society of Chemistry, Letchworth, U.K., 31–47.

Johnson, R.L., 1978. Timber harvests from wetlands. In: P.E. Greeson, J.R. Clark and J.E. Clark, (Editors), *Wetland Functions and Values: The State of Our Understanding*. American Water Resources Association, Minneapolis, MN, 598–605.

Jorgensen, S.E., 1979. Modeling the distribution and effect of heavy metals in aquatic ecosystems. *Ecological Modelling*, 6: 199–222.

Jorgensen, S.E., 1984. Parameter estimation in toxic substance models. *Ecological Modeling*, 22: 1–20.

Jorgensen, S.E., 1986. *Fundamentals of Ecological Modeling, Developments in Environmental Modelling 9*. Elsevier, Amsterdam, 389 pp.

Jorgensen, S.E., 1990. *Modelling in Ecotoxicology*. Elsevier, New York, 340 pp.

Jorgensen, S.E., 1993. Modelling in ecotoxicology. In: A.J. Jakeman, M.B. Beck, and M.J. McAleer (Editors), *Modelling Change in Environmental Systems*, Wiley, New York, 293–316.

Jorgensen, S.E., 1995. Modeling toxic contaminants in an aquatic environment. In: V. Novotny and L. Somlyody (Editors), *Remediation and Management of Degraded River Basins*. Springer, Berlin, 157–195.

Judith, L.B., Wildeman, T.R., and Cohen, R.R., 1991. The use of bench scale parameters for preliminary analysis of metal removal from acid mine drainage by wetlands. Proceedings of the 1991 national meeting of the American Society for Surface Mining and Reclamation, May 14–17, 1991, Durango, CO.

Juste, C. and Mench, M., 1992. Long term application of sewage sludge and its effect on metal uptake in crops. In: D.C. Adriano (Editor), *Biogeochemistry of Trace Metals*, Lewis Publishers, Boca Raton, FL, 159–194.

Kabala-Pendias, A. and Pendias, H., 1993. *Biochemia pierwiastkow sladowych*. PWN, Warszawa.

Kalac, P., Nizanska, M., Berilaqua, D., and Staskova, I., 1996. Concentrations of mercury, copper, cadmium, and lead in fruiting bodies of edible mushrooms in the vicinity of a mercury smelter and a copper smelter. *Science of the Total Environment*, 177(1–3): 251.

Kastens, M.L. and Hutchison, J.C., 1950. Contact sulfuric acid from sulfur. In: W.J. Murphy (Editor), *Modern Chemical Processes*. Reinhold Publishing, New York, 7–16.

Katz, S.A., Jenniss, S.W., and Mount, T., 1981. Comparison of sample preparation methods for the determination of metals in sewage sludges by flame atomic adsorption spectrometry. *International Journal of Environment Analysis Chemistry*, 9: 209–220.

Keller, P.A., 1992. Perspectives on Interfacing Paper Mill Wastewaters and Wetlands. M.S. thesis, University of Florida, Gainesville.

Kelly, J.M., Parker, G.R., and McFee, W.W., 1975. Abstracts with program, International Conference on Heavy Metals in the Environment, Toronto, C167–C169.

Kelly, M., 1988. Mining and the Freshwater Environment. *Elsevier Applied Science*, London.

Kennish, M.J., 1998. Marine pollution. In: Encyclopedia of Environmental Analysis and Remediation. UN Report No. 38, Rome, Italy, 2549–2604.

Kesler, S.E., 1978. Economic lead deposits. In: J.O. Nriagu (Editor), *The Biogeochemistry of Lead in the Environment: Part A. Ecological Cycles*. Elsevier/North-Holland Biomedical Press, Amsterdam, 73–97.

King, H.D., Curtin, G.C., and Shacklette, H.T., 1984. Metal uptake by young conifer trees. U.S. Geological Survey Bulletin, 1617, 23 pp.

Kittle, D.L., McGraw, J.B., and Garbutt, K., 1995. Plant litter decomposition in wetlands receiving acid mine drainage. *Journal of Environmental Quality*, 24: 301–306.

Klein, R.J., 1976. The Fate of Heavy Metals in Sewage Effluent Applied to Cypress Wetlands. M.S. thesis, Department of Environmental Engineering Sciences, University of Florida, Gainesville.

Kneip, T.J. and Hazen, R.E., 1979. Deposit and mobility of cadmium in a marsh–cove ecosystem and the relation to cadmium concentration in biota. *Environmental Health Perspectives*, 28 (February).

Knight, R.L., 1980. Energy Basis of Control in Aquatic Ecosystems. Ph.D. thesis. University of Florida, Gainesville.

Knutson, A.B., Klerks, P.O., and Levinton, J.S., 1987. The fate of heavy metal contaminated sediments in Foundry Cover, New York. *Environmental Pollution (Series B)*, 45: 291.

Krosshavn, M., Steinnes, E., and Varskog, P., 1993. Binding of Cd, Cu, Pb, and Zn in soil organic matter with different vegetational background. *Water, Air, and Soil Pollution*, 71: 185–193.

Kufel, J., 1978. Seasonal changes of Pb, Cu, and Co in above-ground parts of *Phragmites australis* Trin. ex Steudel and *Typha anustifolia* L. *Bulletin de L'Academie Polonaise des Science*, 26(11): C1.II.

Kufel, J., 1989. Akumulacja mikroelementow w trzcinie i palce waskolistnej i drogi ich powrotu do srodowiska. *Polska Akademia Nauk Instytut Ekologii Stacja Hydrobiologiczna*. Praca doktorska, Mikolajki, 71 pp.

Kufel, J., 1991. Lead and molybdenum in reed and cattail-open vs. closed type of metal cycling. *Aquatic Botany*, 40: 275–288.

Kuiters, A.T. and Mulder, W., 1990. Metal complexation by water-soluble organic subtances in forest soils. In: J.W. Patterson and R. Passino (Editors), *Metals Speciation, Separation and Recovery*. Lewis Publishers, Boca Raton, FL, 283–299.

Kuroda, P.K., 1982. *The Origin of the Chemical Elements*. Springer-Verlag, New York, 165 pp.

Kuyucak, N. and Volesky, B., 1990. Biosorption by algal biomass. In: B. Volesky (Editor), *Biosorption of Heavy Metals*. CRC Press, Boca Raton, FL, 173–196.

LaBauve, J.M., Kotuby-Amacher, J., and Gambrell, R.P., 1988. The effect of soil properties and a synthetic municipal landfill leachate on the retention of Cd, Ni, Pb and Zn in soil and sediment materials. *Journal of the Water Pollution Control Federation*, 60: 379–385.

Lake, D.L., Kirk, P.W., and Lester, J.N., 1984. Fractionation, characterization and speciation of heavy metals in sewage sludge and sludge-amended soils: a review. *Journal of Environmental Quality*, 13: 175–183.

Lam, D.C.I. and Simons, T.J., 1976. Computer model for toxicant spills in Lake Ontario. In: J.O. Nriago (Editor), *Environmental Biogeochemistry, Vol. 2, Metals Transfer and Ecological Mass Balances*. Ann Arbor Science, Ann Arbor, MI, 537–549.

Lan, C., Chen, G., Li, L., and Wong, M.H., 1992. Use of cattails in treating wastewater from a Pb/Zn mine. *Environmental Management*, 16(1): 75–80.

Lansdown, R. and Yule, W. (Editors), 1986. *Lead Toxicity, History, and Environmental Impact*. Johns Hopkins University Press, Baltimore, MD, 286 pp.

Lantzy, R.J. and Mackenzie, F.T., 1979. Atmospheric trace metals: global cycles and assessment of man's impact. *Geochimica et Cosmochimica Acta*, 43: 511–525.

Laundre, J.A., 1980. Heavy Metal Distribution in Two Connecticut Wetlands: a Red Maple Swamp and a "Sphagnum" Bog. M.S. thesis, W8005514 OWRTA-073–CONN(2), Institute of Water Resources, University of Connecticut, Storrs.

Lave, L.B. and Hendricksen, C.T., 1997. Clean recycling of lead batteries for electric vehicles. *Industrial Ecology*, 1: 33–37.

Lave, L.B., Hendrickson, C.T., and McMichael, F.C., 1995. Environmental implications of electric cars. *Science*, 268: 993–995.

Lee, C., Mendis, M., and Sullivan, D., 1978. Energy and Environmental Analysis of the Lead-Acid Battery Life Cycle. Contract No. EC-77–C-01–5115. U.S. Department of Energy, Washington, D.C.

Lee, J.A. and Tallis, J.H., 1973. High lead in peats from bogs of Europe. *Nature (London)*, 245: 216–218.

Lee, R., 1980. *Forest Hydrology*. Columbia University Press, New York, 349 pp.

Leenheer, J.A., 1981. Comprehensive approach to preparative isolation and fractionation of dissolved organic carbon from natural waters and wastewaters. *Environmental Science and Technology*, 15: 578–587.

Leeper, G.W., 1978. *Managing the Heavy Metals on the Land*. Marcel Dekker, New York.

Leffler, J.W., 1984. The use of self-selected, generic aquatic microcosms for pollution effects assessment. In: H.H. White (Editor), *Concepts in Marine Pollution Measurements. A Maryland Sea Grant Publication*, University of Maryland, College Park, 139–157.

Lehtonen, J., 1990. *Effects of Acidification on the Metal Levels in Aquatic Macrophytes in Espoo, Finland*. No. 12. Department of Environmental Conservation, University of Helsinki, Helsinki, Poland.

Leitch, J.A. and Ekstrom, B.L., 1989. *Wetland Economics and Assessment: An Annotated Bibliography*. Garland Publishing, New York.

Leitch, J.A. and Shabman, L.A., 1988. Overview of economic assessment methods relevant to wetland evaluation. In: D.D. Hook, W.H. McKee, H.K. Smith, J. Gregory, V.G. Burrell, Jr., M.R. DeVoe, R.E. Sojka, S. Gilbert, R. Banks, L.H. Stolzy, C. Brooks, T.D. Matthews, and T.H. Shear (Editors), *The Ecology and Management of Wetlands*, Vol. 2. Timber Press, Portland, OR, 95–102.

Leitz, W. and Galling, G., 1989. Metals from sediments. *Water Research*, 23: 247–252.

Lester, J.N., 1987. *Heavy Metals in Wastewater and Sludge Treatment Processes*, Vol. 1 and 2. CRC Press, Boca Raton, FL.

Levine, M.B., Hall, A.T., Barrett, G.W., and Taylor, D.H., 1989. Heavy metal concentrations during ten years of sludge treatment in an old field community. *Journal Environmental Quarterly*, 18: 411–418.

Lewis, T.E. and McIntosh, A.W., 1989. Covariation of selected trace elements with binding substrates in cores collected from two contaminated sediments. *Bulletin of Environmental Contamination Toxicology*, 43: 518–528.

Lind, O.T., 1974. *Handbook of Common Methods in Limnology*. C.V. Mosby, St. Louis.

Lion, L.W., Altmann, R.S., and Leckie, J.O., 1982. Trace-metal adsorption characteristics of estuarine particulate matter: evaluation of contributions of Fe/Mn oxide and organic surface coatings. *Environmental Science Technology*, 16: 660–666.

Logan, B.E. and Jiang, Q., 1990. Molecular size distributions of dissolved organic matter. *Journal of Environmental Engineering*, 116: 1046–1062.

Luo, Y. and Christie, P., 1997. Influence of lime stabilized sewage sludge cake on heavy metals and dissolved organic substance in the soil solution. In: M.H.B. Hayes and W.S. Wilson (Editors), *Humic Substances in Soils, Peats, and Sludges: Health and Environmental Aspects*. Royal Society of Chemistry, Cambridge, U.K., 410–437.

Lynch, T.A., 1981. Environmental Economic Damage Assessment for the Sapp Battery (Jackson County) Hazardous Waste Violation Case. Office of Economic Analysis, Florida Department of Environmental Regulation, Tallahassee, 35 pp.

Lyngby, J.E. and Brix, H., 1989. Heavy metals in eelgrass (*Zostera marina L.*) during growth and decomposition. *Hydrobiologia*, 176/177: 189–196.

Lynne, G.D., 1984. Socio-Economic Considerations in Mining While Farming: Depletion of the Histosols. Unpublished report, Food and Resource Economics, University of Florida, Gainesville, 14 pp.

Macnair, M.R., 1990. The genetics of metal tolerance in natural populations. In: A.J. Shaw (editor), *Heavy Metal Tolerance in Plants: Evolutionary Aspects*. CRC Press, Boca Raton, FL, 235–253.

Maher, W.A., 1984. Evaluation of a sequential extraction scheme to study associations of trace elements in estuarine and oceanic sediments. *Bulletin of Environmental Contamination Toxicology*, 32: 339–344.

Manahan, S.E., 1984. *Environmental Chemistry*, 4th ed. Brooks/Cole Publishing, Monterey, CA.

Mance, G., 1987. *Pollution Threats of Heavy Metals in Aquatic Environments*. Elsevier, London, 363 pp.

Margalef, R., 1968. *Perspectives in Ecological Theory*. University of Chicago Press, Chicago.

Marschner, H., 1991. Mechanisms of adaptation of plants to acid soils. *Plant and Soil*, 134: 1–20.

Martin, E.H., 1988. Effectiveness of an urban runoff detention pond-wetlands system. *Journal of Environmental Engineering*, 114(4): 810–827.

Mathews, A.P., 1989. Chemical equilibrium analysis of lead and beryllium speciation in hazardous waste incinerators. In: J.W. Patterson and R. Passino (Editors), *Metals Speciation, Separation and Recovery*, Vol. 2. Lewis Publishers, Boca Raton, FL, 75–84.

Maxim, L.D. and Lesemann, R., 1982. *The International Competitiveness of the U.S. Non-Ferrous Smelting Industry and the Clean Air Act*. Everest Consulting Associates, Princeton Junction, NJ, 451 pp.

McBride, M.B., 1995. Toxic metal accumulation from agricultural use of sludge: are USEPA regulations protective? *Journal of Environmental Quality*, 24: 5–18.

McCafferty, W.P., 1981. *Aquatic Entomology*. Jones and Bartlett Publishers, Boston.

McCloskey, D.N., 1985. *The Applied Theory of Price*, 2nd ed. Macmillan, New York.

McGrane, G., 1993. An Emergy Evaluation of Personal Transportation Alternatives. M.S. thesis, Environmental Engineering Sciences. University of Florida, Gainesville, 105 pp.

McNeil, D., 1989. A systemological approach to hierarchy and duality. *Yearbook of the International Society for the System Sciences*, 32: 307–314.

McNeilly, T., 1987. Evolutionary lessons from degraded ecosystems. In: W.R. Jordan, III, M.E. Gilpin, and J.D. Aber (Editors), *Restoration Ecology — A Synthetic Approach to Ecological Research*. Cambridge University Press, Cambridge, U.K., 271–286.

Meisinger, A.C., 1985. Diatomite. In: U.S. Bureau of Mines (Editor), Mineral Facts and Problems. U.S. Government Printing Office, Washington, D.C., 249–254.

Merritt, R.W. and Cummins, K.W., 1984. *An Introduction to the Aquatic Insects of North America*, 2nd ed. Kendall/Hunt, Dubuque, IA.

Metcalf and Eddy Inc., 1979. *Wastewater Engineering: Treatment, Disposal, Reuse*. McGraw-Hill, New York.

Middleton, G.V., 1970. Generation of the log-normal frequency distribution in sediments. In: M.A. Romanova and O.V. Sarmanov (Editors), *Topics in Mathematical Geology. Special Research Report—Consultants Bureau*, Plenum Press, New York, 34–42.

Miller, E.K. and Friedland, A.J., 1994. Lead migration in forest soils: response to changing atmospheric inputs. *Environmental Science Technology*, 28(4): 662–669.

Miller, J.E., Hassett, J.J., and Koeppe, D.E., 1976. Uptake of cadmium by soybeans as influenced by soil cation exchange capacity, pH, and available phosphorus. *Journal of Environmental Quality*, 5: 157–160.

Miller, R.L. and Goldberg, E.D., 1955. The normal distribution in geochemistry. *Geochimica et Cosmochimica Acta*, 8: 53–62.

Minerals Yearbook, 1968. Bureau of the Superintendent of Documents, U.S. Government Printing Office, Washington, D.C.

Minitab, Inc., 1991. *MINITAB Reference Manual: Release 8*, Macintosh Version. Minitab, Inc., State College, PA.

Mitsch, W.J., 1984. Seasonal patterns of a cypress dome in Florida. In: K.C. Ewel and H.T. Odum (Editors), *Cypress Swamps*. University of Florida Press, Gainesville, 25–33.

Mitsch, W.J., Flanagan, N.E., and Wise, K.M., 1993. Treatment of Acid Mine Drainage at Lick Run Demonstration Wetland: Preconstruction Analysis and Simulation Modeling. Ohio Department of Natural Resources, Columbus, OH, 36 pp.

Mitsch, W.J. and Gosselink, J.G., 1986. *Wetlands*. Van Nostrand Reinhold, New York.

Montgomery, D.C., 1984. *Design and Analysis of Experiments*, 2nd ed. John Wiley, New York, 538 pp.

Moore, J.W. and Ramamoorthy, S., 1984. *Heavy Metals in Natural Waters*. Springer, New York, 268 pp.

Morgan, J.J. and Stumm, W., 1991. Chemical processes in the environment, relevance of chemical speciation. In: E. Merian (Editor), *Metals and Their Compounds in the Environment*. VCH, Weinheim.

Moriarty, F., 1988. *Ecotoxicology*. Academic Press, London.

Mudroch, A. and Capbianco, J.A., 1979. Effects of treated effluent on a natural marsh. *Journal of the Water Pollution Control Federation*, 51: 2243–2256.

Mundrink, D., 1989. Remedial Design Field Investigation Report: Sapp Battery, Jackson County, Florida. U.S. Environmental Protection Agency, Environmental Services Division, Athens, GA.

National Academy of Science, 1980. Lead in the Human Environment. National Academy of Science, Washington, D.C., 524 pp.

National Research Council (U.S.). Committee on Measuring Lead in Critical Populations; Fowler, B.A., National Research Council (U.S.). Board on Environmental Studies and Toxicology; and National Research Council (U.S.). Commission on Life Sciences, 1993. *Measuring Lead Exposure in Infants, Children, and Other Sensitive Populations*. National Academy Press, Washington, D.C.

Needleman, H.L., 1980. *Low Level Lead Exposure: The Clinical Implications of Current Research*. Raven Press, New York.

Needleman, H.L., Gunnoe, C., Leviton, A., Reed, R., Peresie, H., Maher, C., and Barrett, P., 1979. Deficits in psychologic and classroom performance of children with elevated dentine lead levels. *New England Journal of Medicine*, 300: 689–695.

Needleman, H.L., Riess, J., Tobin, M.J., Biesecker, G.E., and Greenhouse, J.B., 1996. Bone lead levels and delinquent behavior. *Journal of the American Medical Association*, 275: 363–369.

Needleman, H.L., Schell, A., Bellinger, D., Leviton, A., and Allred, E.N., 1990. The long-term effects of exposure to low doses of lead in childhood: an 11–year follow-up report. *New England Journal of Medicine*, 322: 83–88.

Neufeld, R.D., Gutierrez, J., and Novak, R.A., 1977. A kinetic model and equilibrium relationship for heavy metal accumulation. *Journal of the Water Pollution Control Federation*, March: 489–497.

New, P.R.C., 1990. *Liposomes: A Practical Approach*. Oirl Press at Oxford University Press, Oxford, U.K.

Newman, M.C. and Jagoe, C.H. (Editors), 1996. *Ecotoxicology, a Hierarchical Treatment*. CRC Press, Boca Raton, FL, 411 pp.

Nixon, S.W. and Lee, V., 1986. Wetlands and Water Quality: A Regional Review of Recent Research in the United States on the Role of Freshwater and Saltwater Wetlands As Sources, Sinks, and Transformers of Nitrogen, Phosphorus, and Various Heavy Metals. Technical Report Y-86-2. U.S. Army Engineer Waterways Experiment Station, Vicksburg, MS.

Norton, G.A., Malaby, K.L., and DeKalb, E.L., 1988. Chemical characterization of ash produced during combustion of refuse-derived fuel with coal. *Environmental Science Technology*, 22(11): 1279–1283.

Nriagu, J.O., 1978a. Lead in soils, sediments, and major rock types. In: J.O. Nriagu (Editor), *The Biogeochemistry of Lead in the Environment, Vol. A: Ecological Cycles*. Elsevier/North-Holland Biomedical Press, Amsterdam, 15–72.

Nriagu, J.O., 1978b. Properties and the biogeochemical cycle of lead. In: J.O. Nriagu (Editor), *The Biogeochemistry of Lead in the Environment, Vol. A: Ecological Cycles*. Elsevier/North-Holland Biomedical Press, Amsterdam, 1–14.

Nriagu, J.O. (Editor), 1978c. *The Biogeochemistry of Lead in the Environment, Part A, Ecological Cycles*. Elsevier Science, Amsterdam.

Nriagu, J.O. (Editor), 1978d. *The Biogeochemistry of Lead in the Environment, Part B, Biological Effects*. Elsevier Science, Amsterdam, 384 pp.

Nriagu, J.O., 1981. Particulate and dissolved trace metals in Lake Ontario. *Journal of Water Resources*, 15: 91–96.

Nriagu, J.O., 1983. *Lead and Lead Poisoning in Antiquity*. John Wiley & Sons, New York, 465 pp.

Nriagu, J.O. (Editor), 1984. *Changing Metal Cycles and Human Health*. Springer-Verlag, Berlin.

Nriagu, J.O. (Editor), 1984. *Environmental Impacts of Smelters*. (*Advances in Environmental Science and Technology* #15). Wiley, New York, 608 pp.

Nriagu, J.O., 1988. A silent epidemic of environmental metal poisoning? *Environmental Pollution*, 50: 139–161.

Nriagu, J.O., 1990. The rise and fall of lead in gasoline. *Science of the Total Environment*, 92: 13–28.

Nriagu, J.O., 1994. Industrial activity and metals emission. In: R.H. Socolow, C. Andrews, F. Berkhart, and V. Thomas (Editors), *Industrial Ecology and Global Change*. Cambridge University Press, New York, 377–286.

Nriagu, J.O., 1998. Tales told in lead. *Science*, 281: 1622–1623.

Nriagu, J.O. and Davidson, C.I. (Editors), 1986. *Toxic Metals in Atmosphere*. Wiley, New York, 634 pp.

Nyholm, N., Nielsen, T.K., and Pedersen, K., 1984. Modeling heavy metals transport in an arctic fjord system polluted from mine tailings. *Ecological Modelling*, 22: 285–324.

Oberts, G. and Osgood, R., 1991. Water-quality effectiveness of a detention/wetland treatment system and its effect on an urban lake. *Environmental Management*, 15(1): 131–138.

Ochiai, E.I., 1987. *General Principles of Biochemistry of the Elements*. Plenum Press, New York.

Ochrona Srodowiska, 1994. Glowny Urzad Statystyczny. Warszawa.

O'Connor, D.J., Connolly, J.P., and Garland, E.J., 1988. Mathematical models — fate, transport and food chain. In: S.A. Levin, M.A. Harwell, J.R. Kelly, and K.D. Kimball (Editors), *Ecotoxicology: Problems and Approaches*. Springer-Verlag, New York, 221–243.

Odum, E.P., 1959. *Fundamentals of Ecology*, 2nd ed. W.B. Saunders, Philadelphia.

Odum, E.P., 1982. *Podstawy Ekologii*. Panstwowe Wydawnictwo Rolnicze i Lesne, Warszawa.

Odum, E.P., 1983. *Basic Ecology*. Saunders College Publishing, Philadelphia.

Odum, H.T., 1956. Primary production in flowing waters. *Limnology and Oceanography*, 1: 102–117.

Odum, H.T., 1971. *Environment, Power, and Society*. Wiley-Interscience, New York.

Odum, H.T., 1983a. Energy Analysis Overview of Nations. Working Paper WP-83–82. International Institute for Applied Systems Analysis, Laxenburg, Austria, 469 pp.

Odum, H.T., 1983b. *Systems Ecology: An Introduction*. John Wiley & Sons, New York.

Odum, H.T., 1984. Summary: cypress swamps and their regional role. In: K.C. Ewel and H.T. Odum (Editors), *Cypress Swamps*. University of Florida Press, Gainesville, 416–443.

Odum, H.T., 1985. *Self-Organization of Ecosystems in Marine Ponds Receiving Treated Sewage*. UNC Sea Grant Publication #UNC-SG-85-04. North Carolina Sea Grant, Chapel Hill.

Odum, H.T., 1986. Emergy in ecosystems. In: N. Polunin (Editor), *Ecosystem Theory and Application*. John Wiley, New York, 337–369.

Odum, H.T., 1988a. Energy, Environment and Public Policy: A Guide to the Analysis of Systems. Regional Seas Reports and Studies No. 95, United Nations Environment Programme, Nairobi, Kenya.

Odum, H.T., 1988b. Self-organization, transformity, and information. *Science*, 242: 1132–1139.

Odum, H.T., 1989. Simulation models of ecological economics developed with energy language methods. *Simulation*, (53): 69–75.

Odum, H.T., 1991. Emergy and biogeochemical cycles. In: C. Rossi and E. Tiezzi (Editors), *Ecological Physical Chemistry: Proceedings of an International Workshop*, November 8–12, 1990, Siena, Italy. Elsevier Science, Amsterdam, 25–56.

Odum, H.T., 1992a. EMERGY and Public Policy, Part I. Text for EES 5306 and EES 6009 classes. Environmental Engineering Sciences, University of Florida, Gainesville.

Odum, H.T., 1992b. EMERGY and Public Policy, Part II. Text for EES 5306 class. Environmental Engineering Sciences, University of Florida, Gainesville.

Odum, H.T., 1994. EMERGY and Public Policy, Part III. Unpublished manuscript.

Odum, H.T., 1996. *Environmental Accounting: EMERGY and Decision Making*. John Wiley, New York. 370 pp.

Odum, H.T. and Arding, J.E., 1991. Emergy Analysis of Shrimp Mariculture in Ecuador. Working Paper. Center for Wetlands, University of Florida, Gainesville.

Odum, H.T. and Hornbeck, D., 1996. Emergy evaluation of Florida salt marsh and its contribution to economic wealth. In: C.L. Coultas and Y.P. Hsieh (Editors), *Ecology and Management of Tidal Marshes*. St. Lucie Press, Boca Raton, FL. 352 pp.

Odum, H.T., Ewel, K.C., Mitsch, W.J., and Ordway, J.W., 1977. Recycling treated sewage through cypress wetlands in Florida. In: F. D'Itri (Editor), *Wastewater Renovation and Reuse*. Marcel Dekker, New York.

Odum, H.T., Odum, E.C., Brown, M.T., LaHart, D., Bersok, C., and Sendzimir, J., 1988. Environmental System and Public Policy. Center for Wetlands, University of Florida, Gainesville.

Odum, H.T., Wang, F.C., Alexander, J.F., Jr., Gilliland, M., Miller, M., and Sendzimir, J., 1987. Energy Analysis of Environmental Value. CFW #78–17. Center for Wetlands, University of Florida, Gainesville.

Oehme, F.W., 1978. *Toxicity of Heavy Metals in the Environment, Part I*. Marcel Dekker, New York.

Olecki, Z., 1991. Promieniowanie Sloneczne w Dorzeczu Gornej Wisly. Monografia UJ, 147 pp.

Onken, B.M. and Hossner, L.R., 1995. Plant uptake and determination of arsenic species in soil solution under flooded conditions. *Journal of Environmental Quality*, 24: 373–381.

O'Shea, T.A. and Mancy, K.H., 1978. The effect of pH and hardness metal ions on the competitive interaction between trace metal ions and inorganic and organic complexing agents found in natural waters. *Water Research*, 12: 703–711.

Overcash, M.R. and Pall, D., 1979. *Design of Land Treatment Systems for Industrial Wastes, Theory and Practice*. Ann Arbor Science, Ann Arbor, MI, 684 pp.

Owens, L.P. and Best, G.R., 1989. Low-energy wastewater recycling through wetland ecosystems: copper and zinc in wetland microcosms. In: R.R. Sharitz and J.W. Gibbons (Editors), Freshwater Wetlands and Wildlife. DOE Symposium Series No. 61, Oak Ridge, TN, 1227–1235.

Pace, C.B. and Di Giulio, R.T., 1987. Lead concentrations in soil sediment and clam samples from the Pungo River peatland area of North Carolina. *Environmental Pollution*, 43: 301–311.

Page, N.J. and Creasy, S.C., 1975. Ore grade, metal production and energy. *Journal of Research, U.S. Geological Survey*, 3: 9–13.

Pahlsson, A.-M.B., 1989. Toxicity of heavy metals (Zn, Cu, Cd, Pb) to vascular plants, a literature review. *Water, Air, and Soil Pollution*, 47: 287–319.

Palm, V. and Ostlund, C., 1996. Lead and zinc flows from technosphere to biosphere in a city region. *Science for the Total Environment*, 192: 95–109.

Pardue, J.H., DeLaune, R.D., and Patrick, W.H., 1992. Metal to aluminum correlation in Louisiana coastal wetlands: identification of elevated metal concentrations. *Journal of Environmental Quality*, 21: 539–545.

Patrick, R., 1949. A proposed biological measure of stream conditions, based on a survey of the Conestoga Basin, Lancaster County, Pennsylvania. *Proceedings of the Academy of Natural Sciences, Philadelphia*, 101: 277–341.

Patterson, C.C., 1965. Contaminated and natural lead environments of man. *Archives of Environmental Health*, 11: 344–363.

Patterson, C.C., 1973. In: E.L. Jernigan (Editor), *Lead Poisoning in Man and the Environment*. MSS Information Corporation, Madison Avenue, New York, 71–87.

Patterson, J.M., Allen, H.E., and Scala, J.J., 1977. Carbonate precipitation for heavy metals pollutants. *Journal of Water Pollution Control Federation*, December: 2397–2410.

Patterson, J.W., 1985. *Industrial Wastewater Treatment Technology*, 2nd ed. Butterworths, Boston.

Patterson, J.W. and Passino, R. (Editors), 1989. *Metals Speciation, Separation, and Recovery*. Lewis Publishers, Boca Raton, FL.

Paul Scherrer Institute, 1998. Annual Report. Zurich, Switzerland, 73 pp.

Paulson, A.J., Feely, R.A., Curl, H.C., Jr., Crecelius, E.A., and Romberg, G.P., 1989. Separate dissolved and particulate trace metal budgets for an estuarine system: an aid for management decisions. *Environmental Pollution*, 57: 317–339.

Pearce, D.W. and Turner, R.K., 1990. *Economics of Natural Resources and the Environment*. Johns Hopkins University Press, Baltimore.

Pedco Environmental, I., 1980. Industrial Process Profiles for Environmental Use: Chapter 27 Primary Lead Industry. EPA-600/2-80-168. U.S. Environmental Protection Agency, Cincinatti, OH.

Pennak, R.W., 1978. *Freshwater Invertebrates of the United States*, 2nd ed. John Wiley & Sons, New York.

Perdue, E.M. and Lytle, C.R., 1983. A critical examination of metal-ligand complexation models: application to defined multiligand mixtures. In: R.F. Christman and E.T. Gjessing (Editors), *Aquatic and Terrestrial Humic Materials*. Ann Arbor Science, Ann Arbor, MI, 295–313.

Perez, K.T., Morison, G.M., Lackie, N.F., Oviatt, C.A., Nixon, S.W., Buckley, B.A., and Heltsche, J.F., 1977. The importance of physical and biotic scaling to the experimental simulation of a coastal marine ecosystem. *Helgol Wiss. Meeresunters*, 30: 144–162.

Perez, K., 1995. Role and significance of scale to ecotoxicology. In: *Ecological Toxicity Testing*. Lewis Publishers, Boca Raton, FL, 49–72.

Peterson, P.J., 1978. Lead and vegetation. In: J.O. Nriagu (Editor), *The Biogeochemistry of Lead in the Environment, Vol. B: Biological Effects*. Elsevier/North-Holland Biomedical Press, Amsterdam, 355–384.

Pierzynski, G.M., Schnoor, J.L., Banks, M.K., Tracey, J.C., Licht, L.A., and Erikson, L.E., 1994. In: R.E. Hester and R.M. Harrison (Editors), *Mining and Its Environmental Impact. Issues in Environmental Science and Technology*. Royal Society of Chemistry, Letchworth, U.K., 49–69.

Pimental, D. (Editor), 1980. *Handbook of Energy Utilization in Agriculture*. CRC Press, Boca Raton, FL.

Posner, H.S., Damstra, T., and Nriagu, J.O., 1978. Human health effects of lead. In: J.O. Nriagu (Editor), *Biogeochemistry of Lead, Part B, Biological Effects*. Elsevier, Amsterdam, 172–223.

Pourbaix, M., 1966. *Atlas Electrochemical Equilibria: In Aqueous Solutions*. Pergamon Press, New York.

Pritchard, L., Jr., 1992. The Ecological Economics of Natural Wetland Retention of Lead. M.S. thesis, University of Florida, Gainesville.

Pritchard, P.H. and Bourquin, A.W., 1984. A perspective on the role of microcosms in environmental fate and effects assessments. In: H.H. White (Editor), *Concepts in Marine Pollution Measurements*. Maryland Sea Grant Publication, University of Maryland, College Park, 117–138.

Purchase, N.G. and Fergusson, J.E., 1986. The distribution and geochemistry of lead in river sediments, Christchurch, New Zealand. *Environmental Pollution (Series B)*, 12: 203–216.

Putnam Hayes and Bartlett Inc., 1986. The Impacts of Lead Industry Economics on Battery Recycling. Prepared for Office of Policy Analysis, Environmental Protection Agency. Putnam Hayes and Bartlett, Cambridge, MA.

Putnam Hayes and Bartlett Inc., 1987. The Impacts of Lead Industry Economics and Hazardous Waste Regulations on Lead-Acid Battery Recycling: Revision and Update. Prepared for Office of Policy Analysis, Environmental Protection Agency. Putnam Hayes and Bartlett, Cambridge, MA.

Rabinowitch, G., Wetherall, W., and Kopple, J.D., 1976. Kinetics of lead metabolism in healthy humans. *Journal of Clinical Investigations*, 58: 260–270.

Radzicki, M.J., 1988. Institutional dynamics: an extension of the institutionalist approach to socioeconomic analysis. *Journal of Economic Issues*, 22: 633–665.

Radzicki, M.J., 1990. Institutional dynamics, deterministic chaos, and self-organizing systems. *Journal of Economic Issues*, 24: 57–102.

Raghi-Atri, F. and Segebade, C., 1979. Einfluss von Schwermetallen (PB, Hg) in Substrat auf Glyceria Maxima Angew. *Botanic*, 53: 175–178.

Randall, A., 1987. *Resource Economics: An Economic Approach to Natural Resource and Environmental Policy*, 2nd ed. John Wiley & Sons, New York.

Rankama, K. and Sahama, T.G., 1950. *Geochemistry*. Interscience, New York, 591 pp.

Rea, B.A., Kent, D.B., LeBlanc, D.R., and Davis, J.A., 1991. Mobility of zinc in a sewage-contaminated aquifer, Cape Cod, Massachusetts. In: G.E. Mallard and D.A. Aronson (Editors), Water Resources Investigations Report 91–4034. U.S. Geological Survey Toxic Substances Hydrology Program. Proceedings of the Technical Meeting, March 11–15, 1991, Monterey, CA.

Reddy, C.N. and Patrick, W.H., Jr., 1977. Effect of redox potential and pH on the uptake of Cd and Pb by rice plants. *Journal of Environmental Quality*, 6: 259–262.

Reddy, K.R. and Patrick, W.H., 1983. Effects of aeration on reactivity and mobility of soil constituents. In: *ASA SSSA, Chemical Mobility and Reactivity in Soil System*. ASA Soil Science Society of America, Madison, WI, 11–33.

Reed, S.C., Middlebrooks, E.J., and Crites, R.W., 1988. *Natural Systems for Waste Management and Treatment*. McGraw-Hill, New York.

Reuter, J.H. and Perdue, E.M., 1977. Importance of heavy metal-organic matter interactions in natural waters. *Geochimica et Cosmochimica Acta*, 41: 325–334.

Richardson, J.R., Straub, P., Ewel, K.C., and Odum, H.T., 1983. Sulfate enriched water effects on a floodplain forest. *Environmental Management* 7(4): 321–326.

Rickard, D.T. and Nriagu, J.O., 1978. Aqueous environmental chemistry of lead. In: J.O. Nriagu (Editor), *The Biogeochemistry of Lead in the Environment, Vol. A. Ecological Cycles*. Elsevier/North-Holland Biomedical Press, Amsterdam, 219–284.

Roberts, S., Sanderson, D.J., and Gumiel, P., 1998. Fractal analysis of SN-W mineralization from central Iberia: insights into the role of fracture connectivity in the formation of an ore deposit. *Economic Geology*, 300–365.

Rolfe, G.L., Chaker, A., Melin, J., and Ewing, B.B., 1972. Modeling lead pollution in a watershed ecosystem. *Journal of Environmental Systems*, 2(4): 339–349.

Rolfe, G.L. and Haney, A., 1975. *An Ecosystem Analysis of Environmental Contamination by Lead*. Institute for Environmental Studies, University of Illinois, Urbana, 113 pp.

Rosen, J.F. and Sorell, M., 1978. The metabolism and sub-clinical effects of lead in children. In: J.O. Nriagu (Editor), *Biogeochemistry of Lead, Part B, Biological Effects*. Elsevier, Amsterdam, 151–172.

Rubin, A.J. (Editor), 1974. *Aqueous-Environmental Chemistry of Metals*. Ann Arbor Science, Ann Arbor, MI.

Ruby, M.V., Davis, A., Kempton, J.H., Drexler, J.W., and Bergstrom, P.D., 1992. Lead bioavailability: dissolution kinetics under simulated gastric conditions. *Environmental Science Technology*, 26(6): 1242–1248.

Rudd, T., Campbell, J.A., and Lester, J.N., 1988. The use of model compounds to elucidate metal forms in sewage sludge. *Environmental Pollution*, 225–243.

Rudd, T., Lake, D.L., Mehrotra, I., Sterritt, R.M., Kirk, P.W.W., Campbell, J.A., and Lester, J.N., 1988. Characterization of metal forms in sewage sludge by chemical extraction and progressive acidification. *Science of the Total Environment*, 74: 149–175.

Russell, L.H., 1978. Heavy metals in foods of animal origin. In: F.W. Oehme (Editor), *Toxicity of Heavy Metals in the Environment, Part 1*. Marcel Dekker, New York, 3–23.

Rygg, B., 1985. Effect of sediment copper on benthic fauna. *Marine Ecological Progress Series*, 25: 83.

Saar, R.A. and Weber, J.M., 1980. Lead(II)-fulvic acid complexes: conditional stability constants, solubility, and implications for lead(II) mobility. *Environmental Science and Technology*, 14: 877–880.

Saar, R.A. and Weber, J.M., 1982. Fulvic acid: modifier of metal-ion chemistry. *Environmental Science and Technology*, 16: 510A-517A.

Salomons, W. and Förstner, U., 1984. *Metals in the Hydrocycle*. Springer-Verlag, New York.

Salomons, W., Förstner, U., Mader, P. (Editors), 1995. *Heavy Metals, Problems and Solutions*. Springer, Berlin, 412 pp.

Samant, H.S., Doe, K.G., and Vaidya, O.C., 1990. An integrated chemical and biological study of the bioavailability of metals in sediments from two contaminated harbors in New Brunswick, Canada. *Science of the Total Environment*, 96: 253–268.

Sather, J.H. and Smith, R.D., 1984. An overview of major wetland functions and values, Western Energy and Land Use Term, U.S. Fish and Wildlife Service, FWS/OBS-84/18, Washington, D.C.

Scanferlato, V.S. and Cairns, J., Jr., 1990. Effect of sediment-associated copper on ecological structure and function of aquatic microcosms. *Aquatic Toxicology*, 18: 23–34.

Schalscha, E.B., Morales, M., Vergara, I., and Chang, A.C., 1982. Chemical fractionation of heavy metals in wastewater-affected soils. *Journal of the Water Pollution Control Federation*, 54: 175–180.

Schlesinger, W.H., 1991. *Biogeochemistry: An Analysis of Global Change*. Academic Press, San Diego, CA.

Schnitzer, M. and Khan, S.U., 1972. Humic Substances in the Environment. Marcel Dekker, New York.

Schwartz, J., 1994a. Low-level lead exposure and children's IQ: a meta-analysis and search for a threshold. *Environmental Research*, 65: 42–55.

Schwartz, J., 1994b. Societal benefits of reducing lead exposure. *Environmental Research*, 66: 105–124.

Scienceman, D.M., 1987. Energy and emergy. In: G. Pillet and T. Murota (Editors), *Environmental Economics: The Analysis of a Major Interface*. Leimgruber, Geneva, Switzerland, 257–276.

Scodari, P.F., 1990. *Wetlands Protection: The Role of Economics*. Environmental Law Institute, Washington, D.C.

Seaward, M.R.D. and Richardson, D.H.S., 1990. Atmospheric sources of metal pollution and effects on vegetation. In: A.J. Shaw (Editor), *Heavy Metal Tolerance in Plants: Evolutionary Aspects*. CRC Press, Boca Raton, FL, 75–92.

Seidel, K., 1966. Reinigung von Gewassern durch höhere Pflanzen. *Die Naturwissenschaften*, Heft 12, 53 Jahrgang.

Seidel, K., 1969. Höhere Wasserplflanzen in ihrer Umwelt-eine Neuorienterung. *Revue Roumaine De Biologie Serie De Zoologie*, (2)14.

Seip, K.L., 1979. A mathematical model for the uptake of heavy metals in benthic algae. *Ecological Modelling*, 6: 183–197.

Senesi, N., 1992. Metal humic substance complexes in the environment. Molecular and medical aspects by multiple spectroscopic approach. In: D.C. Adriano (Editor), *Biogeochemistry of Trace Metals*, Lewis Publishers, Boca Raton, FL, 429–496.

Senesi, N. and Miano, T.M., 1994. *Humic Substances in the Global Environment and Implications on Human Health*. Elsevier, Amsterdam, 1368 pp.

Shabman, L.A. and Batie, S.S., 1978. Economic value of natural coastal wetlands: a critique. *Coastal Zone Management Journal*, 4: 231–247.

Shacklette, H.T., Erdman, J.A., Harms, T.F., and Papp, C.S.E., 1978. Trace elements in plant foodstuffs. In: F.W. Ohme (Editor), *Toxicity of Heavy Metals in the Environment, Part I*. Marcel Dekker Inc., New York, 25–68.

Shaw, A.J. (Editor), 1990. *Heavy Metal Tolerance in Plants: Evolutionary Aspects*. CRC Press, Boca Raton, FL.

Sheehan, P.J. and Winner, R.W., 1984. Comparison of gradient studies in heavy-metal-polluted streams. In: P.J. Sheehan, D.R. Miller, G.C. Butler, and P. Bourdeau (Editors), *Effect of Pollutants at the Ecosystem Level*. John Wiley & Sons, New York, 255–271.

Sheppard, M.I. and Thibault, D.H., 1992. Desorption and extraction of selected heavy metals from soils. *Soil Science Society of America Journal*, 56: 415–423.

Shotyk, W., Weiss, D., Appleby, P.G., Cheburkin, A.K., Frei, R., Gloor, M., Kramers, J.D., Reese, S., and VanDer Knaap, W.O., 1998. History of atmospheric lead disposition since 12,370 14C yr BP from a peat bog, Jura Mountains, Switzerland. *Science*, 281: 1635–1618.

Sidle, R.C., Hook, J.E., and Kardos, L.T., 1977. Accumulation of heavy metals in soils from extended wastewater irrigation. *Journal of the Water Pollution Control Federation*, 311–318.

Siegel, S.M., Golan, M., and Siegel, B.Z., 1990. Filamentous fungi and metal biosorbents. *Water, Air, and Soil Pollution*, 53: 339–344.

Simkiss, K. and Taylor, M.G., 1995. In: A. Tessier and D.R. Turner (Editors), *Metal Speciation and Bioavailability*. John Wiley & Sons, New York, 679 pp.

Simpson, J.A. and Weiner, E.S.C. (Editors), 1989. *The Oxford English Dictionary*, 2nd ed. Oxford University Press, New York.

Simpson, J.C., 1985. Estimation of spatial patterns and inventories of environmental contaminants using Kriging (geostatistics). In: J.J. Breen and P.E. Robinson (Editors), *Chemometrics*. American Chemical Society, Washington, D.C., 203–242.

Simpson, R.L., 1983. The role of Delaware River freshwater tidal wetlands in the retention of nutrients and heavy metals. *Journal of Environmental Quality*, 12(1).

Simpson, R.L., Good, R.E., Dubinski, B.J., Pasquale, J.J., and Philip, K.R., 1983. *Fluxes of Heavy Metals in Delaware River Freshwater Tidal Wetlands*. Center for Coastal and Environmental Studies and Biology Department, Rutgers University, Camden, NJ.

Simpson, R.L., Good, R.E., Walker, R., and Frasco, B.R., 1983. Role of Delaware River freshwater tidal wetlands in the retention of nutrients and heavy metals. *Journal of Environmental Quality*, 12: 41–48.

Singer, P.C. (Editor), 1973. *Trace Metals and Metal-Organic Interactions in Natural Waters*. Ann Arbor Science, Ann Arbor, MI.

Skinner, B.J., 1979. The many origins of hydrothermal mineral deposits. In: H.L. Barnes (Editor), *Geochemistry of Hydrothermal Ore Deposits*, 2nd ed. John Wiley & Sons, New York, 1–21.

Skinner, B.J., 1986. Earth Resources, 3rd ed. Prentice-Hall, Englewood Cliffs, NJ.

Smith, R., 1983. Evaluation of various techniques for the pretreatment of sewage sludge prior to trace metal analysis by atomic absorption spectrophotometry. *Water South Africa*, 9: 31–36.

Smith, R.A., Alexander, R.B., and Wolman, M.G., 1987. Water-quality trends in the nation's rivers. *Science*, 235: 1607–1615.

Snoeyink, V.L. and Jenkins, D., 1980. *Water Chemistry*. John Wiley & Sons, New York.

Socolow, R. and Thomas, V., 1997. The industrial ecology of lead and electric vehicles. *Journal of Industrial Ecology*, 1(1): 13–36.

Socolow, R.H., Andrews, C., Berkhart, F., and Thomas, V. (Editors), 1994. *Industrial Ecology and Global Change*. Cambridge University Press, New York, 500 pp.

Sposito, G., Holtzclaw, K.M., and LeVesque-Madore, C.S., 1981. Trace metal complexation by fulvic acid extracted from sewage sludge. I. Determination of stability constants and linear correlation analysis. *Soil Science Society of America Journal*, 45: 465–468.

Staley, L., 1996. Lead reclamation from superfund waste. *Tech Trends*, EPA 542–N-96–002: No. 23, p. 1.

Steelink, C., 1985. Implications of elemental characteristics of humic substances. In: G.R. Aiken et al. (Editors), *Humic Substances in Soil, Sediment, and Water*. John Wiley & Sons, New York, 457–476.

Stephenson, T., 1987. Sources of heavy metals in wastewater. In: J.N. Lester (Editor), *Heavy Metals in Wastewater and Sludge Treatment Processes*, Vol. 1. CRC Press, Boca Raton, FL, 31–64.

Sterritt, R.M. and Lester, J.N., 1980a. Determination of silver, cobalt, manganese, molybdenum and tin in sewage sludge by a rapid electrothermal atomic-absorption spectroscopic method. *Analyst*, 105: 616–620.

Sterritt, R.M. and Lester, J.N., 1980b. Atomic absorption spectrophotometric analysis of the metal content of waste water samples. *Environmental Technology Letters*, 1: 402–417.

Stevenson, F.J., 1976. Stability constants of Cu^{2+}, Pb^{2+} and Cd^{2+} complexes with humic acids. *Soil Science Society of America Journal*, 40: 665–672.

Stevenson, F.J., 1982. *Humis Chemistry*. Wiley-Interscience, New York.

Stevenson, F.J., 1985. Geochemistry of soil humic substances. In: G.R. Aiken et al. (Editors), *Humic Substances in Soil, Sediment, and Water*. John Wiley & Sons, New York, 13–52.

Stevenson, F.J., 1986. Cycles of soil carbon, nitrogen, phosphorus, sulfur and micronutrient. Wiley, New York, 279 pp.

Stockdale, F.C., 1991. *Freshwater Wetlands, Urban Stormwater, and Non Point Pollution Control: A Literature Review and Annotated Bibliography*, 2nd ed. Washington State University Department of Ecology, Olympia.

Stover, R.C., Sommers, L.E., and Silviera, D.J., 1976. Evaluation of metals in wastewater sludge. *Journal of Water Pollution Control Federation*, 48: 2165–2175.

Stumm, W. and Morgan, J.J., 1981. *Aquatic Chemistry*, 2nd ed. Wiley-Interscience, New York.

Summers, R.S., Cornel, P.K., and Roberts, P.V., 1987. Molecular size distribution and spectroscopic characterization of humic substances. *Science of the Total Environment*, 62: 27–37.

Suter, G.W., II, 1993. A critique of ecosystem health concepts and indexes. *Environmental Toxicology Chemistry*, 12: 1533–1539.

Swiderska-Broz, M., 1993. Mikrozanieczyszczenia w srodowisku wodnym. Wroclaw.

Szewczynski, J. and Skrodzka, Z., 1997. *Higiena Zywienia*. Panstwowy Zaklad Wydawnictw Lekarskich, Warszawa.

Tatsumoto, M. and Patterson, C.C., 1963. Concentrations of common lead in some Atlantic and Mediterranean waters and in snow. *Nature*, 199: 350–352.

Taub, F.B., 1984. Laboratory microcosms, introduction. In: H.H. White (Editor), *Concepts in Marine Pollution Measurements*. Maryland Sea Grant Publication, University of Maryland, College Park, 113–116.

Taylor, J.K., 1981. Quality assurance of chemical measurements. *Analytical Chemistry*, 53: 1589A–1596A.

Taylor, G. and Crowder, A.A., 1983. Uptake and accumulation of heavy metals by *Typha latifolia* in wetlands. *Canadian Journal of Botany*, 61: 63–73.

Terhune, E.C., 1980. Energy used in the United States for agricultural liming materials. In: D. Pimental (Editor), *Handbook of Energy Utilization in Agriculture*. CRC Press, Boca Raton, FL, 23.

Tessier, A. and Turner, D.R. (Editors), 1995. *Metal Speciation and Bioavailability*. John Wiley & Sons, New York, 679 pp.

Thibodeau, F.R. and Ostro, B.D., 1981. An economic analysis of wetland protection. *Journal of Environmental Management*, 12: 19–30.

Thoman, R.V., 1984. Physico-chemical and ecological modeling the fate of toxic substances in natural water systems. *Ecological Modeling*, 22(1): 45–70.

Thoman, R.V., 1988. Deterministic and statistical models of chemical fate in aquatic systems. In: S.A. Levin, M.A. Harwell, J.R. Kelly, and K.D. Kimball (Editors), *Ecotoxicology: Problems and Approaches*. Springer-Verlag, New York, 245–278.

Thomas, V. and Spiro, T., 1994. Emissions and exposure to metals: cadmium and lead. In: R.H. Socolow, C. Andrews, F. Berkhart, and V. Thomas (Editors), *Industrial Ecology and Global Change*. Cambridge University Press, New York, 297–318

Thompson, D.M., 1981. Distribution of Heavy Metals in Selected Florida Sediments. M.S. thesis. Environmental Engineering Sciences, University of Florida, Gainesville, 159 pp.

Thompson, K.C. and Waghstaff, K., 1980. Simplified method for the determination of cadmium, chromium, copper, nickel, lead and zinc in sewage sludge using atomic-absorption spectrophotometry. *Analyst*, 105: 883–896.

Thurman, E.M., 1985. Humic substances in groundwater. In: G.R. Aiken et al. (Editors), *Humic Substances in Soil, Sediment and Water*. John Wiley & Sons, New York, 87–103.

Tipping, E., Backes, C.A., and Hurley, M.A., 1988. The complexation of protons, aluminum and calcium by aquatic humic substances: a model incorporating binding-site heterogeneity and macroionic effects. *Water Resources*, 5: 579–611.

Ton, S., 1990. Natural Wetland Retention of Lead from a Hazardous Waste Site. M.S. thesis. University of Florida, Gainesville.

Ton, S., 1993. Lead Cycling Through a Hazardous Waste-Impacted Wetland. Ph.D. dissertation, University of Florida, Gainesville.

Ton, S., Delfino, J.J., and Odum, H.T., 1993. Wetland retention of lead from a hazardous waste site. *Bulletin of Environmental Contamination and Toxicology*, 51: 430–437.

Ton, S., Odum, H.T., and Delfino, J.J., 1998. Ecological-economic evaluation of wetland management alternatives. *Ecological Engineering*, 11: 291–302.

Trnovsky, M., Oxer, J.P., Rudy, R.J., Hanchak, M.J., and Hartsfield, B., 1988. Site remediation of heavy metals contaminated soils and groundwater at a former battery reclamation site in Florida. In: R. Abbou (Editor), *Hazardous Waste: Detection, Control, Treatment*. Elsevier Science, Amsterdam, 1581–1590.

Truitt, R.E. and Weber, J.H., 1981a. Determination of complexing capacity of fulvic acid for copper(II) and cadmium(II) by dialysis titration. *Analytical Chemistry*, 53: 337–342.

Truitt, R.E. and Weber, J.H., 1981b. Copper (II)- and cadmium(II)-binding abilities of some New Hampshire freshwaters determined by dialysis titration. *Environmental Science and Technology*, 15: 1024–1028.

Turekian, K.K., 1969. The oceans, streams, and atmosphere. In: K.H. Wedepohl (Editor), *The Oceans, Streams, and Atmosphere, Handbook of Geochemistry*, Vol. 1. Springer-Verlag, New York, 297–323.

Turner, R.S., Johnson, A.H., and Wang, D., 1985. Biogeochemistry of lead in McDonalds Branch watershed, New Jersey pine barrens. *Journal of Environmental Quality*, 15: 305–314.

Tuschall, J.R., Jr., 1981. Heavy Metal Complexation with Naturally Occurring Organic Ligands in Wetland Ecosystems. Ph.D. dissertation, Environmental Engineering Sciences, University of Florida, Gainesville, 212 pp.

Tuschall, J.R., Jr., 1983. *Heavy Metal Binding by Dissolved Aquatic Organic Matter.* Center for Wetlands Publication #83-20, University of Florida, Gainesville.

Tuschall, J.R., Jr. and Brezonik, P., 1983. *Aquatic Humus.* Center for Wetlands Publication #83-20, University of Florida, Gainesville.

Tyler, G., 1989. Uptake, retention and toxicity of heavy metals in lichens. *Water, Air, and Soil Pollution*, 47: 321–333.

Tyler, G., Pahlsson, A.-M.B., Bengtsson, G., Baath, E., and Tranvik, L., 1989. Heavy-metal ecology of terrestrial plants, microorganisms and invertebrates. *Water, Air, and Soil Pollution*, 47: 189–215.

U.S. Army Corps of Engineers, 1972. Charles River watershed, Massachusetts. New England Division, Waltham, MA.

U.S. Congress, 1991. Lead Pollution Prevention: Hearing before the Subcommittee on Transportation and Hazardous Materials of the Committee on Energy and Commerce, House of Representatives, One Hundred First Congress, Second Session, On HR 5372. Serial No. 101-198. U.S. Government Printing Office, Washington, D.C.

U.S. Department of Commerce and Bureau of the Census, 1990. Statistical Abstract of the United States: 1991, 111th ed. U.S. Government Printing Office, Washington, D.C.

U.S. Environmental Protection Agency, 1979. Methods for Chemical Analysis of Water and Wastes. EPA-600/4–79–020, Environmental Monitoring and Support Laboratory, Cincinnati, OH.

U.S. Environmental Protection Agency, 1979. Lead-Acid Battery Manufacture: Background Information for Proposed Standards. Draft EIS: EPA 450/3–79–028. U.S. Environmental Protection Agency, Springfield, VA.

U.S. Environmental Protection Agency, 1986. Air Quality Criteria for Lead, Vols. I to IV. EPA-600/8–83/028, Triangle Park, NC.

U.S. Environmental Protection Agency, 1989. The Toxics Release Inventory: A National Perspective, 1987. U.S. Government Printing Office, Washington, D.C.

U.S. Environmental Protection Agency, 1991. The Superfund Innovative Technology Evaluation Program: Technology Profiles, 4th ed. Report No. EPA/540/5–91/008. U.S. EPA, Washington, D.C.

U.S. Soil Conservation Service, 1979. Soil Survey of Jackson County, Florida. Department of Agriculture, Washington, D.C.

Valiela, I., Bnus, M.D., and Teal, J.M., 1974. Response of salt marsh bivalves to enrichment with metal-containing sewage sludge and retention of lead, zinc and cadmium by marsh sediments. *Environmental Pollution*, 7: 149–157.

van der Leeden, F., Troise, F.L., and Todd, D.K., 1990. *The Water Encyclopedia*, 2nd ed. Lewis Publishers, Chelsea, MI.

Vedagiri, U. and Ehrenfeld, J., 1991. Effects of Sphagnum moss and urban runoff on bioavailability of lead and zinc from acidic wetlands of the New Jersey pinelands. *Environmental Pollution*, 72: 317–330.

Vedagiri, U. and Ehrenfeld, J., 1992. Partitioning of lead in pristine and runoff-impacted waters from acidic wetlands in the New Jersey pinelands. *Water, Air, and Soil Pollution*, 64: 511–524.

Veizer, J., Laznicka, P., and Jansen, S.L., 1989. Mineralization through geologic time: recycling perspective. *American Journal of Science*, 289: 484–524.

Venugopal, B. and Luckey, T.D., 1978. *Metal Toxicity in Mammals — 2: Chemical Toxicity of Metals and Metalloids.* Plenum Press, New York.

Vernet, J.P. (Editor), 1992. *Impact of Heavy Metals on the Environment.* Elsevier, Amsterdam, 444 pp.

Volesky, B. (Editor), 1990. *Biosorption of Heavy Metals.* CRC Press, Boca Raton, FL, 396 pp.

Vuceta, J. and Morgan, J.J., 1978. Chemical modeling of trace metals in fresh waters: role of complexation and adsorption. *Environmental Science and Technology*, 12: 1302–1309.

Vymazul, J., 1995. *Algae and Element Cycling in Wetlands.* Lewis Publishers, Boca Raton, FL, 704 pp.

Ward, N.I., Roberts, R., and Brooks, R.R., 1977. Lead pollution from a New Zealand battery factory and smelter compared with that from motor-vehicle exhaust emissions. *New Zealand Journal of Science*, 20: 407–412.

Watts, G.B., 1984. Remedial Investigation: Sapp Battery Site, Jackson County, Florida. Florida Department of Environmental Regulation, Groundwater Section, Tallahassee, FL.

Webb, J.S., 1978. *The Wolfson Geochemical Atlas of England and Wales*. Clarendon Press, Oxford, U.K.

Weber, C.I. (Editor), 1973. Biological Field and Laboratory Methods for Measuring the Quality of Surface Waters and Effluents. EPA-670/4–73–001. U.S. Environmental Protection Agency, Cincinnati, OH.

Weber, J.H., 1983. Metal ion speciation studies in the presence of humic materials. In: R.F. Christman and E.T. Gjessing (Editors), Aquatic and Terrestrial Humic Materials. Ann Arbor Science, Ann Arbor, MI, 315–331.

Webster's Encyclopedic Unabridged Dictionary, 1989. Dilithium Press, New York.

Wehle, M., 1974. Alkalies and chlorine. In: J.G. Myers, et al. (Editors), *Energy Consumption in Manufacturing*. Ballinger, Cambridge, MA, 181–203.

Weinstein, D.A. and Buk, E.M., 1994. The effects of chemicals on the structure of terrestrial ecosystems, mechanisms and patterns of change. In: S.A. Levin, M.A. Howell, J.R. Kelley, and K.D. Kimboll (Editors), *Ecotoxicology Problems and Approaches*. Springer, New York, 181–209.

Whitfield, M. and Turner, D.R., 1982. Ultimate removal mechanism of elements from the ocean — a comment. *Geochimica et Cosmochimica Acta*, 46: 1989–1992.

Whitton, B.A., Say, P.J., and Jupp, B.P., 1982. Accumulation of zinc, cadmium and lead by the aquatic liverwort *Scapania*. *Environmental Pollution (Series B)*, 3: 299–316.

Wickland, D.E., 1990. Vegetation of heavy metal-contaminated soils in North America. In: A.J. Shaw (Editor), *Heavy Metal Tolerance in Plants: Evolutionary Aspects*. CRC Press, Boca Raton, FL, 39–51.

Wieder, R.K., 1990. Metal cation binding to Sphagnum peat and sawdust relation to wetland treatment of metal-polluted waters. *Water, Air, and Soil Pollution*, 53: 391–400.

Wildeman, T., Gusek, J., Cevaal, J., Whiting, K., and Scheuering, J., 1996. Biotreatment of acid rock drainage at a gold mining operation. In: R.E. Hichee, J.L. Means, and D. Burns (Editors), *International In-Situ-On-Site Bioremediation Symposium*. Batelle Press, Columbus, OH, 141–148.

Wildeman, T., Gusek, J., Miller, A., and Ficke, J., 1997. Metals, sulfur, and carbon budget in a pilot reactor treating lead in wastes. In: *3rd International In-Situ-On-Site Bioremediation Symposium*. Batelle Press, Columbus, OH, 401–406.

Williams, J.D. and Dodd, C.K., Jr., 1979. Importance of wetlands to endangered and threatened species. In: P.E. Greeson, J.R. Clark, and J.E. Clark (Editors), *Wetland Functions and Values: the State of Our Understanding*. American Water Resources Association, Minneapolis, 565–575.

Windom, H.L., Schropp, S.L., Calder, F.D., Ryan, J.D., Smith, R.G., Jr., Burney, L.C., Lewis, F.G., and Rawlinson, C.H., 1989. Natural trace metal concentratrions in estuarine and coastal marine sediments of the southeastern United States. *Environmental Science and Technology*, 23: 314–320.

Windom, H., Smith, R., and Rawlinson, C., 1988. Trace metal transport in a tropical estuary. *Marine Chemistry*, 24: 293–305.

Wixson, B.G., 1978. Biogeochemical cycling of lead in the New Lead Belt of Missouri. In: J.O. Nriagu (Editor), *The Biogeochemistry of Lead in the Environment, Vol. A: Ecological Cycles*. Elsevier/North-Holland Biomedical Press, Amsterdam, 119–136.

Wixson, B.R. and Davies, B.E., 1993. *Lead in Soil, Recommended Guidelines*. Science Reviews, Society for Environmental Geochemistry and Health, Northwood, U.K., 117 pp.

Wójcik, W., 1993. Application of emergy analysis for assessment of optional wastewater treatment methods. In: L. Preisner (Editor), *Economics and Environment #12, Institutions and Environmental Protection Vol. 3*. Krakow Academy of Economics, Krakow, Poland, 360–375.

Wójcik, W., Leszczynski, S., and Wójcik, M., 1990. Heavy Metals Interactions with the Biala River Wetland. Technical Report from the Research Sponsored by the Sendzimir Family. Center for Research and Development, Krakow, Poland.

Wójcik, W. and Wójcik, M., 1989. Heavy Metals Interactions with the Biala River Wetland. Technical Report to the Subcontract between the University of Florida and the Academy of Mining and Metallurgy, to the D.T. Sendzimir Family Research Contract. Academy of Mining and Metallurgy, Krakow, Poland.

Wolverton, B.C. and Bounds, B.K., 1988. Aquatic plants for pH adjustment and removal of toxic chemicals and dissolved minerals from water supplies. *Journal of the Mississippi Academy of Science*, 32: 71–80.

Wolverton, B.C. and McDonald, R.C., 1975. Water Hyacinths and Alligator Weeds for Removal of Lead and Mercury from Polluted Waters. NASA Technical Memorandum TM-X-72723. NASA, National Space Technology Laboratories, Bay St. Louis, Hancock, MS.

Wolverton, B.C. and McDonald, R.C., 1977. Wastewater treatment utilizing water hyacinths (*Eichhornia crassipes*). Proceedings of the 1977 National Conference on Treatment and Disposal of Industrial Wastewater and Residue, April 26–29, 1977, Houston, TX.

Wolverton, B.C. and McDonald, R.C., 1978a. Water Hyacinth Sorption Rates of Lead, Mercury, and Cadmium. Earth Resources Laboratory Report No. 170. NASA, National Space Technology Laboratories, Bay St. Louis, Hancock, MS.

Wolverton, B.C. and McDonald, R.C., 1978b. Bioaccumulation and detection of trace levels of cadmium in aquatic systems by *Eichhornia crassipes*. *Environmental Health Perspectives*, 27: 161–164.

Wong, P.T.S., Siverberg, B.A., Chau, Y.K., and Hodson, P.V., 1978. Lead and the aquatic biota. In: J.O. Nriagu (Editor), *Biogeochemistry of Lead, Part B, Biological Effects*. Elsevier, Amsterdam, 395 pp.

Woodbury, W.D., 1988. Lead. In: U.S. Bureau of Mines (Editor), Minerals Yearbook. U.S. Department of the Interior, Washington, D.C., 581–618.

Wu, L., 1990. Colonization and establishment of plants in contaminated environments. In: A.J. Shaw (Editor), *Heavy Metal Tolerance in Plants: Evolutionary Aspects*. CRC Press, Boca Raton, FL, 269–284.

Wyandotte Chemicals, 1961. *Caustic Soda*. Wyandotte Chemicals Corporation, Wyandotte, MI.

Zar, J.H., 1984. *Biostatistical Analysis*, 2nd ed. Prentice-Hall, Englewood Cliffs, NJ.

Zhigalovskaya, T.N., Makhan'ko, E.P., Shilina, A.I., Egorov, V.V., Malakhov, S.G., and Pervunina, R.I., 1974. Trace Elements in Natural Waters and Atmosphere. Translation of the Institute for Experimental Meteorology, Issue 2. Gidrometeoizdat, Moscow.

Zhulidov, A.V., Headley, J.V., Roberts, R.D., Nikanorov, A.M., and Ischenko, A.A. (Editors), 1996. Atlas of Russian Wetlands, Biogeography and Metal Concentrations. National Hydrology Research Institute, Environment Canada, 11 Innovation Blvd., Saskatoon, Sk Canada S7N 3H5.

INDEX

"Biala R." refers to study sites in Poland; "Sapp Swamp" to sites in Florida

A

H

9 780367 398620